T0252610

Wireless Ad Hoc and Sensor Networks

Protocols, Performance, and Control

AUTOMATION AND CONTROL ENGINEERING

A Series of Reference Books and Textbooks

Editor

FRANK L. LEWIS, Ph.D.

Professor
Applied Control Engineering
University of Texas at Arlington
Fort Worth, Texas

Wireless Ad Hoc and Sensor Networks

Protocols, Performance, and Control

Jagannathan Sarangapani

The University of Missouri-Rolla
Rolla, Missouri, U.S.A.

CRC Press
Taylor & Francis Group
Boca Raton   London   New York

CRC Press is an imprint of the
Taylor & Francis Group, an informa business

Dedication

This book is dedicated to my parents, Sarangapani Jagannathan

and Janaki Sarangapani; my wife, Sandhya; my daughter, Sadhika,

and my son, Anish Seshadri

Table of Contents

Preface

The purpose of this book is to initiate the newcomer into the control of computer and wireless communication networks, one of the fastest growing fields in the engineering world. Technical concepts, which are at the core of the design, implementation, research and invention of computer network and wireless communication network control protocols, are presented in an order that is conducive to understanding general concepts, as well as those specific to particular wired, cellular, wireless ad hoc, and sensor networks.

The unprecedented growth of Internet traffic and the huge commercial success of wireless communications, along with the emerging popularity of Internet Protocol (IP)-based multimedia applications are the major driving forces behind the current and next-generation network evolution wherein data, voice, and video are brought into wired, ad hoc, and sensor networks that require diverse quality of service (QoS). Wired and cellular networks have certain infrastructure — for instance, a router in the case of computer networks and a base station in the case of wireless cellular networks — to route and to perform certain functionality. By contrast, an ad hoc network, which does not need any fixed infrastructure, also has many applications including home and personal area networking, sensor networking, search-and-rescue missions, weather prediction, and so on. Sensor networks are currently being addressed to monitor the health of industrial machinery, civil and military infrastructure, forest fire, earthquake monitoring, tsunami alerts, homeland security, and many other applications. Given the promising applications, this book addresses the basic theory, architectures, and technologies that are needed in order to implement QoS control in wired and wireless networks.

The success of telecommunications in supporting a wide variety of Internet services, such as multimedia conferencing and video-on-demand, depends on (in addition to high-speed transmission and switching technologies), reliable control in the underlying high-speed networks to provide guaranteed QoS. Computer networks require sophisticated, real-time controllers to manage traffic and to ensure QoS because Internet traffic is dominated by multimedia services, which have bursty traffic characteristics and various quality of service (QoS) and bandwidth requirements. As the line speed is increasing towards 100 Gbit/sec, and the number of connections in each line increases to several hundreds of thousands, implementing QoS control, under the constraints of timing and memory requirements, becomes extremely difficult. Moreover, as the traffic pattern and bandwidth required for establishing a connection are typically difficult to determine beforehand, adaptive techniques hold out the promise of improved learning under uncertainty. The QoS control techniques in wired networks include flow and congestion control, admission control, buffer management, fair scheduling, and so on. This book provides an overview of the existing QoS control techniques and describes in detail the most recent adaptive ones, along with practical approaches to implement the techniques in high-speed networks.

On the other hand, in modern wireless networks, distributed power control (DPC) for transmitters along with rate adaptation allows interfering communications sharing the same channel to achieve the required QoS levels. Moreover, unlike in wired networks, the channel state affects the power control, rate adaptation, and routing protocols. Further, attaining QoS goals of the users requires a unified approach to protocol development across physical, transport, and network layers so that cross-layer optimization must be taken into consideration. This book covers existing, and the most recent, protocol designs for cellular, ad hoc, and sensor networks. Most important, the book presents certain underlying QoS control techniques, which were developed using Lyapunov-based design so that the controller performance can be demonstrated. Thorough development, rigorous stability proofs, and simulation, examples are presented in each case.

Chapter 1 lays the foundation of QoS control in wired and wireless networks and presents a systematic overview of QoS control methods including admission control, traffic rate and congestion control, and QoS routing. Background information on the Internet, asynchronous transfer networks (ATM), as well as cellular, ad hoc, and sensor networks is discussed. The QoS parameters for networks in general are introduced. In Chapter 2, background on dynamical systems, Lyapunov analysis, and controllability of dynamical systems is given. Chapter 3 focuses on congestion control of high-speed networks — Internet and ATM using Lyapunov-based analysis and protocol design. The congestion control techniques address the regulation of traffic loading across a network for congestion avoidance and recovery.

In Chapter 4, admission control is discussed in detail using hybrid system theory in order to make suitable network decisions as to whether or not any new connections based on QoS requirements should be allowed. Chapter 5 explores the distributed power control (DPC) of CDMA-based cellular and peer-to-peer networks in the presence of channel uncertainties. Active link protection schemes for existing cellular users are discussed to meet a desired QoS level. In Chapter 6, the DPC is extended to wireless adhoc and wireless sensor networks. The medium access control protocol design to implement the DPC is presented as well. An important aspect of any QoS control system is its implementation on actual hardware using motes. Therefore, in Chapter 6, we develop the framework needed to implement QoS control at each network node using UMR mote hardware. A benchmarking test-bed that is developed for testing and validating the wireless ad hoc and sensor network protocols is also presented. Thorough overhead analysis is described and hardware implementation of DPC is covered.

Chapter 7 presents a historical overview of different packet scheduling schemes and how to regulate the flow of packets among contending flows at a network node so that the QoS requirements of each flow can be met. Additionally, the framework needed to implement QoS control at each network node using embedded computer hardware is also discussed. Using the benchmarking test-bed, the effect of channel uncertainties in the case of wireless networks on packet scheduling is detailed and performance of the scheduling protocol is evaluated. Chapter 8 covers link state routing using the QoS parameters for each connection. The decision of selecting one path from many may depend upon the resource availability along the path. The link state routing protocol is extended to wireless networks where resources, transmitter powers, and expected delays are taken into account. A dynamic optimal link state routing protocol is detailed. Hardware implementation of the routing protocol is also covered.

Chapter 9 describes predictive congestion protocol for wireless sensor networks. Analytical proofs are provided. In Chapter 10, the DPC is extended to a different type of wireless network referred to as radio frequency identification networks. Using DPC, improvement in read rates and coverage is presented.

The book surveys the most recent technical works that readers can refer to for the most up-to-date development of control techniques, both in wired and wireless networks. The book should be useful as a senior undergraduate or a graduate level course in electrical and computer engineering and computer science departments. This book is also useful to software, hardware, and system engineers in the networking design and operation.

Special thanks to my students — in particular, Maciej Zawodniok, Sarat Dontula, Niranjan Regatte, Kainan Cha, Anil Ramachandran, and James Fonda who forced me to take the work seriously and be a part of it. Without monumental efforts of proof checking by Atmika Singh, typing by Maheswaran Thiagarajan, and tireless help by Regina Kohout, this book would not be a reality. Finally, thanks to Kyle Mitchell of St. Louis University for helping us with the hardware.

This research work was supported by the National Science Foundation under grants EcS-9985739, EcS-0216191, NSF Industry/University Cooperative Research Center grant IIp 0639182, CISE #0325613, Department of Education GAANN Program, Missouri Research Board and Intelligent Systems Center.

Jagannathan Sarangapani
Rolla, Missouri

1

Background on Networking

Computer or communication networks can be broadly classified based on whether they are physically connected or are intermittently connected using radio signals. Computer networks that are connected by a piece of wiring, such as a coaxial cable are known as *wired networks*. Wireless networks use radio signals as their physical layer. The ability to connect two or more computers without the need of cumbersome wiring and the flexibility to adapt to mobile environments have been fueling the widespread acceptance and popularity of wireless networks. For an overview of computer networks refer to (Stallings 2002, Chao and Guo 2002, Walrand and Varaiya 1996, Tanenbaum 1996).

1.1 Computer Networks

Common examples of wired networks are the Internet and asynchronous transfer mode (ATM) networks. The broadband integrated services digital network (B-ISDN) with standardized ATM is envisaged to support new services with varying traffic characteristics and quality of service (QoS) requirements that are imposed by the users. In simple terms, ATM is a connection-oriented packet switching and multiplexing technique that uses short fixed-size cells to transfer information over a B-ISDN network. The short cell size of ATM at high transmission rates is expected to offer full bandwidth flexibility and provide the basic framework for guaranteeing QoS requirements of applications with a wide range of performance metrics, such as delay and loss.

Meanwhile, the advent of broadband networking technology has dramatically increased the capacity of packet-switched networks, from a few megabits per second to hundreds or even thousands of megabits per second. This increased data communication capacity allows new applications such as videoconferencing and Internet telephony. These applications have diverse QoS requirements. Some require stringent end-to-end delay bounds; some require a minimal transmission rate whereas others simply require high throughput. With the use of the Internet diversifying and expanding at an exceptional rate, the issue of how to provide necessary QoS for a wide variety of different user applications is also gaining importance. This book attempts to clarify the QoS issues and examines the effectiveness of some proposed network solutions mainly by using congestion, scheduling, and admission control.

In short, QoS depends on the statistical nature of traffic. An appropriate service model should be defined, and some network QoS control methods should be engineered to meet a range of QoS performance requirements (e.g., throughput, delay, and loss), which are usually represented as a set of QoS parameters associated with the service model. There are two main traffic types: delay-sensitive and loss-sensitive traffic. Delay-sensitive traffic is characterized by rate and duration and may need transmission in real-time. Examples include videoconferencing, telephone, and audio/ video on demand, which usually have stringent delay requirements but can accept a certain data loss. Loss-sensitive traffic is characterized by the amount of information transmitted. Examples are web pages, files, and mail. It usually has stringent data-loss requirements but no deadline for completing a transmission.

There are other traffic types, such as multicast traffic [e.g., conferences, distributed interactive simulation (DIS), and games], and traffic aggregation [e.g., from local area network (LAN) interconnection]. Observations of LAN traffic (Stallings 2002) reveal that traffic on the network displays self-similar or long-range dependent behavior. The rate is variable at all time scales; it is not possible to define a duration over which the traffic intensity is approximately constant. These observations have been confirmed repeatedly for every type of network. A plausible explanation for self-similarity is that LAN traffic results from a superposition of bursts whose duration has a heavy-tailed distribution.

The networks have been evolving to provide QoS guarantees to users. For instance, ATM, widely adopted in the backbone network, can reserve the bandwidth and buffer for each virtual connection. Similarly, the Internet integrated service (Intserv) can also provide QoS for each flow in the Internet Protocol (IP) network. Internet differentiated service (Diffserv) provides different treatment options for packets of different classes, instead of on a flow basis, so that it has better scalability than Intserv. Multiprotocol label switching (MPLS), a recent technology trend for the Internet, allows the network providers to have better control and provision of QoS through traffic-engineering policies.

The ATM forum summarizes the traffic parameters used in ATM networks. The constant bit rate (CBR) service category applies to connections that require cell loss and delay guarantees. The bandwidth resource provided to the connection is always available during the connection lifetime, and the source can send at or below the peak cell rate (PCR) or not at all. A CBR connection must specify the parameters, including PCR or peak emission interval denoted as T, which is given by $T = 1/PCR$, cell delay variation tolerance (CDVT), maximum cell transfer delay, and cell loss ratio. The standard defines the rate in terms of an algorithm making necessary allowances for jitter (cell delay variation) introduced between a terminal and the network interface. The chosen algorithm, called the virtual scheduling algorithm or the continuous state leaky bucket, is not standardized as the generic cell rate algorithm (GCRA).

The variable bit rate (VBR) service category is intended for a wide range of connections; it includes real-time constrained connections (rt-VBR), as well as connections that do not need timing constraints (nrt-VBR). (Note that CBR is normally used for a real-time service.) The VBR is basically defined by its PCR, sustainable cell rate (SCR), and maximum burst size (MBS). The SCR indicates the upper bound for the mean data rate, and the MBS indicates the number of consecutive cells sent at peak rate.

The available bit rate (ABR) standard specifies how users should behave in sending data and resource management (RM) cells in response to network feedback in the form of explicit rate or congestion indications. An application using ABR specifies a PCR that it will use and a minimum cell rate (MCR) that it requires. The network allocates resources so that all ABR applications receive at least their MCR capacity. Any unused capacity is then shared in a fair and controlled fashion among all ABR sources. The ABR mechanism uses explicit feedback to sources to ensure that capacity is fairly allocated.

At any given time, a certain amount of the capacity of an ATM network is consumed in carrying CBR and the two types of VBR traffic. Additional capacity may be available for one or both of the following reasons: (1) not all of the total resources have been committed to CBR and VBR traffic, and (2) the bursty nature of VBR traffic means that sometimes less than the committed capacity is used. Any capacity not used by ABR sources remains available for unspecified bit rate (UBR) traffic as explained as follows.

The UBR service is suitable for applications that can tolerate variable delays and some cell losses, which is typically true of transmission control protocol (TCP) traffic. With UBR, cells are forwarded on a first-in, first-out (FIFO) basis using the capacity not consumed by other services; delays and variable losses are possible. No initial commitment is made to a UBR source, and no feedback concerning congestion is provided; this is referred to as a best-effort service.

The guaranteed frame rate (GFR) is intended to refine UBR by adding some form of QoS guarantee. The GFR user must specify a maximum packet size that he or she will submit to the ATM network and a minimum throughput that he or she would like to have guaranteed, that is, a MCR. The user may send packets in excess of the MCR, but they will be delivered on a best-effort basis. If the user remains within the throughput and packet size limitations, he or she can expect that the packet loss rate will be very low. If the user sends in excess of the MCR, he or she can expect that if resources are available, they will be shared equally among all competing users.

1.1.1 Integrated Services (Intserv)

The Internet, as originally conceived, offers only point-to-point best-effort data delivery. Routers use a simple FIFO service policy and reply on buffer management and packet discarding as a means to control network congestion. Typically, an application does not have knowledge of when, if at all, its data will be delivered to the other end unless explicitly notified by the network. New service architecture is needed to support real-time applications, such as remote video, multimedia conferencing, with various QoS requirements. It is currently referred to as the Intserv.

The concept of flow is introduced as a simplex, distinguishable stream of related datagrams that result from a single-user activity and require the same QoS. The support of different service classes requires the network and routers specifically, to explicitly mange their bandwidth and buffer resources to provide QoS for specific flows. This implies that resource reservation, admission control, packet scheduling, and buffer management are also key building blocks of Intserv.

Furthermore, this requires a flow-specific state in the routers, which represents an important and fundamental change to the Internet model. And because the Internet is connectionless, a soft-state approach is adopted to refresh flow rates periodically using a signaling system, such as resource reservation protocol (RSVP). Because ATM is connection-oriented, it can simply use a hard-state mechanism, in that each connection state established during call setup remains active until the connection is torn down. Because it implies that some users are getting privileged service, resource reservation will also need enforcement of policy and administrative controls.

There are two service classes currently defined within the Intserv: guaranteed service (GS) and controlled-load service (CLS) (Stallings 2002). GS is a service characterized by a perfectly reliable upper bound on end-to-end packet delay. The GS traffic is characterized by peak rate, token bucket parameters (token rate and bucket size), and maximum packet size. GS needs traffic access control (using the token bucket) at the user side and fair packet queuing (FPQ) at routers to provide a minimum bandwidth. Because this upper bound is based on worst-case assumptions on the behavior of other flows, proper buffering can be provided at each router to guarantee no packet loss.

The CLS provides the client data flow with a level of QoS closely approximating what the same flow would receive from a router that was not heavily loaded or congested. In other words, it is designed for applications that can tolerate variance in packet delays, as well as a minimal loss rate that must closely approximate the basic packet error rate of the transmission medium. CLS traffic is also characterized by (optional) peak rate, token bucket parameters, and maximum packet size. The CLS does not accept or make use of specific target values for control parameters such as delay or loss. It uses loose admission control and simple queue mechanisms, and is essentially for adaptive real-time communications. Thus, it does not provide a worst-case delay bound like the GS.

Intserv requires packet scheduling and buffer management on a per flow basis. As the number of flows and line rate increase, it becomes very difficult and costly for the routers to provide Intserv. A solution called Diffserv can provide QoS control on a service class basis. It is more feasible and cost-effective than the Intserv.

1.1.2   Differentiated Services (Diffserv)

Service differentiation is desired to accommodate heterogeneous application requirements and user expectations, and to permit differentiated pricing of Internet services. DS, or Diffserv (Stallings 2002), are intended to provide scalable service discrimination in the Internet without the need for per flow state and signaling at every hop, as with the Intserv. The DS approach to providing QoS in networks employs a small, well-defined set of building blocks from which a variety of services may be built. The services may be either end-to-end or intradomain. A wide range of services can be provided by a combination of the following:

- Setting bits in the type of service (ToS) byte at network edges and administrative boundaries
- Using those bits to determine how packets are treated by the routers inside the network
- Conditioning the marked packets at network boundaries in accordance with the requirements of each service

According to this model, network traffic is classified and conditioned at the entry to a network and assigned to different behavior aggregates. Each such aggregate is assigned a single DS codepoint (i.e., one of the markups possible with the DS bits). Different DS codepoints signify that the packet should be handled differently by the interior routers. Each different type of processing that can be provided to the packets is called a different per-hop behavior (PHB). In the core of the network, packets are forwarded according to the PHBs associated with the codepoints. The PHB to be applied is indicated by a Diffserv codepoint (DSCP) in the IP header of each packet. The DSCP markings are applied either by a trusted customer or by the boundary routers on entry to the DS network.

The advantage of such a scheme is that many traffic streams can be aggregated to one of a small number of behavior aggregates (BAs), which are each forwarded using the same PHB at the routers, thereby simplifying the processing and associated storage. Because QoS is invoked on a packet-by-packet basis, there is no signaling, other than what is carried in the DSCP of each packet, and no other related processing is required in the core of the DS network. Details about DS can be found in (Stallings 2002).

1.1.3 Multiprotocol Label Switching (MPLS)

MPLS has emerged as an important new technology for the Internet. It represents the convergence of two fundamentally different approaches in data networking: datagram and virtual circuit. Traditionally, each IP packet is forwarded independently by each router hop by hop, based on its destination address, and each router updates its routing table by exchanging routing information with the others. On the other hand, ATM and frame relay (FR) are connection-oriented technologies — a virtual circuit must be set up explicitly by a signaling protocol before packets can be sent into the network.

MPLS uses a short, fixed-length label inserted into the packet header to forward packets. An MPLS-capable router, termed the label-switching router (LSR), uses the label in the packet header as an index to find the next hop and the corresponding new label. The LSR forwards the packet to its next hop after it replaces the existing label with a new one assigned for the next hop. The path that the packet traverses through an MPLS domain is called a label-switched path (LSP). Because the mapping between labels is fixed at each LSR, an LSP is determined by the initial label value at the first LSR of the LSP.

The key idea behind MPLS is the use of a forwarding paradigm based on label swapping that can be combined with a range of different control modules. Each control module is responsible for assigning and distributing a set of labels, as well as for maintaining other relevant control information. Because MPLS allows different modules to assign labels to packets using a variety of criteria, it decouples the forwarding of a packet from the contents of the packet's IP header. This property is essential for such features as traffic engineering and virtual private network (VPN) support.

1.1.4 QoS Parameters for Internet and ATM Networks

In general, QoS is a networking term that specifies a guaranteed throughput level, which allows network providers to guarantee their customers that end-to-end delay will not exceed a specified level. For an Internet and ATM network, the QoS parameters include peak-to-peak packet or cell

TABLE 1.1

ATM Service Category Attributes

	ATM Layer Service Category				
Attribute	Constant Bit Rate	real time-Variable Bit Rate	non real time-Variable Bit Rate	Unapplied Bit Rate	Available Bit Rate
Traffic parameters	Peak cell rate	Peak cell rate Sustainable cell rate Mean Bit Rate	Peak cell rate Sustainable cell rate Mean Bit Rate	Peak cell rate	Peak cell rate
	Cell delay variable tolerance	Cell delay variable tolerance	Cell delay variable tolerance		Cell delay variable tolerance
QoS parameters	Cell delay variation maximum Cell transfer delay	Cell delay variation maximum Cell transfer delay			
	Cell loss ratio	Cell loss ratio	Cell loss ratio		Cell loss ratio
Conformance definitions	Generic cell rate algorithm	Generic cell rate algorithm	Generic cell rate algorithm		Dynamic GCRA
Feedback	Unspecified---				Specified

delay variation (CDV), maximum packet or cell transfer delay (maxCTD), and packet loss rate or cell loss ratio (CLR). For instance, Table 1.1 presents the overview of traffic and QoS parameters.

The goal of traffic management in an Internet or ATM network is to maximize network resource utilization while satisfying each individual user's QoS. For example, the offered loading to a network should be kept below a certain level to avoid congestion, which, in turn, causes throughput decrease and delay increase. Next we highlight a set of QoS control methods for traffic management.

1.2 QoS Control

Consider the data communication between two users in a network who are separated by a network of routers or packet switches, referred to as nodes for brevity. If the source has a message that is longer than the maximum packet size, it usually breaks the message up into packets and sends these packets, one at a time, to the network. Each packet contains a portion of the message plus some control information in the packet header. The control information, at a minimum, includes the routing information (IP destination address for the Internet, or virtual channel identifier for FR and ATM networks) that the network requires to be able to deliver the packet to the intended destination.

The packets are initially sent to the first-hop node to which the source end system attaches. As each packet arrives at this node, it stores the packet briefly in the input buffer of the corresponding node, determines the next hop of the route by searching the routing table (created through a routing protocol) with the routing control information in the packet header, and then moves the packet to the appropriate output buffer associated with that outgoing link. When the link is available, each packet is transmitted to the next node *en route* as rapidly as possible; this is, in effect, statistical time-division multiplexing. All of the packets eventually work their way through the network and are delivered to the intended destination.

Routing is essential to the operations of a packet switching network. Some sort of adaptive or dynamic routing technique is usually necessary. The routing decisions that are made change as conditions on the network change. For example, when a node or trunk fails, it can no longer be used as part of a route; when a portion of the network is heavily congested, it is desirable to route packets around the area of congestion rather than through it.

To maximize network resource (e.g., bandwidth and buffer) utilization while satisfying the individual user's QoS requirements, special QoS control mechanisms should be provided to prioritize access to resources at network nodes. For example, real-time queuing systems are the core of any implementation of QoS-controlled network services. The provision of a single class of QoS-controlled service requires the coordinated use of admission control, traffic access control, packet scheduling, and buffer management. Other techniques include flow and congestion control and QoS routing, as briefly explained in the following subsections. Each of them will be further explained with detailed references later in the book some for computer networks and other wireless for wireless networks.

1.2.1 Admission Control

Admission control limits the traffic on the queuing system by determining if an incoming request for a new user can be met without disrupting the service guarantees to established data flows. Basically, when a new request is received, the admission control (AC) or call admission control (CAC) in the case of an ATM network is executed to decide whether to accept or reject the request. The user provides a source traffic descriptor — the set of traffic parameters of the request or ATM source (e.g., PCR, SCR, MBS, and MCR), QoS requirements (such as delay, delay variation, and cell loss rate), and conformance definition (e.g., generic cell rate algorithm (GCRA) or dynamic GCRA (DGCRA) for ABR). The network then tries to see whether there are sufficient network resources (buffer and bandwidth) to meet the QoS requirement.

Given that most real-time queuing systems cannot provide QoS-controlled services at arbitrarily high loads, admission control determines when to generate a busy signal. Proper resources may be reserved for an accepted request based on its QoS specification, such as minimum bandwidth and buffer space. Chapter 4 introduces a recent algorithm for admission control for packet-switched networks.

1.2.2 Traffic Access Control

Traffic Access control (e.g., GCRA) shapes the behavior of data flows at the entry and at specific points within the network. Once the connection is accepted to the network, traffic to the network should comply with the traffic descriptor. If not, the excess traffic can either be dropped or tagged to a lower priority, or delayed (i.e., shaped). Different scheduling and admission control schemes have different limitations on the characteristics (e.g., rate, burstiness) of traffic that may enter the network. Traffic access control algorithms filter data flows to make them conform to the expectations of the scheduling algorithms. For details on traffic access control, refer to Stallings (2002).

1.2.3 Packet Scheduling

Packet scheduling specifies the queue service discipline at a node — that is, the order in which queued packets are actually transmitted. Because packets of many users may depart from the same outgoing node, packet scheduling also enforces a set of rules in sharing the link bandwidth. For example, if a user is given the highest priority to access the link, his or her packets can always go first, whereas packets from others will be delayed; and this privileged user can have his or her packets marked through some traffic access control algorithm when they enter the network.

In other words, packet scheduling prioritizes a user's traffic into two categories: delay priority for real-time traffic and loss priority for data-type traffic. One major concern is to ensure that the link bandwidth is fairly shared between connections and to protect the individual user's share from being corrupted by malicious users (i.e., put a firewall between connections). In this respect, FPQ is very promising. Chapter 7 introduces various kinds of scheduling algorithms for wireless networks targeted at different goals. Scheduling schemes for the wireless networks are quite similar to a wired network although unpredictable channel state becomes an issue in the design of the scheduling schemes for wireless networks.

There is a challenging design issue in that FPQ's packet reordering and queue management impose increased computational overhead and forwarding burden on networks with large volumes of data and very high-speed links. Chapter 7 presents the development and implementation of a weighted fair scheduler and how to assess the overhead due to the scheduler for wireless ad hoc networks.

1.2.4 Buffer Management

The problem of buffer sharing arises naturally in the design of high-speed communication devices such as packet switches, routers, and multiplexers, where several flows of packets may share a common pool of buffers. Buffer management sets the buffer-sharing policy and decides which packet should be discarded when the buffer overflows. Thus, the design of buffer-sharing strategies is also very critical to the performance of the networks. Because there are variable-length packets in routers and switches, the per-time slot processing imposes a difficulty in handling large volumes of data at high-speed links for both buffer management and PFQ, as mentioned earlier. Buffer management is very critical even for congestion control. For details refer to Stallings (2002).

1.2.5 Flow and Congestion Control

In all networks, there are scenarios where the externally offered load is higher than can be handled by the network. If no measures are taken to limit the traffic entering into the network, queue sizes at bottleneck links will grow fast and packet delays will increase. Eventually, the buffer space may be exhausted, and then some of the incoming packets are discarded, possibly violating maximum-delay loss specifications. Flow control and congestion control are necessary to regulate the packet population within the network. Flow control is also sometimes necessary between two users for speed matching, that is, for ensuring that a fast transmitter does not overwhelm a slow receiver with more packets than the latter can handle. Chapter 3 presents the schemes for ATM and Internet, whereas Chapter 9 details the congestion control for wireless networks.

1.2.6 QoS Routing

The current routing protocols used in IP networks are typically transparent to any QoS requirements that different packets or flows may have. As a result, routing decisions are made by neglecting the resource availability and requirements. This means that flows are often routed over paths that are unable to support their requirements although alternate paths with sufficient resources are available. This may result in significant deterioration in performance, such as high call-blocking probability in the case of ATM.

To meet the QoS requirements of the applications and improve the network performance, strict resource constraints may have to be imposed on the paths being used. QoS routing refers to a set of routing algorithms that can identify paths that have sufficient residual (unused) resources to satisfy the QoS constraints of a given connection (flow). Such a path is called a *feasible path*. In addition, most QoS routing algorithms also consider the optimization of resource utilization measured by metrics, such as delay, hop count, reliability, and bandwidth. Further details are provided in Chapter 8 for wireless ad hoc and sensor networks. Next we present an overview of wireless networking.

1.3 Overview of Wireless Networking

Wireless networking in recent years is becoming an integral part of residential, commercial, and military computing applications. The elimination of unsightly and cumbersome wiring, and the increase in mobility made possible by wireless networks are only some of the advantages that have led to the widespread acceptance and popularity of such networks. The rapid worldwide growth in cellular telephone industry and the number of subscribers demonstrate that wireless communication is a robust and viable data and voice transport mechanism.

Recent developments in ad hoc wireless networking have eliminated the requirement of fixed infrastructure (central base station as required in cellular networking) for communication between users in a network and expanded the horizon of wireless networking. These networks termed as mobile ad hoc networks (MANET) are a collection of autonomous terminals that communicate with each other by forming a multihop radio network and maintaining connectivity in a decentralized manner. MANET and, in particular, wireless sensor networks (WSNs) are finding increasing applications in communication between soldiers in a battlefield, emergency-relief-personnel coordinating efforts, earthquake aftermath, natural disaster relief, wired homes, and in today's highly mobile business environment.

Wireless communication is much more difficult to achieve than wired communication because the surrounding environment interacts with the signal, blocking signal paths while introducing noise and echoes. Consequently, poor quality connections are observed as compared to wired connections: lower bandwidth, high error rates, and more frequent spurious disconnections. Wireless communications can also be degraded because of mobility and inefficient battery power control. Users may enter regions of high interference or out-step the coverage of network transceivers. Unlike typical wired networks, the number of devices in a cell varies dynamically, and large concentrations of mobile users, such as at conventions and public events, may overload network capacity. MANET and WSNs offer many challenging problems to solve, such as routing protocols in the presence of topology changes, QoS constraints, and so on. Separate standards and protocols are being developed for cellular, ad hoc wireless and sensor networks.

1.3.1   Cellular Wireless Networks

Mobile radio systems were designed with the main objective of providing a large coverage area to users by using a single, high-powered transmitter (Rappaport 1999). This approach achieved good coverage, but it was impossible to increase the number of users in the system beyond a certain limit. Attempts to increase the number of users in the network resulted in greater interference due to simultaneous transmissions.

The cellular networking concept was a major breakthrough in solving the problem of spectral congestion and user capacity. The cellular concept replaced the existing high-power transmitters with many low-power trans-mitters (base stations serving smaller cells) to combat interference and to increase capacity within a limited spectrum. Each of the low-power trans-mitters served only a small service area, defined as a *cell*. Each cell was provided with different groups of channel frequencies so that all the avail-able bandwidth is assigned to a relatively small number of neighboring cells. Neighboring base stations or cells are assigned different groups of frequencies so that the interference between base stations is minimized.

Success of the Internet has initiated the search for technologies that can provide high-speed data access even to mobile users. The third-generation (3G) in cellular systems is aimed at realizing this vision. The 3G cellular systems are expected to provide the ability to communicate using the voice over Internet protocol (VoIP), unparalleled network capacity, data access with multiple users and high-speed data access even when sub-scribers are on the move.

Researchers are now looking at the technologies that may constitute a fourth-generation (4G) system. Different modulation technologies, such as orthogonal frequency division modulation (OFDM), which is used in other wireless systems, for instance WirelessMAN™ (IEEE 802.16) and digital video broadcast (DVB), are being investigated. Research is also taking place on mesh network technology, where each handset or terminal can be used as a repeater, with the links routed and maintained in a dynamic way. New standards and technologies are being implemented to replace the existing copper cable lines. Most of today's cellular com-munication systems utilize second-generation (2G) cellular standards commonly known as 2G technologies. A brief description of the wireless cellular standards is provided below. Later, the IEEE 802.11 standard for ad hoc networks is described in detail.

1.3.1.1   *Wireless Cellular Standards and Protocols*

First-generation cellular networks: The first-generation cellular systems relied on FDMA/FDD (frequency division multiple access/fre-quency division duplexing) multiple access techniques to effi-ciently utilize the available bandwidth resources. Analog FM, an analog modulation scheme, was used in transmission of signals between a user and a mobile.

Second-generation (2G) cellular networks: Present cellular networks conform to the 2G cellular standards. The 2G cellular standards employ digital modulation formats and TDMA/FDD (time division multiple access/frequency division duplexing) and CDMA/FDD (code division multiple access/frequency division duplexing) multiple access schemes. Of the many cellular standards proposed, four popular 2G standards have gained wide acceptance. These include three TDMA standards and one CDMA standard. A brief description of these standards is provided in the following paragraphs.

GSM (global system mobile): GSM supports eight time-slotted users for each 200-kHz radio channel and has been deployed widely in the cellular and personal communication systems (PCS) bands by cellular service providers. It has been effectively deployed in Asia, Europe, Australia, South America, and in some parts of U.S.

Interim standard 136 (IS-136): This cellular networking standard is also commonly known as North American digital cellular (NADC) or U.S. digital cellular (USDC), which supports three time-slotted users for each 30-kHz radio channel and is a popular choice for carriers in North America, South America, and Australia.

Pacific digital cellular (PDC): It is a Japanese TDMA standard that is similar to IS-136 with more than 50 million users.

Interim standard 95 code division multiple access (IS-95): This popular 2G CDMA standard supports up to 64 users that are orthogonally coded or multiplexed. The signal generated from these 64 users is simultaneously transmitted each on a 1.25-MHz channel.

Mobile radio networks (2.5G): 2G cellular standards contain circuit-switched data modems that limit data users to a single circuit-switched data voice channel. Data transmissions in 2G are thus generally limited to the data throughput rate of an individual user. These systems support only single-user data rates of the order of 10 kbps.

The need for high-speed data access rates called for improvement in the 2G cellular system standard. Three different upgrade paths have been developed for GSM carriers and two of these solutions support IS-136. The three TDMA upgrade options available include: (1) high-speed circuit switched data (HSCSD), (2) general packet radio service (GPRS), and (3) enhanced data rates for GSM evolution (EDGE).

The upgrade options for 2G cellular networks have provided significant improvement in Internet access speed over GSM and IS-136 standards. HSCSD, the standard for 2.5G GSM, was able to provide a raw transmission rate of up to 57.6 kbps to individual users. GPRS and IS-136 standard was able to achieve data rates to the tune of 171.2 kbps. EDGE, the standard for 2.5GSM and IS-136, uses 8-PSK digital modulation in addition to the GSM's standard Global Mobile shift keying (GMSK) modulation, without any error protection. All 8 time slots of a GSM radio channel are dedicated to a single user and a raw peak throughput data rate of 547.2 kbps can be provided. By combining the capacity of different radio channels (e.g., using multicarrier transmissions), EDGE can provide up to several megabits per second of data throughput to the individual data user.

Third-generation (3G) wireless networks: Third-generation standards for wireless cellular networks promise unparalleled wireless access, multimegabit Internet access, communications using VoIP, voice-activated calls, and ubiquitous "always-on" access and many other distinguishable wireless access features. Several 3G standards, many of which provide backward compatibility, are being followed around the world. The various upgrade paths for the 2G technologies to the 3G standards are shown in Figure 1.1. Among the available 3G standards the following cellular standards have gained greater acceptance.

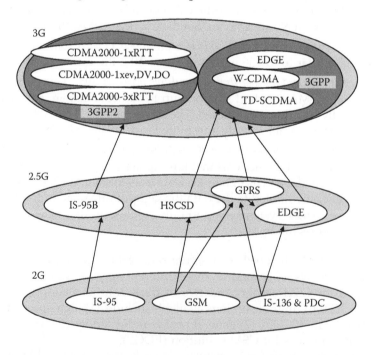

FIGURE 1.1
Upgrade paths for 2G technologies. (From Rappaport, T.S., *Wireless Communications: Principles and Practice*, Prentice Hall, Upper Saddle River, NJ, 1999. With permission.)

3G W-CDMA (UMTS): Universal mobile telecommunications systems (UMTS) is a visionary air interface standard that was developed by European Telecommunications Standards Institute (ETSI). UMTS was designed to provide a high-capacity upgrade path for GSM. Several other competing wideband CDMA (W-CDMA) proposals agreed to merge into a single W-CDMA standard. The resulting W-CDMA standard is now called UMTS.

UMTS: UMTS assures backward compatibility with the second generation GSM, IS-136 technologies. The 3G W-CDMA air interface standard had been designed for "always-on" packet-based wireless service. It supports packet data rates up to 2.048 Mbps per user. Future versions of W-CDMA will support stationary user data rates in excess of 8 Mbps.

3G CDMA2000: The cdma2000 vision is to provide a seamless and evolutionary high data-rate upgrade path for current users of 2G and 2.5G CDMA technology, using a building block approach that centers on the original 2G CDMA channel bandwidth of 1.25 MHz per radio channel. The cdma2000 standard is being developed under the auspices of working group 45 of the Telecommunications Industry Association (TIA) of the U.S. Cdma2000 1X supports an instantaneous data rate of up to 307 kbps for a user in packet mode, and yields typical throughput rates of up to 144 kbps per user, depending on the number of users, the velocity of a user, and the propagation conditions.

3G TD-SCDMA: The China Academy of Telecommunications Technology (CATT) and Siemens Corporation jointly submitted an IMT-2000 3G standard proposal in 1998, based on time division synchronous code division multiple access (TD-SCDMA). This proposal was accepted by ITU as one of the 3G options in 1999. TD-SCDMA relies on the existing core GSM infrastructure. It combines TDMA and TDD techniques to provide a data-only overlay in an existing GSM network. Up to 384 kbps of packet data is provided to data users in TD-SCDMA.

Fourth-generation (4G) cellular networks: Broadband applications in future wireless networks may require data rates that are many times the maximum data rate as promised for UMTS. Broadband services like wireless high-quality videoconferencing (up to 100 Mbps) or wireless virtual reality (up to 500 Mbps, when allowing free body movements) are envisioned. The goal of the next generation of wireless systems — the 4G — is to provide data rates yet higher than the ones of 3G while granting the same degree of user mobility.

TABLE 1.2

Summary of Cellular Standards

	EDGE	GERAN	W-CDMA	TD-CDMA	UMTS TD-CDMA	HSDPA	CDMA2000 1×RTT	1×EV-DO 1×EV-DV
Carrier bandwidth [MHz]	0.2		5		1.6	According to base technology	1.25	
Minimum spectrum required [MHz]	2 × 2.4 (due to BCCH for 4/12)		2 × 5	1 × 5	1 × 1.6		2 × 1.25	
Multiple access principle	Time and frequency		code	code and time			code	UL: code DL: code and time
Chip rate [Mcps]	Not applicable		3.84		1.28		1.2288	
Modulation	GMSK, 8-PSK		QPSK	QPSK, 8-PSK	QPSK, 8-PSK	QPSK, 16QAM	BPSK, QPSK	BPSK, QPSK, B-PSK, 16QAM
Peak user data rate [kbps][a]	473		384 [2048b]	2048	2048	10000c	307 [625d]	2400 3100
System asymmetry (UL:DL)	1:1		1:1	2:13–14:1	1:6–6:1	1:1–5:1	1:1	1:1–4:1
QoS classes	3 and 4		1 ... 4				None	3 classes of service only
Transport network	PCM (CS), FR (PO)		PCM, FR, ATM for both CS and PO service domains				Sonet for CS domain, IP-network (PPP and SDLC) for PO domain IS-41, IP protocols for data	
Mobility support	MAP		ATM					

[a] According to presently defined framing, coding and modulation schemes and assuming ideal radio conditions.
[b] For pico cells.
[c] Present assumptions.
[d] Second phase.

Source: Reprinted with permission from Siemens preprint.

1.3.1.2   Cellular Network Design

The process of frequency allocation to the base stations to decrease the interference between the neighboring base stations is called *frequency reuse*, and is explained in detail in the following section. Frequency reuse concept is explained by considering the cell shape to be hexagonal. The hexagonal cell is conceptual and is a simplistic model of the radio coverage for the base station. This shape, which has been universally adopted, permits easy and manageable analysis of a cellular system. The actual radio coverage of a cell, known as the *footprint*, differs from the assumed hexagonal structure and can actually be determined from field measurements and propagation prediction models. The concept of frequency reuse is explained in the next paragraph.

Frequency Reuse

Intelligent allocation and reuse of channels throughout a coverage region is important for a cellular radio system. The design process of selecting and allocating channel groups for all of the cellular base stations within a system is called *frequency reuse*. The concept of cellular frequency reuse is clearly illustrated in Figure 1.2. In the figure, the seven shaded cells named A to G form a cell cluster. This cell cluster is repeated over the entire coverage area. Total available channels are equally distributed among the cells in a cell cluster. The cells with same name signify that they use the same set of channels.

The total number of cells in a cluster defined as cluster size is equal to 7 for the scenario shown in Figure 1.2. Because each cell contains one-seventh of the total number of available channels, the frequency reuse factor is 1/7. The figure clearly shows the frequency reuse plan overlaid

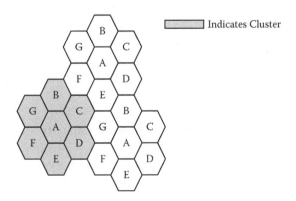

FIGURE 1.2
Illustration of cellular frequency reuse. (From Rappaport, T.S., *Wireless Communications: Principles and Practice*, Prentice Hall, Upper Saddle River, NJ, 1999. With permission.)

upon the map, which indicates the different frequency channels used. Frequency channels are assigned to the cells in the entire coverage area in such a way that the interference between the cells using same set of frequencies is reduced to a minimum.

In the cellular model, when considering the model coverage areas as hexagonal, base station transmitters are assumed to be either in the center of the cell or on three of the six-cell vertices (edge-excited cells). The frequency reuse concept can be explained below. The cellular system is considered to have total of S duplex channels available for use. Each cell is assumed to be allocated k channels ($k < S$). If the S channels are divided among N cells equally, then the total number of available radio channels can be expressed as (Rappaport 1999)

$$S = kN \qquad (1.1)$$

The N cells, which collectively use the complete set of available frequencies is called a *cluster*. If a cluster is repeated M times within the system, the total number of duplex channels, C, can be used as a measure of network capacity and it is given by

$$C = MkN = MS \qquad (1.2)$$

From Equation 1.2, we see that the capacity of a cellular network is directly proportional to the number of times a cluster is replicated in a fixed service area. The factor N is called the cluster size. If the cluster size, N defined in terms of number of cells per cluster, is reduced while the cell size is kept constant, more clusters are required to cover a given area, and hence more capacity is achieved. A larger cluster size causes the ratio between the cell radius and the distance between cochannel cells to decrease, leading to weaker cochannel interference. Conversely, a small cluster size indicates that cochannel cells are located much closer together. The frequency reuse factor of a cellular system is given by $1/N$, because each cell within a cluster is only assigned $1/N$ of the total available channels in the system.

1.3.1.3   Effects of Interference on Network Capacity

Interference is the major limiting factor in the performance of cellular radio systems. Sources of interference include another mobile in the same cell, a call in progress in a neighboring cell, other base stations operating in the same frequency band, or any noncellular system that inadvertently leaks energy into the cellular frequency band. Interference on voice channels causes cross talk, where the subscriber hears interference in the background because of an undesired transmission. Interference is more severe in urban areas, owing to the greater RF noise floor and the large number of base stations and mobiles. Interference has been recognized as a major bottleneck in increasing capacity and is often responsible for dropped calls. The two major types of system-generated cellular interference are cochannel interference and adjacent channel interference. Interfering signals often generated within the cellular network are difficult to control in practice. Even more difficult to control is interference due to out-of-band users, which arises without warning owing to front end overload of subscriber equipment or intermittent intermodulation products.

1.3.1.4 Cochannel Interference and Network Capacity

The concept of frequency reuse assigns the same set of frequencies to several cells in a given coverage area to increase the capacity of the cellular network. These cells, which share a common set of frequencies, are called *cochannel cells*, and the interference between signals from these cells is known as *cochannel interference*. Interference cannot simply be kept in check by raising the power of the transmitter, as is the case with thermal noise. The increase in power increases the interference to neighboring cells, decreasing the system capacity. Interference can only be reduced by physical separation of cochannel cells by a minimum distance.

When the size of each cell is considered to be equal and the base stations are considered to transmit with the same power, the cochannel interference ratio is independent of the transmitted power and becomes a function of the radius of the cell (R) and the distance between the centers of the nearest cochannel cells (D). By increasing the ratio of D/R, the spatial separation between cochannel cells relative to the coverage distance of a cell is increased. Thus, interference is reduced from improved isolation of RF energy from the cochannel cell. The parameter Q, called the cochannel reuse ratio, is related to the cluster size. For a hexagonal geometry (Rappaport 1999),

$$Q = \frac{D}{R} = \sqrt{3N} \tag{1.3}$$

A small value of Q provides larger capacity because the cluster size N is small, whereas a large value of Q improves the transmission quality due to a smaller level of cochannel interference. A trade-off must be made between these two objectives in actual cellular design. Let i_0 be the number of cochannel interfering cells. Then, the signal-to-interference ratio (SIR) or referred as signal-to-noise ratio (SNR) for a mobile receiver, which monitors a forward channel, can be expressed as

$$\frac{S}{I} = \frac{S}{\sum_{i=1}^{i_0} I_i}, \tag{1.4}$$

where S is the desired signal power from the desired base station and I_i is the interference power caused by the ith interfering cochannel cell base station. If the signal levels of cochannel cells are known, the SIR ratio for the forward link can be found using Equation 1.4.

Propagation measurements in a mobile radio channel show that the average received signal strength at any point decays as a power law of the distance of separation between a transmitter and receiver. The average received power P_r at a distance, d, from the transmitting antenna is approximated by (Rappaport 1999)

$$P_r = P_0 \left(\frac{d}{d_0} \right)^{-n},$$ (1.5)

or

$$P_r = P_0(dBm) - 10n \log \left(\frac{d}{d_0} \right),$$ (1.6)

where P_0 is the power received at a close-in reference point in the far region of the antenna at a small distance d_0 from the transmitting antenna, and n is the path loss exponent. Now, consider the forward link where the desired signal is the serving base station and the interference is due to cochannel base stations. If D_i is the distance of the ith interferer from the mobile, the received power at a given mobile due to the ith interfering cell will be proportional to $(D_i)^{-n}$. The path loss exponent typically ranges between two and four in urban cellular systems. When the transmit power of each base station is equal and the path loss exponent is the same throughout the coverage area, SIR for a mobile can be approximated as

$$\frac{S}{I} = \frac{R^{-n}}{\sum_{i=1}^{i_0} (D_i)^{-n}}.$$ (1.7)

1.3.1.5 Capacity of Cellular CDMA Network

In TDMA and FDMA systems, the capacity of the network is bandwidth-limited. But in the case of cellular systems, as described in the preceding sections, their capacity is interference-limited. It can be inferred that when the number of users decreases, the link performance increases. Several ways have been proposed to reduce the interference in the channel. They are (Rappaport 1999):

Multisectorized antenna: The directional antennas receive signals from only a fraction of the current users, thus leading to the reduction of interference.

Discontinuous transmission mode: Another way of increasing CDMA capacity is to operate the antenna in discontinuous transmission (DTX) mode. In DTX, the transmitter is turned off during the periods of silence in speech. It has been observed that voice signals have a duty factor of about 3/8 in landline networks and 1/2 for mobile systems, where background noise and vibration can trigger voice activity detectors. Thus, the average capacity of a CDMA system can be increased by a factor inversely proportional to the duty factor.

Effective power control schemes: An effective power control scheme would choose an appropriate power value that causes decreased interference but still maintains the power required for transmission. A good power control scheme would help accommodate an increased number of users in the network, thereby increasing system capacity. Higher spatial reuse can be obtained because the cochannel interference is reduced to a minimum with efficient power control.

1.3.1.6 Evaluation of Capacity of a Cellular CDMA Network

The cellular network consists of a large number of mobile users communicating with a base station within a cell. The cell-site transmitter consists of a linear combiner, which adds the spread signals of the individual users and also uses a weighting factor for each signal for forward-link power control purposes. For a single-cell system under consideration, these weighting factors can be assumed to be equal. A pilot signal is also included in the cell-site transmitter and is used by each mobile to set its own power control for the reverse link. For a single-cell system with power control, all the signals on the reverse channel are received at the same power level at the base station. Let the number of users be N. Then, each demodulator at the cell site receives a composite waveform containing the desired signal of power S and $(N - 1)$ interfering users, each of which has power, S. Thus the SNR is

$$SNR = \frac{S}{(N-1)S} = \frac{1}{N-1} \qquad (1.8)$$

In addition to SNR, bit energy-to-noise ratio is an important parameter in communication systems. It is obtained by dividing the signal power by the baseband information bit rate, R, and the interference power by the total RF bandwidth, W. The SNR at the base station receiver can be represented in terms of $\frac{E_b}{N_0}$ given by (Rappaport 1999)

$$\frac{E_b}{N_0} = \frac{S/R}{(N-1)(S/W)} = \frac{W/R}{N-1} \qquad (1.9)$$

Equation 1.9 does not take into account the background thermal noise, and spread bandwidth. To take this noise into consideration, $\frac{E_b}{N_0}$ can be expressed as

$$\frac{E_b}{N_0} = \frac{W/R}{(N-1)+(\eta/S)} \qquad (1.10)$$

where η is the background thermal noise. The number of users that can access the system is thus given by simplifying Equation 1.10 as

$$\frac{E_b}{N_0} = 1 + \frac{W/R}{N-1} - (\eta/S) \qquad (1.11)$$

where W/R is called the processing gain. The background noise determines the cell radius for a given transmitter power.

1.3.2   Channel Assignment

Several channel assignment strategies have been developed to achieve certain objectives like increased capacity and channel reuse. In a fixed channel assignment strategy, each cell is allocated a predetermined set of channels. A new user requesting to call is provided with a channel in the set of unused channels available. If no unused channel is available, the call may be blocked. To minimize call blocking, borrowing strategies are used to assign channels, if available, from neighboring cells to serve the new user. These borrowing processes are supervised by a mobile switching center (MSC). Another kind of channel assignment is dynamic channel assignment. Here a serving base station requests a channel from MSC every time a call is made. The channel is then allocated depending upon several constraints which increase the channel reuse.

1.3.3   Handoff Strategies

When a mobile moves into a different cell while a conversation is in progress, the MSC automatically transfers the call to a new channel belonging to the new base station. This handoff operation requires the knowledge for identifying the new base station and the voice and control signal channels that need to be associated with the new base station.

There are two kinds of handoff strategies that are usually employed. They are hard handoff and soft handoff. In a hard handoff procedure, the MSC assigns different radio channels to service the user in the event of handoff. The IS-95 CDMA system provides a more efficient handoff, called soft handoff, which cannot be provided with other wireless systems. In soft handoff, there will not be any physical change in the assigned channel, but a different base station handles the radio communication task. An MSC actually decides the servicing base station depending upon the received signals from a user at several neighboring base stations. Soft handoff effectively reduces the inadvertent dropping of calls during handoff.

1.3.4 Near–Far Problem

All the signals in CDMA systems are transmitted on the same frequency band at the same time. When all the mobiles are transmitting to the base station at the same power level, then the power of a nearby (unwanted) mobile arriving at the listening base station overwhelms the signal from a distant (wanted) mobile. This is the primary hurdle in implementing the CDMA cellular systems and is defined as the near–far problem (Mohammed 1993). Efficient power control schemes are useful to combat this near–far problem.

1.3.5 CDMA Power Control

The limit on the system performance and the number of simultaneous users available with CDMA is a function of the system's ability to overcome the near–far problem. To achieve the maximum number of simultaneous users, the transmitter power of each mobile should be controlled such that its signal arrives at the cell site with the minimum required SIR.

If the signal power of all mobile transmitters within an area covered by a cell site are controlled, then the total signal power received at the cell site from all mobiles will be equal to the average received power times the number of mobiles operating in the region of coverage. A trade-off must be made; if a strong signal from a mobile user arrives at the cell site along with a weak signal from another user, the weak user will be dropped. If the received power from a mobile user is significantly high, the performance may be acceptable, but it will add undesired interference to all other users in the cell, wasting energy and causing a drop in network capacity. Efficient power control schemes can be employed to satisfy this trade-off. Chapter 5 presents an overview of power control in cellular networks and describes in detail how to design a distributed power control (DPC) scheme. Next we provide an overview of mobile ad hoc networks.

1.4   Mobile Ad hoc Networks (MANET)

The major drawback for cellular networks is the need for a centralized infrastructure. Recent technological advancements enable portable computers to be equipped with wireless interfaces, allowing networked communication even while mobile. Wireless networking greatly enhances the utility of carrying a computing device. It provides mobile users with versatile and flexible communication between people and continuous access to networked services with much more flexibility than cellular phones or pagers.

A MANET is an autonomous collection of mobile users communicating over a relatively bandwidth-constrained wireless link with limited battery power in highly dynamic environments. The network topology, due to the mobility in the network is, in general, dynamic and may change rapidly and unpredictably over time. Hence, the connectivity among the nodes may vary with time because of node departures, new node arrivals, and the possibility of having mobile nodes. To maintain communication between the nodes in the network, each node in a wireless ad hoc network functions as a transmitter, host, and a router. The management and control functions of the network are also distributed among the nodes. Moreover, as the network is highly decentralized, all network activity, including discovering the topology, transmitting information, and efficient use of the battery power, must be executed by the nodes themselves.

As the users in the MANET communicate over wireless links, they have to contend with the effects of radio communication, such as noise, fading, shadowing, and interference. Regardless of the application, a MANET needs efficient distributed algorithms to determine network organization, link scheduling, power control, and routing.

A MANET is formed by a cluster of mobile hosts and can be rapidly deployed without any established infrastructure or centralized administration. Because of the transmission range constraint of transceivers, two mobile hosts can communicate with each other either directly if they are close enough (peer-to-peer) or (multihop) indirectly by having other intermediate mobile hosts relay their packets. The combination of networking and mobility will engender new services, such as collaborative software to support impromptu meetings, self-adjusting lighting and heating, natural disaster relief operations, and navigation software to guide users in unfamiliar places and on tours. Protocol standards that were developed for cellular networks are no longer useful for ad hoc networks because of lack of centralized infrastructure. The standards for ad hoc networks, which are different from those for cellular networks, are presented in the following subsection.

1.4.1 IEEE 802.11 Standard

The IEEE 802.11 standard (IEEE standard 1999) is proposed to develop a medium access control (MAC) and physical layer (PHY) specification for wireless connectivity for fixed, portable, and moving users within a local area.

This standard also offers regulatory bodies a means of standardizing access to one or more frequency bands for the purpose of local area communication. This standard has great significance because of the following features:

1. Describes the functions and services required by an IEEE 802.11 compliant device to operate within ad hoc and infrastructure networks as well as the aspects of user mobility (transition) within the network

2. Defines the MAC procedures to support the asynchronous MAC service data unit (MSDU) delivery services

3. Defines several physical layer signaling techniques and interface functions that are controlled by the IEEE 802.11 MAC

4. Permits the operation of an IEEE 802.11 conformant device within a wireless LAN that may coexist with multiple overlapping IEEE 802.11 wireless LANs

5. Describes the requirements and procedures to provide privacy of user information being transferred over the wireless medium (WM) and authentication of IEEE 802.11 conformant devices

1.4.2 IEEE 802.11 Physical Layer Specifications

IEEE, in May 1997, published the initial standard, "Wireless LAN Medium Access Control (MAC) and Physical Layer (PHY) Specifications" also known as IEEE 802.11. The standard provides three physical layer specifications for radio, operating in the 2400- to 2483.5-MHz band, 902 to 928 MHz, 5.725- to 5.85-GHz regions. The standard (IEEE Standard 1999) allowed three different transmissions.

Frequency hopping spread spectrum radio physical layer: This physical layer provides for 1 Mbps (with 2 Mbps optional) operation. The 1-Mbps version uses 2-level Gaussian frequency shift keying (GFSK) modulation and the 2-Mbps version uses 4-level GFSK.

Direct sequence spread spectrum radio physical layer: This physical layer provides both 1- and 2-Mbps operation. The 1-Mbps version uses differential binary phase shift keying (DBPSK) and the 2-Mbps version uses differential quadrature phase shift keying (DQPSK).

Infrared physical layer: This PHY provides 1 Mbps with optional 2 Mbps. The 1-Mbps version uses pulse position modulation (PPM) with 16 positions (16-PPM) and the 2-Mbps version uses 4-PPM.

1.4.3   IEEE 802.11 Versions

IEEE 802.11b: An appendix to the existing 802.11 protocol "Higher Speed Physical Layer Extension in the 2.4GHz Band" was published in 1999 (IEEE 802.11b 1999). It is known as 802.11b standard. This standard working in 2.4-GHz frequency provides 5.5-Mbps and 11-Mbps data rate, in addition to the existing 1 Mbps and 2 Mbps provided. This extension uses 8-chip complementary code keying (CCK) as the modulation scheme. The IEEE ratified the IEEE 802.11 specification in 1997 as a standard for WLAN. Current versions of 802.11 (i.e., 802.11b) support transmission up to 11 Mbps. Wi-Fi, as it is known, is useful for fast and easy networking of PCs, printers, and other devices in a local environment, for example, the home. Current PCs and laptops as purchased have the hardware to support Wi-Fi. Purchasing and installing a Wi-Fi router and receivers is within the budget and capability of home PC enthusiasts.

IEEE 802.11a: Another appendix to the existing 802.11 (IEEE 802.11a standard 1999) known as IEEE 802.11a was published in 1999. The standard was called "High-Speed Physical Layer in the 5GHz Band." This standard allows up to 54 Mbps using 5-GHz frequency range. At these high speeds, multipath delay is a major problem. A new modulation method, called coded-orthogonal frequency-division multiplexing (COFDM), must be used to overcome this multipath delay problem.

IEEE 802.11g: In February 2003, IEEE 802.11g, 54-Mbps extension to 802.11b wireless LANs, gained working group approval. IEEE 802.11g, which is called "Wireless LAN Medium Access Control (MAC) and Physical Layer (PHY) Specifications: Higher Speed Physical Layer (PHY) Extension to IEEE 802.11b," will boost wireless LAN speed to 54 Mbps by using OFDM (orthogonal frequency division multiplexing). The IEEE 802.11g specification is backward compatible with the widely deployed IEEE 802.11b standard.

1.4.4   IEEE 802.11 Network Types

IEEE 802.11 recognizes two types of networks. They are ad hoc networks and extended set networks. An ad hoc network doesn't require any existing

TABLE 1.3
Wireless LAN Technical Specifications

Technology	Multiple Access Techniques	Modulation Technique	User Data Rate	Key Performance Parameters and Target Values			Frequency Band
				Typical BER	Typical Delay	Connectivity	
IEEE 802.11	DSSS, FHSS	2GFDK, 4GFSK, DBPSK, DQPSK	Up to 2 Mbps Type 1 Mbps	$\approx 10^5$	10–50 msec	Connectionless	2.4-GHz unlicensed band (ISM band)
IEEE 802.11b	CCK-DSSS	DBPSK, DQPSK	Up to 11 Mbps Type 5 Mbps	Same as wired IP or ATM			
IEEE 802.11g	CCK-DSSS, OFDM	DBPSK, DQPSK 16-QAM, 64-QAM	Up to 54 Mbps Type 25 Mbps				
IEEE 802.11a ETSI HiperLAN2	OFDM			$< 5 \times 10^{14}$	<5 msec	Connectionless and connection oriented	5-GHz unlicensed band (RLAN band)

infrastructure. Each node in the network must act both as a host and a router in addition to a transmitter. Every node in the network should forward packets that are intended for other nodes that are not in direct transmission range.

1.4.4.1   Ad Hoc Network

According to the IEEE 802.11 standard, an *ad hoc network* is defined as a network composed solely of stations within mutual communication range of each other via the WM. An ad hoc network is typically created in a spontaneous manner. The principal distinguishing characteristic of an ad hoc network is its limited temporal and spatial extent. These limitations allow the act of creating and dissolving the ad hoc network to be sufficiently straightforward and convenient so as to be achievable by nontechnical users of the network facilities; i.e., no specialized technical skills are required and little or no investment of time or additional resources is required beyond the stations that are to participate in the ad hoc network. The term ad hoc is often used as a slang for an independent basic service set (IBSS).

The extended set network on the other hand relies on existing infrastructure for the nodes to communicate with other nodes including the wired network. The network infrastructure consists of one or more access points (AP), which provide the routing for the nodes in the network.

Similar to the case of cellular networks, transmitter power of a node in an ad hoc network affects the channel capacity. Hence, power control schemes are necessary. Efficient power control in ad hoc wireless networks, which can increase the channel reuse and the number of nodes in the network, is considered in this book. To understand the operation of power control, knowledge of how the nodes communicate in the network is required. In the next section, network simulator (NS) (Fall and Varadhan 2002) implementation and MAC protocol for ad hoc networks are given.

1.4.5   IEEE 802.11 MAC Protocol

IEEE 802.11 specifications provide information about two MAC protocols, PCF (point coordination function) and DCF (distributed coordination function). PCF is a centralized scheme and DCF is a distributed scheme. The specifications of DCF are considered in this thesis. Important definitions required to understand the specifications are given next.

> *Transmission range:* This is the distance range around the transmitter within which any user in the network can receive, and correctly decode, packets sent by the transmitter. When highest transmitter power is used, the transmission range is 250 m.

Carrier sensing range: This is the distance range around the transmitter within which any user in the network can sense the information sent by the transmitter. When highest transmitter power is used, the carrier sensing range is 500 m.

Carrier sensing zone: This is the distance range around the transmitter within which any user in the network can sense the information sent by the transmitter but cannot decode it correctly. When highest transmitter power is used, the carrier sensing zone is between 250 and 500 m.

Figure 1.3 shows transmission range, carrier sensing range, and carrier sensing zone for the user C. Here, users B and D are in the transmission range of C as per the definition of transmission range. Users A and E are in the carrier sensing zone. All the nodes shown in the figure are in the carrier sensing range.

The DCF in IEEE 802.11 works on the principle of carrier sense multiple access with collision avoidance (CSMA/CD). Carrier sensing is performed using physical carrier sensing through air as an interface, and also by virtual carrier sensing. Virtual carrier sensing uses the duration of the packet transmission, which is included in the header of RTS, CTS, and DATA frames. The duration included in each of these frames can be used to infer the time when the source node would receive an ACK frame from the transmitter. The duration in RTS signifies the time needed for CTS,

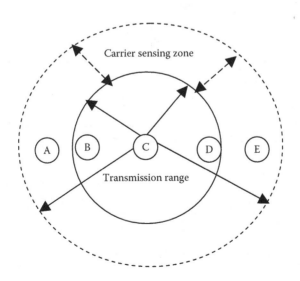

FIGURE 1.3
An IEEE 802.11 wireless LAN.

DATA, and ACK transmissions. Similar is the case with duration fields in CTS and DATA frames.

The users in the IEEE 802.11 network maintain a NAV (network allocation vector), which indicates the remaining time of the ongoing transmission sessions. The NAVs are updated using the duration field information from RTS, CTS, and DATA frames. The channel is considered busy if either physical or virtual carrier sensing indicates that the channel is busy.

Figure 1.4 shows the way users in the network in the carrier sensing range adjust their NAV during the RTS, CTS, DATA-ACK exchange. There are four IFS (interframe space) and they are: SIFS (short interframe space), DIFS (DCF interframe space), PIFS (PCF interframe space), and EIFS (extended interframe space). The IFS provides priority levels for accessing the channel. The SIFS is used after RTS, CTS, and DATA frames to give the highest priority to CTS, DATA, ACK, respectively. In DCF, when the channel is idle, a node waits for the DIFS duration before transmitting any packet.

As shown in Figure 1.4, nodes in the transmission range correctly set their NAVs when receiving RTS or CTS. However, because nodes in the carrier sensing zone cannot decode the packet, they do not know the duration of the packet transmission. To prevent a collision with the ACK

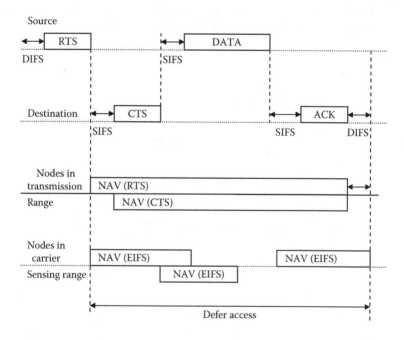

FIGURE 1.4
NAVs during RTS-CTS-DATA-ACK transmission.

reception at the source node, when nodes detect a transmission and cannot decode it, they set their NAVs for the EIFS duration. The main purpose of the EIFS is to provide enough time for a transmitting user to receive the ACK frame, so that the duration of EIFS is longer than that of an ACK transmission. As per the standard IEEE 802.11, the EIFS is obtained using the SIFS, the DIFS, and the length of time to transmit an ACK frame at the physical layer's lowest mandatory rate, as in the following equation EIFS = SIFS + DIFS + [(8 × ACK size) + Preamble length + PLCP header length]/Bit rate, where ACK size is the length (in bytes) of an ACK frame, and Bit rate is the physical layer's lowest mandatory rate. Preamble length is 144 b and PCLP header length is 48 b.

1.4.6 Need for Power Sensitive Schemes and Protocols

Wireless cellular, ad hoc, and sensor networks operate with limited bandwidth and battery resources. Unlike computer networks, wireless networks are bandwidth and interference limited. Interference due to the users in the network decreases the network's capacity. Hence, efficient ways to combat interference to increase the capacity and spatial reuse of the network are of great importance. A solution to this problem is to efficiently control signals from randomly distributed users in the network, which would increase the network capacity and battery lifetime. Therefore, centralized power control schemes (Dontula and Jagannathan 2004) were proposed in the case of cellular networks. A major disadvantage of these schemes is that they are computationally expensive, consuming large base station computation resources. On the other hand, DPC schemes are a more realistic approach for minimizing the interference. DPC schemes adjust the power levels of the users in the network just by sensing the local environment. These DPC schemes (Dontula and Jagannathan 2004) are straightforward, consume minimal mobile user resources, and minimize the transmission of control packet information.

In ad hoc wireless networks, where there is no centralized governing agent, a DPC scheme is a natural choice. Determining the power with which a mobile user in the network should transmit, by keeping the interference in check, in a highly dynamic environment is a challenging problem. In a military environment, where the networks are required to maintain a low probability of intercept and a low probability of detection, as they operate in remote and unstructured environments, nodes need to radiate as little power as necessary and transmit as infrequently as possible, thus decreasing the probability of detection or interception. A lapse in any of these requirements may degrade the performance and dependability of the network.

With the advent of ad hoc networks of geographically distributed sensors in remote site environments (e.g., sensors dropped from aircraft for personnel/vehicle surveillance), there is a focus on increasing the lifetimes of sensor nodes through power generation, conservation, and management. Current research is in designing small MEMS (microelectromechanical systems), RF components for transceivers, including capacitors, inductors, etc. The limiting factor now is in fabricating micro-sized inductors. Another thrust is in designing MEMS power generators using technologies including solar, vibration (electromagnetic and electrostatic), thermal, etc.

Meanwhile, software power management techniques can greatly decrease the power consumed by RF sensor nodes. TDMA is especially useful for power conservation, because a node can power down or "sleep" between its assigned time slots, waking up in time to receive and transmit messages.

The required transmission power increases as the square of the distance between source and destination. Therefore, multiple short message transmission hops require less power than one long hop. In fact, if the distance between source and destination is R, the power required for single-hop transmission is proportional to R^2. If nodes between source and destination are taken advantage of to transmit n short hops instead, the power required by each node is proportional to R^2/n^2. This is a strong argument in favor of distributed networks with multiple nodes, that is, nets of the mesh variety.

A current topic of research is active transmission power control, whereby each node cooperates with all other nodes in selecting its individual transmission power level. This is a decentralized feedback control problem. Congestion is increased if any node uses too much power, but each node must select a large-enough transmission range, so that the network remains connected. For n nodes randomly distributed in a disk, the network is asymptotically connected with probability one, if the transmission range r of all nodes is selected using (Kumar 2001)

$$r \geq \sqrt{\log n + \gamma(n) \Big/ \pi n} \qquad (1.12)$$

where, $\gamma(n)$ is a function that goes to infinity as n becomes large.

1.4.7 Network Simulator

NS-2 (Fall and Varadhan 2002) is a discrete event simulator targeted at networking research. NS provides substantial support for simulation of wired and wireless networks, satellite networks, TCP, and routing. NS began as a variant of the REAL network simulator in 1989 and has evolved substantially over the past few years. In 1995, NS development was supported by DARPA through the VINT project at LBL, Xerox PARC, UCB, and USC/ISI. Currently, NS development is supported through DARPA with SAMAN and through NSF with CONSER, both in collaboration with other researchers, including ACIRI. NS has always included substantial contributions from other researchers, including wireless code from the UCB Daedelus and CMU Monarch projects and Sun Microsystems.

NS-2 provides a split-programming model. OTcl is interpreted and is used to define the composition of the objects in the simulation (nodes, links, etc.) to allow changing scenarios without having to recompile. C++ focuses on the mechanisms and internals of the objects and protocols to allow packet simulation to be efficient. This split-programming approach can benefit research productivity. Also, NS-2 can produce a detailed trace file and an animation file for each network simulation, which is very convenient for analyzing the behavior. It is open source and can be downloaded from the Internet. For additional functionality, existing protocols can be extended and new protocols can also be implemented. Complex traffic patterns, topologies, and dynamic events can be automatically generated to test the created network topologies. There are several related protocols for both wired and wireless communications that have been already implemented. Thus, NS-2 simulator was normally chosen to implement the computer network and ad hoc and sensor networking schemes and protocols.

1.4.8 Power Control Implementation in MAC 802.11 Using NS

Efficient control of transmission power would combat interference and increase channel reuse. At present, in IEEE 802.11 protocol, a transmitter is allowed to transmit only at a single power level. But, incorporation of the power control scheme requires the users in the network to transmit at different power levels. Hence, modifications are to be made in the 802.11 protocol to incorporate power control.

Modifications to the IEEE 802.11 protocol can be made such that when a receiver receives an RTS message, it will encode the ratio of the received signal strength of the RTS message to the interference felt by the receiver in the header of the CTS reply message. Similarly, when transmitting the DATA message, the transmitter will encode onto it the ratio with respect to the received CTS. Thus, during one RTS-CTS-DATA-ACK exchange, both the transmitter and the receiver inform each other about the quality of their transmitted signals. Both nodes can now alter their transmit power levels according to the power control algorithms for further communication between each other. The sequence of events that usually take place between a transmitter and receiver in a network is shown in the Figure 1.5.

The performance of a power control scheme is measured by its ability to preserve the accuracy of the encoded data. In mobile wireless networks,

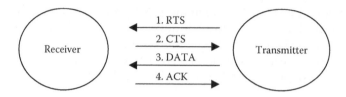

FIGURE 1.5
IEEE 802.11 signaling for addressed messages.

path loss, fading, and interference cause variations in the received SIR. Such variations also cause variations in the bit error rate (BER), because the lower the SIR or SNR, the more difficult it is for the receiver to decode the received signal. Hence, SIR is an important metric, which needs to be maintained even when there are fluctuations in the wireless network.

The block diagram representation of the DPC is illustrated in Figure 1.6. The receiver (as shown in the block diagram), after receiving the signal from the transmitter, measures the SIR value and compares it against the required target SIR threshold. The difference between the desired and the received signal SIR is sent to the power update block, which then calculates the optimal power level with which the transmitter has to send the next packet to maintain the required SIR. This power level is sent as feedback to the transmitter, which then transmits with the power level as requested by the receiver during the next packet transmission.

Chapter 5 and Chapter 6, respectively, present DPC schemes for cellular and ad hoc networks, which satisfy the stringent requirements placed by the wireless networking environment. Outage probability, total power consumed, and rate of convergence of the new links seeking admission into the network are taken as metrics to evaluate the proposed DPC scheme's performance and to compare various power control schemes. In fact, a few DPC schemes (Dontula and Jagannathan 2004) have been proposed in the past. Convergence of these schemes has not been proven

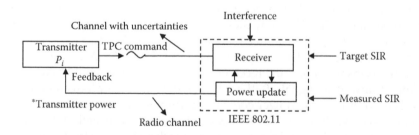

FIGURE 1.6
Block diagram representation of DPC.

mathematically, and the performance requirements, like low-outage probability and high spatial reuse, an important requirement in wireless networks, are not satisfied. A system theory approach for solving a networking problem has been taken advantage of in the proposed scheme.

When new users try to access the channel, active users may be inadvertently dropped because of fluctuation in maintaining the QoS metrics. An active user may be dropped even if there is a possibility of new users eventually getting accommodated at the steady state. Moreover, the admission of new users on to the wireless network is essential to maintain high system utilization and to ensure certain QoS metrics like low end-to-end delay and low packet losses in the network. An efficient power control scheme, given in Chapter 5 in this book, provides protection for links that are currently operational, by maintaining the QoS of the users at all times even as new users try to enter the network.

Constant topology changes, due to mobile users in the wireless network, require that the active wireless links are protected during a session in progress. Very few DPC schemes proposed in the past have considered addressing this issue. Link between the users in the network should be protected by satisfying minimum QoS requirements for example, mechanisms to maintain a minimum SNR when a call is initiated and is in progress. The proposed power control scheme is aimed at providing active link protection for the users in the network by satisfying the QoS requirements.

Modifications are needed for the existing protocols to accommodate the proposed power control schemes. Existing standards and protocols in place for ad hoc wireless networks do not allow users to transmit at different power levels. So, changes to the packet formats in control messages, and minor changes in the functionality in physical and MAC layers, will enable the implementation of the proposed DPC schemes with active link protection. Next, we present an overview of WSNs.

1.5   Wireless Sensor Networks

Smart environments represent the next evolutionary step in building, utilities, industrial, home, shipboard, and transportation system automation. Like any organism, the smart environment relies on real world sensory data, which comes from multiple sensors of different modalities in distributed locations. The smart environment needs information about its surroundings as well as about its internal workings.

Sensor networks are the key to gathering the information needed by smart environments, whether in buildings, utilities, industrial, home, shipboard, transportation systems, automation, or elsewhere. Recent terrorist and guerrilla warfare countermeasures require distributed networks of sensors that can be deployed using aircraft and have self-organizing capabilities. In such applications, running wires or cabling is usually impractical. A sensor network is required that is fast and easy to install and maintain.

The challenges in the hierarchy of detecting the relevant quantities, monitoring, collecting and aggregating the data, assessing and evaluating the information, formulating meaningful user displays, and performing decision-making and alarm functions are enormous. The information needed by smart environments is provided by distributed WSNs, which are responsible for sensing as well as for the first stages of the processing hierarchy.

The study of WSNs is challenging in that it requires an enormous breadth of knowledge from a variety of disciplines. In this chapter, we outline communication networks, WSNs and smart sensors, commercially available wireless sensor systems, self-organization, and finally, some concepts for home automation.

Routing tables for distributed networks increase exponentially as nodes are added. An $n \times m$ mesh network has nm links, and there are multiple paths from each source to each destination. Hierarchical network structures simplify routing, and also are amenable to distributed signal processing and decision making, because some processing can be done at each hierarchical layer. It is important to note that a fully connected network has NP-hard complexity, while imposing routing protocols by restricting the allowed paths to obtain a reentrant flow topology results in polynomial complexity. Such streamlined protocols are natural for hierarchical networks.

As nodes are added, the number of links increases exponentially. This makes for NP-complexity problems in routing and failure recovery. To simplify network structure, we can use hierarchical clustering techniques. The hierarchical structure must be consistent, that is, it must have the same structure at each level. Hierarchical structure is quite common in WSNs, where geographically located nodes are clustered, and a cluster head is elected during self-organization (see Chapter 8). The data from the nodes are aggregated at each cluster head and transmitted to a central base station over the multihop network constructed from different cluster heads.

There are many fundamental differences between ad hoc and sensor networks. Ad hoc networks may not have stringent memory, power and processing constraints as much as a sensor network. Also, WSN may not have an IP address, whereas an ad hoc network node has an IP address. Typically, an ad hoc network has mobility, whereas a WSN can have stationary or mobile nodes. A WSN has to transmit large quantities of data and, therefore, hierarchical structure is imposed and data aggregation is performed. By contrast, in MANET, data is not always abundant, and it is information that is transmitted. Moreover, the density of nodes in a WSN is significantly higher compared to an ad hoc network. Sensor nodes are prone to failure in comparison with an ad hoc network. Finally, a sensor node is more energy-constrained compared to an ad hoc network. Although traditional networks aim to achieve high QoS provisions, WSN protocols must focus primarily on power conservation. Therefore, WSNs must have an inbuilt power or energy monitor and smartness that will give an end user the option of prolonging network lifetime at the cost of lower throughput or higher transmission delay. Many research works in the literature, which are covered in the book, are engaged in developing schemes that fulfill these requirements.

WSNs are used to monitor a wide variety of ambient conditions that include temperature, humidity, vehicular movement, pressure, soil makeup, and so on, besides many military applications, such as battlefield surveillance, reconnaissance mission and terrain, ammunition targeting, battle damage assessment, nuclear, biological, and chemical attack detection as well as water quality monitoring etc. There are a number of home applications such as home automation and smart environment. Sensor nodes in a WSN can be used for continuous assessment of process/system health, event detection, location sensing, and control of actuators.

A WSN design is influenced by many factors that include hardware constraints, transmission media, power consumption, network topology, scalability, and fault tolerance. Though many researchers have addressed many of these factors, none of these studies has a full integrated view of all factors (Akyildiz et al. 2002).

Sensor nodes can fail or be blocked because of lack of power, causing problems to the functioning of the network. The reliability $R_i(k)$ or fault tolerance of a sensor node is modeled (Hoblos et al. 2000) using the Poisson distribution to capture the probability of not having a failure within the time interval $(0, k)$ as

$$R_i(k) = e^{-\lambda_i k} \tag{1.14}$$

where λ_i and k are the failure rate of sensor node i and the time instant, respectively. Protocols and algorithms have to be designed to address the level of fault tolerance required by the sensor networks.

A WSN is supposed to have several hundred or thousand sensor nodes. Therefore, the protocols and algorithms should be scalable, which is a major problem observed in ad hoc networks. The density can range from a few sensor nodes in a region, which can be less than 10 m in diameter. The density can be calculated according to (Bulusu et al. 2001) as

$$\gamma(R) = (N\pi R^2) / A \tag{1.15}$$

where N is the number of sensor nodes in a geographical region A, and R, the radio transmission range. The term $\gamma(R)$ provides the number of nodes within the transmission radius of each node in region A.

Topology maintenance in a WSN is a challenging task due to the sheer numbers of inaccessible and unattended sensor nodes, which are prone to failures. Due to their sheer numbers, it may not be possible to place the nodes according to a carefully engineered deployment plan; the schemes for initial deployment must eliminate the need for any pre-organization and preplanning and should promote self-organization and fault tolerance.

Because a WSN employs a multihop routing protocol for transmitting information at the cluster head level, communicating nodes are linked by a wireless medium formed by either radio, infrared, or optical media. Wireless media uses a standard ISM frequency band. Besides radio links, nodes use infrared communication because it is license-free and robust to interference from electrical devices.

A sensor node expends more energy in communication, which normally involves transmission of data and reception. In Shih et al. (2001), formulation of the radio power consumption (P_c) is

$$P_c = N_T(P_T(T_{on} + T_{st}) + P_{out}T_{on}) + N_R(P_R(R_{on} + R_{st}))$$ (1.16)

where P_T, P_R is the power consumed by the transmitter/receiver; P_{out} is the output power of the transmitter; T_{on}/R_{on} is the transmitter/receiver on time, T_{st}/R_{st} is the transmitter/receiver start-up time, and N_T/R is the number of times transmitter/receiver is switched on per unit time, which depends on the task and MAC scheme used. The term T_{on} can be further written as L/R, where L is the packet size and R is the data rate.

Energy expended in data processing is much less compared to data communication. Hence, data aggregation by local processing is very important to minimize power consumption. Research is being done to minimize energy expended during processing as well. Computing energy consumed, during data processing, is a difficult task because it is dependent upon the processor clock cycle, number of memory operations, and so on. However the power consumption in data processing is given by (Shih et al. 2001)

$$P_d = CV_{dd}^2 f + V_{dd}I_0 e^{\frac{V_{dd}}{\zeta' V_T}}$$ (1.17)

where C is the total switching capacitance, V_{dd} is the voltage swing, and f is the switching frequency.

Now the sensor network communication architecture is covered briefly. Data from the sensor node is routed to a base station via a multihop path. The protocol stack for communication used by the base station and all sensor nodes is depicted in Figure 1.7 (Akyildiz et al. 2002).

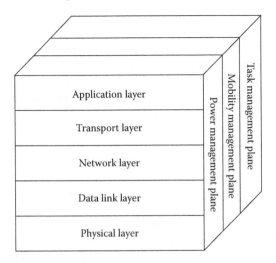

FIGURE 1.7
Sensor network protocol stack. (Reprinted from Akyildiz, I.F. et al., *Computer Networks*, Vol. 38, 393–422, 2002. With permission.)

WINS (Pottie and Kaiser 2000), smart dust motes (Kahn et al. 1999), $\mu AMPS$ (Akyildiz et al. 2002) use the protocol stack. This protocol stack combines power and routing awareness, integrates data with networking protocols, communicates power efficiently through the wireless medium even in the presence of mobility, and cooperating efforts among the sensor nodes.

Depending on the sensing tasks, a variety of application software can be built and used in the application layer. Common application management protocols include sensor management protocol (SMP), task assignment and data advertisement protocol (TADAP), and sensor query and dissemination protocol (SQDDP) (Akyildiz et al. 2002). Application level protocols perform administrative tasks, such as rules for data aggregation, clustering of nodes, exchanging location determination information, time synchronization of the sensor nodes, mobility, ON/OFF feature, querying the sensor node configuration, and authentication, key distribution, and security.

The transport layer helps maintain the flow of data, whereas the networking layer oversees the routing of data supplied by the transport layer. There are not many transport layer protocols for WSNs. Communication between base station and sensor nodes are purely through UDP type protocols because each sensor node has limited memory. Protocols such as TCP from computer networks cannot be used because sensor nodes do not have a global address. Instead, end-to-end schemes must consider attribute-based naming to indicate the destination. The power consumption, scalability, and data-centric routing makes sensor networks different and they demand new types of transport layer protocols.

The MAC protocol must be power-aware and able to minimize collisions due to the CSMA/CA paradigm with the neighbors' broadcast. Protocols for ad hoc networks cannot be ported to WSNs because, in WSN, power efficiency is critical. In cellular networks, a mobile node is a single hop away and QoS provisioning is given greatest importance for a MAC protocol whereas power consumption is given secondary importance. This makes MAC protocols for cellular networks to be impractical for WSNs. Besides being energy efficient, WSNs are data-centric, have attribute-based address and location awareness, aggregate data and, therefore, MAC protocols must work under these constraints. The MAC layer protocols must allow self-organization while fairly and efficiently sharing communication resources between sensor nodes. Chapter 8 presents the self-organization and routing protocols for WSNs. Bluetooth is closest peer to a WSN. Regardless of which type of medium access control scheme is used for WSNs, it certainly must support power saving modes for a sensor node. The most simple form of power saving is to turn the transceiver off when it is not required and turning it on only when desired.

The data link layer is responsible for error control in communication networks. Bit error rate (BER) is used as a performance parameter for assessing link reliability due to unpredictable wireless channel. A good choice of error correcting code such as forward error correction (FCC) or Hamming codes can result in several orders of magnitude reduction in BER without additional energy consumption and is therefore preferred in WSNs.

The physical layer addresses the need of simple but robust modulation, transmission and receiving techniques. Direct-sequence spread spectrum architecture, ultrawideband (UWB) and impulse radio (IR), have been proposed for WSN applications such as indoor location determination. Low transmission power and simple transceiver circuitry make UWB an attractive candidate for WSNs. Strategies to overcome unpredictable channels have to be dealt at this layer.

UWB radio is of great interest (recently) for communications in distributed sensor networks. This is because UWB is a short-range technology that can penetrate walls, it is suitable for multinode transmissions, and it has built in time-of-flight properties that make it very easy to measure ranges down to 1 cm with a range of 40 m. This means that the same medium, UWB, can be used for communications, localization, and target tracking in a distributed surveillance network. Moreover, UWB transceivers can be made very small and are amenable for MEMS technology; because PPM is used, no carrier is needed, meaning that antennas are not inductive. Also, the receiver is based on a rake detector and correlator bank so that no IF stage is needed.

UWB uses signals such as (Ray 2001)

$$s(t) = \sum_j w(t - jT_f - c_j T_c - \delta d_{[j/N_s]}) \qquad (1.18)$$

where $w(t)$ is the basic pulse of duration approximately 1 nsec, often a wavelet or a Gaussian monocycle, and T_f is the frame or pulse repetition time. In a multinode environment, catastrophic collisions are avoided by using a pseudorandom sequence c_j to shift pulses within the frame to different compartments, and the compartment size is T_c sec. One may have, for instance $T_f = 1$ μ sec and $T_c = 5$ nsec. Data is transmitted using digital PPM, where if the data bit is 0 the pulse is not shifted, and if the data bit is 1 the pulse is shifted by δ. The modulation shift is selected to make the correlation of $w(t)$ and $w(t - δ)$ as negative as possible. The meaning of $d_{[j/N_s]}$ is that the same data bit is transmitted Ns times, allowing for very reliable communications with low probability of error.

In addition, the power, mobility, and task management planes monitor the power, movement, and task distribution among the sensor nodes and help the sensor nodes coordinate the sensing aspect while lowering the overall power consumption. Many protocols and algorithms for WSN are designed by taking into account the physical layer requirements such as the microprocessor and receiver types. This is essential to minimize the power consumption even though incorporating the physical layer needs in the protocol design may result in a custom solution.

1.5.1 Closest Peers

The following are the closest peers to a WSN. They are:

Bluetooth was initiated in 1998 and standardized by the IEEE as Wireless Personal Area Network (WPAN) specification IEEE 802.15. Bluetooth is a short-range RF technology aimed at facilitating communication of electronic devices between each other and with the Internet, allowing for data synchronization that is transparent to the user. Supported devices include PCs, laptops, printers, joysticks, keyboards, mice, cell phones, PDAs, and consumer products. Mobile devices are also supported. Discovery protocols allow new devices to be hooked up easily to the network. Bluetooth uses the unlicensed 2.4-GHz band and can transmit data up to 1 Mbps, can penetrate solid nonmetal barriers, and has nominal range of 10 m that can be extended to 100 m. A master station can service up to seven simultaneous slave links. Forming a network of these networks, for example, a piconet, can allow one master to service up to 200 slaves. Currently, Bluetooth development kits can be purchased from a variety of suppliers, but the systems generally require a great deal of time, effort, and knowledge for programming and debugging. Forming piconets has not yet been streamlined and is unduly difficult.

Home RF was initiated in 1998 and has similar goals to Bluetooth for WPAN. Its goal is shared data/voice transmission. It interfaces with the Internet as well as the public switched telephone network. It uses the 2.4-GHz band and has a range of 50 m, suitable for home and yard. A maximum of 127 nodes can be accommodated in a single network.

1.5.2   IEEE 1451 and Smart Sensors

Desirable functionality of sensor nodes in a WSN include: ease of installation, self-indication, self-diagnosis, reliability, time awareness for coordination with other nodes, some software functions and DSP, and standard control protocols and network interfaces (IEEE 1451 Expo 2001).

There are many sensor manufacturers and many networks in the market today. It is too costly for manufacturers to make special transducers for every network on the market. Different components made by different manufacturers should be compatible. Therefore, in 1993, the IEEE and the National Institute of Standards and Technology (NIST) began work on a standard for smart sensor networks. IEEE 1451, the standard for smart sensor networks was the result. The objective of this standard is to make it easier for different manufacturers to develop smart sensors and to interface those devices to networks.

> *Smart sensor, virtual sensor:* A major outcome of IEEE 1451 studies is the formalized concept of a smart sensor. A smart sensor is one that provides extra functions beyond those necessary for generating a correct representation of the sensed quality. Included might be signal-conditioning, signal-processing, and decision-making/alarm functions. Objectives of a smart sensor include moving the intelligence closer to the point of measurement; making it cost-effective to integrate and maintain distributed sensor systems; creating a confluence of transducers, control, computation, and communications towards a common goal; and seamlessly interfacing numerous sensors of different types. The concept of virtual sensor also needs to be discussed. A virtual sensor is the physical sensor/transducer, plus the associated signal conditioning and digital signal processing (DSP) that is required to obtain reliable estimates of the required sensory information. The virtual sensor is a component of the smart sensor.

1.5.3   Sensors for Smart Environments

Many vendors now produce commercially available sensors of many types that are suitable for wireless network applications. See, for instance,

TABLE 1.4

Physical Principles and Measurands

Measurements of Wireless Sensors Networks		
	Measurand	**Principle**
Physical properties	Pressure	Piezoresistive, capacitive
	Temperature	Thermistor, thermomechanical, thermocouple
	Humidity	Resistive, capacitive
	Flow	Pressure change, thermistor
Motion properties	Position	E-mag, GPS, contact sensor
	Velocity	Doppler, Hall effect, optoelectronic
	Angular velocity	Optical encoder
	Acceleration	Piezoresistive, piezoelectric, optical fiber
Contact properties	Strain	Piezoresistive
	Force	Piezoelectric, piezoresistive
	Torque	Piezoresistive, optoelectronic
	Slip	Dual torque
	Vibration	Piezoresistive, piezoelectric, optical fiber, sound, ultrasound
Presence	Tactile/contact	Contact switch, capacitive
	Proximity	Hall effect, capacitive, magnetic, seismic, acoustic, RF
	Distance/range	E-mag (sonar, radar, lidar), magnetic, tunneling
	Motion	E-mag, IR, acoustic, seismic (vibration)
Biochemical	Biochemical agents	Biochemical transduction
Identification	Personal features	Vision
	Personal ID	Fingerprints, retinal scan, voice

the Web sites of SUNX Sensors, Schaevitz, Keyence, Turck, Pepperl & Fuchs, National Instruments, UE Systems (ultrasonic), Leake (IR), CSI (vibration). Table 1.4 shows which physical principles may be used to measure various quantities. MEMS sensors are by now available for most of these measurands.

1.5.4 Commercially Available Wireless Sensor Systems

Many wireless communication nodes are commercially available, including Lynx technologies, and various Bluetooth kits, including the Casira devices from Cambridge Silicon Radio, CSR.

Crossbow Berkley Motes: They may be the most versatile WSN devices on the market for prototyping purposes. Crossbow (http://www.xbow.com/) makes three Mote processor radio module families — MICA (MPR300) (first generation), MICA2 (MPR400) and MICA2-DOT (MPR500) (second generation). Nodes come with five sensors installed — temperature, light, acoustic (microphone), acceleration/seismic, and magnetic. These are especially suitable for surveillance networks for personnel and vehicles. Different sensors can be installed, if desired. Low power and small physical size enable placement virtually anywhere. Because all sensor nodes in a network can act as base stations, the network can self-configure and has multihop routing capabilities. The operating frequency of the ISM band is either 916 MHz or 433 MHz, with a data rate of 40 Kbps and a range of 30 to 100 ft. Each node has a low-power microcontroller processor with speed of 4 MHz, a flash memory with 128 kB, and SRAM and EEPROM of 4 kB each. The operating system is Tiny-OS, a tiny microthreading distributed operating system developed by UC-Berkeley, with a NES-C (nested C) source code language (similar to C). Installation of these devices requires a great deal of programming. A comparison of the crossbow motes with that of the UMR/SLU motes is included in several chapters of this book.

Mircrostrain's X-Link measurement system (http://www.microstrain.com/): It may be the easiest system to get up and running and to program. The frequency used is 916 MHz, which lies in the U.S. license-free ISM band. The sensor nodes are multichannel, with a maximum of eight sensors supported by a single wireless node. There are three types of sensor nodes — S-link (strain gauge), G-link (accelerometer), and V-link (supports any sensors generating voltage differences). The sensor nodes have a preprogrammed EPROM, so a great deal of programming by the user is not needed. Onboard data storage is 2 MB. Sensor nodes use a 3.6-V lithium ion internal battery (9-V rechargeable external battery is supported). A single receiver (base station) addresses multiple nodes. Each node has a unique 16-bit address, so a maximum of 2^{16} nodes can be addressed. The RF link between base station and nodes is bidirectional, and the sensor nodes have a programmable data logging sample rate. The RF link has a 30-m range with a 19200-baud rate. The baud rate on the serial RS-232 link between the base station and a terminal PC is 38400. Lab VIEW interface is supported.

Radio frequency identification devices (RFID): RFID tags are transponder microcircuits having an inductor and capacitor (L-C) tank circuit that stores power from received interrogation signals, and then uses the power to transmit a response. Passive tags have no onboard power source and limited onboard data storage, whereas active tags have a battery and up to 1 MB of data storage. RFID devices operate in a low-frequency range of 100 kHz to 1.5 MHz or a high-frequency range of 900 MHz to 2.4 GHz, which has an operating range up to 30 m. RFID tags are very inexpensive, and are used in manufacturing and sales inventory control, container shipping control, and for homeland security applications. RFID tags are installed on water meters in some cities, allowing a metering vehicle to simply drive by and remotely read the current readings. They are also being used in automobiles for automatic toll collection. RFID networks can be viewed as mobile WSNs and therefore the DPC scheme developed for cellular and ad hoc networks is shown to be extended to a dense RFID network for the improvement of read rates and coverage in Chapter 10.

1.5.5 Self-Organization and Localization

Ad hoc networks of nodes may be developed using, for example, aircrafts or ships. Self-organization of ad hoc networks includes both communications self-organization and positioning self-organization. In the former, the nodes must wake up, detect each other, and form a communication network. Technologies for this are by now standard, by and large developed within a mobile phone industry. Distributed surveillance sensor networks require information about the relative positions of the nodes for distributed signal processing, as well as absolute positioning information for reporting data related to detected targets. Chapter 8 presents a self-organization scheme for WSNs.

Relative layout positioning (localization): Relative positioning or localization requires internode communications and a TDMA message header frame that includes both communications and localization fields. There are various means for a node to measure distance to its neighbors, mostly based on RF time-of-flight information. In air, the propagation speed is known, so time differences can be converted to distances. Given the relative distances between nodes, we want to organize the web into a grid specified in terms of relative positions.

Absolute geographical positioning: A network is said to be relatively calibrated if the relative positions of all nodes are known. Now, it is necessary to determine the absolute geographic position of the network. For the net to be known as a (fully) calibrated flat two-dimensional (2-D) net, at least three nodes in the net must determine their absolute positions. There are many ways for a node to determine its absolute position, including GPS and techniques based on stored maps, landmarks, or beacons (Bulusu et al. 2002).

1.6 Summary

Guaranteeing performance in a network whether it is a computer or wireless network is extremely important. Although QoS issues have been explicitly addressed in the case of wired networks, energy efficiency is of paramount importance for wireless networks. The QoS issues can be met by using congestion and admission control, scheduling and routing protocols in a network. Many of the wired networking schemes do not offer analytical performance except simulation results. Though wireless networks provide mobility and low cost and rapid deployment of a WSN create many new opportunities in application areas, a WSN has to satisfy hard constraints introduced by factors, such as fault tolerance, scalability, cost, hardware, topology change, environment, and power consumption. Because these constraints are stringent, new protocols and algorithms are necessary that ensure performance. This book intends to address these issues and offer analytical framework on how to ensure performance regardless of the type of network.

References

Akyildiz, I.F., Su, W., Sankarasubramaniam, Y., and Cayirci, E., Wireless sensor networks: a survey, _Computer Networks_, Vol. 38, 393–422, 2002.

Bulusu, N., Estrin, D., Girod, J., and Heidemann, J., Scalable coordination for wireless sensor networks: self-configuring localization systems, _Proceedings of the International Symposium on Communication Theory and Applications (IS-CTA 2001)_, Ambleside, U.K., July 2001.

Bulusu, N., Heidemann, J., Estrin, D., and Tran, T., Self-configuring localization systems: design and experimental evaluation, _ACM TECS Special Issue on Networked Embedded Computing_, 1–31, August 2002.

Chao, H.J. and Guo, X., _Quality of Service Control in High-Speed Networks_, Wiley-Interscience, New York, 2002.

Dontula, S. and Jagannathan, S., Active link protection with distributed power control of wireless networks, _Proceedings of the World Wireless Congress_, May 2004, pp. 612–617.

Fall, K. and Varadhan, K., ns Notes and Documentation, Technical report UC Berkley LBNL USC/IS Xerox PARC, 2002.

Hoblos, G., Staroswiecki, M., and Aitouche, A., Optimal design of fault tolerant sensor networks, _Proceedings of the IEEE International Conference on Control Applications_, Anchorage, AK, September 2000, pp. 467–472.

IEEE 802.11a-1999, Amendment to IEEE 802.11: High-speed Physical Layer in the 5 GHz band.

IEEE 802.11b-1999, Supplement to 802.11-1999, Wireless LAN MAC and PHY specifications: Higher speed Physical Layer (PHY) extension in the 2.4 GHz band, _NS-2 manual_, http://www.isi.edu/nsnam/ns/ns-documentation.html.

Kahn, J.M., Katz, R.H., and Pister, K.S.J., Next century challenges: mobile networking for smart dust, *Proceedings of the ACM MOBICOM*, Washington, D.C., 1999, pp. 271–278.

Mohammed, A.F., Near-far problem in direct-sequence code-division multiple-access systems, *Proceedings of the Seventh IEE European Conference on Mobile and Personal Communications*, December 1993, pp. 151–154.

Pottie, G.J. and Kaiser, W.J., Wireless Integrated Network Sensors, *Communications of the ACM*, Vol. 43, No. 5, 2000, pp. 551–558.

Rappaport, T.S., *Wireless Communications: Principles and Practice*, Prentice Hall, Upper Saddle River, NJ, 1999.

Ray, S. "an introduction to ultra wide band (impulse ratio)", Internal Report Elec. and Comp. Eng. Dept., Boston University, Oct. 2001.

Shih, E., Cho, S., Ickes, N., Min, R., and Sinha, A., Wang, A., and Chandrahasan, A., Physical layer driven protocols and algorithm design for energy-efficient wireless sensor networks, *Proceedings of the ACM MOBICOM*, Rome, Italy, July 2001, pp. 272–286.

Siemens, 3G Wireless Standards for Cellular Mobile Services, 2002.

Sinha, A. and Chandrahasan, A., Dynamic power management in wireless sensor networks, *Proceedings of the IEEE International Conference on Communications*, Helsinki, Finland, June 2001.

Stallings, W., *High-Speed Networks and Internets: Performance and Quality of Service*, 2nd ed., Prentice Hall, Upper Saddle River, NJ, 2002.

Tanenbaum, A.S., *Computer Networks*, 3rd ed., Prentice Hall, Upper Saddle River, NJ, 1996.

Walrand, J. and Varaiya, P., *High-Performance Communication Networks*, Morgan Kaufmann Publishers, San Francisco, CA, 1996.

Wireless LAN Medium Access Control (MAC) and Physical Layer (PHY) Specifications, *ANSI/IEEE Standard 802.11*, 1999 Edition.

2

Background

In this chapter, we provide a brief background on dynamical systems, mainly covering the topics that will be important in a discussion of protocol and algorithm development for networking such as congestion control, admission control, scheduling, and neural network (NN) applications in closed-loop control of networks. It is quite common for non-control engineers working in wireless networking systems and control applications to have little understanding of feedback control and dynamical systems. Many of the phenomena they observe are not attributable to properties of NN but to the properties of feedback control systems. Control applications of dynamical systems are a complex area with several facets. An incomplete understanding of any one of these can lead to incorrect conclusions being drawn, with inaccurate attributions of causes — many are convinced that often the exploratory, regulatory, and behavioral phenomena observed in NN control systems are completely because of the NN; in fact, most are due to the rather remarkable nature of feedback itself. Included in this chapter are discrete-time systems, computer simulation, norms, and stability.

2.1 Dynamical Systems

Many systems in nature, including neurobiological systems, are dynamical in nature, in the sense that they are acted upon by external inputs, have internal memory, and behave in certain ways that are captured by the notion of the development of activities through time. According to the notion of systems defined by Whitehead (1953), it is an entity distinct from its environment, whose interactions with the environment can be characterized through input and output signals. An intuitive feel for dynamic systems is provided by Luenberger (1979), which includes many examples.

2.1.1 Discrete-Time Systems

If the time index is an integer k instead of a real number t, the system is said to be of discrete-time. A general class of discrete-time systems can be described by the nonlinear ordinary difference equation in discrete-time state space form

$$x(k+1) = f(x(k), u(k)), \quad y(k) = h(x(k), u(k)) \qquad (2.1)$$

where $x(k) \in \Re^n$ is the internal state vector, $u(k) \in \Re^m$ is the contol input, and $y(k) \in \Re^p$ is the system output.

These equations may be derived directly from an analysis of the dynamical system or process being studied, or they may be sampled or discretized versions of continuous-time dynamics of a nonlinear system. Today, controllers are implemented in digital form by using embedded hardware making it necessary to have a discrete-time description of the controller. This may be determined by design, based on the discrete-time system dynamics. Sampling of linear systems is well understood with many design techniques available. However, sampling of nonlinear systems is not an easy topic. In fact, the exact discretization of nonlinear continuous dynamics is based on the Lie derivatives and leads to an infinite series representation (e.g., Kalkkuhl and Hunt 1996). Various approximation and discretization techniques use truncated versions of the exact series.

2.1.2 Brunovsky Canonical Form

Letting $x(k) = [x_1(k) \ldots x_n(k)]^T$, a special form of nonlinear dynamics is given by the class of systems in discrete Brunovsky canonical form

$$x_1(k+1) = x_2(k)$$
$$x_2(k+1) = x_3(k)$$
$$\vdots \qquad\qquad\qquad\qquad\qquad (2.2)$$
$$x_n(k+1) = f(x(k)) + g(x(k))u(k)$$
$$y(k) \quad = h(x(k))$$

As seen from Figure 2.1, this is a chain or cascade of unit delay elements z^{-1}, that is, a shift register. Each delay element stores information and requires an initial condition. The measured output $y(k)$ can be a general function of the states as shown, or can have more specialized forms such as

$$y(k) = h(x_1(k)) \qquad (2.3)$$

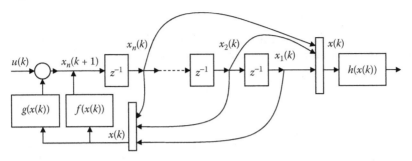

FIGURE 2.1
Discrete-time single-input Brunovsky form.

The discrete Brunovsky canonical form may equivalently be written as

$$x(k+1) = Ax(k) + bf(x(k)) + bg(x(k))u(k) \qquad (2.4)$$

where,

$$A = \begin{bmatrix} 0 & 1 & 0 & . & . & . & 0 \\ 0 & 0 & 1 & . & . & . & 0 \\ & & & . & & & \\ & & & . & & & \\ & & & . & & & \\ 0 & 0 & . & . & . & 1 & 0 \\ 0 & 0 & 0 & . & . & . & 0 \end{bmatrix}, \quad b = \begin{bmatrix} 0 \\ 0 \\ . \\ . \\ . \\ 0 \\ 1 \end{bmatrix} \qquad (2.5)$$

A discrete-time form of the more general version may also be written. It is a system with m parallel chains of delay elements of lengths $n_1, n_2, \ldots$ (e.g., m shift registers), each driven by one of the control inputs.

Many practical systems occur in the continuous-time Brunovsky form. However, if a system of the continuous Brunovsky form (Lewis, Jagannathan, and Yesilderek 1999) is sampled, the result is not the general form as in Equation 2.2. Under certain conditions, general discrete-time systems of the form having the form of Equation 2.1 can be converted to discrete Brunovsky canonical form systems (e.g., Kalkkuhl and Hunt 1996).

2.1.3   Linear Systems

A special and important class of dynamical systems is the discrete-time linear time invariant (LTI) system

$$x(k+1) = Ax(k) + Bu(k)$$
$$y(k) = Cx(k)$$

(2.6)

with A, B, and C constant matrices of general form (e.g., not restricted to Equation 2.5). An LTI is denoted by (A,B,C). Given an initial state $x(0)$, the solution to the LTI system can be explicitly written as

$$x(k) = A^k x(0) + \sum_{j=0}^{k-1} A^{k-j-1} Bu(j)$$

(2.7)

The next example shows the relevance of these solutions and demonstrates that the general discrete-time nonlinear systems are even easier to simulate on a computer than continuous-time systems, as no integration routine is needed.

Example 2.1.1: Discrete-Time System — Savings Account (Lewis, Jagannathan, and Yesilderek 1999)

Discrete-time descriptions can be derived from continuous-time systems by using Euler's approximation or system discretization theory. However, many phenomena are naturally modeled, using discrete-time dynamics, including population growth and decline, epidemic spread, economic systems, and so on. The dynamics of the savings account using compound interest are given by the first-order system

$$x(k+1) = (1+i)x(k) + u(k)$$

where i represents the interest rate over each interval, k is the interval iteration number, and $u(k)$ is the amount of the deposit at the beginning of the kth period. The state $x(k)$ represents the account balance at the beginning of interval k.

ANALYSIS

According to Equation 2.7, if equal annual deposits are made of $u(k) = d$, the account balance is

$$x(k) = (1+i)^k x(0) + \sum_{j=0}^{k-1} (1+i)^{k-j-1} d$$

with $x(0)$ being the initial amount in the account. Using the standard series summation formula,

$$\sum_{j=0}^{k-1} a^j = \frac{1-a^k}{1-a}$$

one derives the standard formula for complex interest with constant annuities of d.

$$x(k) = (1+i)^k x(0) + d(1+i)^{k-1} \sum_{j=0}^{k-1} \frac{1}{(1+i)^j}$$

$$= (1+i)^k x(0) + d(1+i)^{k-1} \left[\frac{1 - \dfrac{1}{(1+i)^k}}{1 - \dfrac{1}{(1+i)}} \right]$$

$$= (1+i)^k x(0) + d \left[\frac{(1+i)^{k-1} - 1}{i} \right]$$

SIMULATION

It is very easy to simulate a discrete-time system. No numerical integration driver program is needed in contrast to the continuous-time case. Instead, a simple "do loop" can be used. A complete MATLAB program that simulates the compound interest dynamics is given by

```
%Discrete-Time Simulation program for Compound
Interest Dynamics
d =100; i =0.08; % 8% interest rate
x(1)=1000;
for k =1:100
x(k+1)=(1+i)*x(k)
end
k =[1:101];
plot(k,x);
```

2.2 Mathematical Background

2.2.1 Vector and Matrix Norms

We assume the reader is familiar with norms, both vector and induced matrix norms (Lewis, Abdallah, and Dawson 1993). We denote any suitable vector norm by $\|\cdot\|$. When required to be specific, we denote the p-norm by $\|\cdot\|_p$. Recall that for any vector $x \in \Re^n$,

$$\| x \|_1 = \sum_{i=1}^{n} |x_i| \tag{2.8}$$

$$\|x\|_p = \left(\sum_{i=1}^{n} |x_i|^p \right)^{\frac{1}{p}} \tag{2.9}$$

$$\|x\|_\infty = \max_i |x_i| \tag{2.10}$$

The 2 -norm is the standard Euclidean norm.

Given a matrix A, its induced p-norm is denoted by $\|A\|_p$. Letting $A = [a_{ij}]$, recall that the induced 1-norm is the maximum absolute column sum

$$\|A\|_1 = \sum_i^{max} |a_{ij}| \tag{2.11}$$

and the induced ∞-norm is the maximum absolute row sum

$$\|A\|_\infty = \max_i \sum_i |a_{ij}| \tag{2.12}$$

The induced matrix p-norm satisfies the inequality, for any vector x,

$$\|A\|_p \leq \|A\|_p \| x \|_p \tag{2.13}$$

and for any two matrices A, B one also has

$$\|AB\|_p \leq \|A\|_p \|B\|_p \tag{2.14}$$

Given a matrix $A = [a_{ij}]$, the Frobenius norm is defined as the root of the sum of the squares of all the elements:

$$\left\|A\right\|_F^2 \equiv \sum a_{ij}^2 = tr(A^T A) \tag{2.15}$$

with tr(.) the matrix trace (i.e., sum of diagonal elements). Though the Frobenius norm is not an induced norm, it is compatible with the vector 2-norm so that

$$\|Ax\|_2 \le \|A\|_F \|x\|_2 \qquad (2.15a)$$

2.2.1.1 Singular Value Decomposition

The matrix norm $\|A\|_2$ induced by the vector 2-norm is the maximum singular value of A. For a general $m \times n$ matrix A, one may write the singular value decomposition (SVD)

$$A = U\Sigma V^T \qquad (2.16)$$

where U is $m \times n$, V is $n \times n$, and both are orthogonal; that is,

$$U^T U = U U^T = I_m$$
$$V^T V = V V^T = I_n \qquad (2.17)$$

where I_n is the $n \times n$ identity matrix. The $m \times n$ singular value matrix has the structure

$$\Sigma = diag\{\sigma_1, \sigma_2, \ldots, \sigma_r, 0, \ldots, 0\} \qquad (2.18)$$

where r is the rank of A and σ_i are the singular values of A. It is conventional to arrange the singular values in a nonincreasing order, so that the largest singular value is $\sigma_{max}(A) = \sigma_1$. If A is full rank, then r is equal to either m or n, whichever is smaller. Then the minimum singular value is $\sigma_{min}(A) = \sigma_r$ (otherwise, the minimum singular value is equal to zero).

The SVD generalizes the notion of eigenvalues to general nonsquare matrices. The singular values of A are the (positive) square roots of the nonzero eigenvalues of AA^T, or equivalently $A^T A$.

2.2.1.2 Quadratic Forms and Definiteness

Given an $n \times n$ matrix, Q the *quadratic form* $x^T Q x$, with x being an n-vector, will be important for stability analysis in this book. The quadratic form can, in some cases, have certain properties that are independent of the vector x selected. Four important definitions are:

Q is positive definite, denoted $Q > 0$, if $x^T Q x > 0$, $\forall x \ne 0$.

Q is positive semidefinite, denoted $Q \ge 0$, if $x^T Q x \ge 0$, $\forall x$.

Q is negative definite, denoted $Q < 0$, if $x^T Q x < 0$, $\forall x \ne 0$.

Q is negative semidefinite, denoted $Q \le 0$, if $x^T Q x \le 0$, $\forall x$. $\qquad (2.19)$

If Q is symmetric, then it is positive definite if and only if all its eigenvalues are positive and positive semidefinite if and only if all its eigenvalues are nonnegative. If Q is not symmetric, tests are more complicated and involve determining the minors of the matrix. Tests for negative definiteness and semidefiniteness may be found by noting that Q is negative (semi) definite if and only if Q is positive (semi) definite.

If Q is a symmetric matrix, its singular values are the magnitudes of its eigenvalues. If Q is a symmetric positive semidefinite matrix, its singular values and its eigenvalues are the same. If Q is positive semidefinite then, for any vector x, one has the useful inequality

$$\sigma_{\min}(Q)\,\|x\|^2 \leq x^T Q x \leq \sigma_{\max}(Q)\,\|x\|^2 \tag{2.20}$$

2.2.2   Continuity and Function Norms

Given a subset $S \subset \Re^n$, a function $f(x): S \to \Re^m$ is *continuous* on $x_0 \in S$ if, for every $\varepsilon > 0$, there exists a $\delta(\varepsilon, x_0) > 0$ such that $\|x - x_0\| < \delta(\varepsilon, x_0)$ implies that $\|f(x) - f(x_0)\| < \varepsilon$. If δ is independent of x_0, then the function is said to be uniformly continuous. Uniform continuity is often difficult to test. However, if $f(x)$ is continuous and its derivative $f'(x)$ is bounded, then it is uniformly continuous.

A function $f(x): \Re^n \to \Re^m$ is differentiable if its derivative $f'(x)$ exists. It is continuously differentiable, if its derivative exists and is continuous. $f(x)$ is said to be locally Lipschitz if, for all $x, z \in S \subset \Re^n$, one has

$$\|f(x) - f(z)\| < L\,\|x - z\| \tag{2.21}$$

for some finite constant $L(S)$, where L is known as a *Lipschitz constant*. If $S = \Re^n$, then the function is globally Lipschitz.

If $f(x)$ is globally Lipschitz, then it is uniformly continuous. If it is continuously differentiable, it is locally Lipschitz. If it is differentiable, it is continuous. For example, $f(x) = x^2$ is continuously differentiable. It is locally, but not globally, Lipschitz. It is continuous but not uniformly continuous.

Given a function $f(t): [0, \infty) \to \Re^n$, according to *Barbalat's Lemma*, if

$$\int_0^\infty f(t)dt \leq \infty \tag{2.22}$$

and $f(t)$ is uniformly continuous, then $f(t) \to 0$ as $t \to \infty$.

Given a function $f(t):[0,\infty)\to\Re^n$, its L_p (function) norm is given in terms of the vector norm $\|f(t)\|_p$ at each value of t by

$$\|f(\cdot)\|_p = \left(\int_0^\infty \|f(t)\|_p^p\,dt\right)^{\frac{1}{p}} \tag{2.23}$$

and if $p = \infty$,

$$\|f(\cdot)\|_\infty = \sup_t \|f(t)\|_\infty \tag{2.24}$$

If the L_p norm is finite, we say $f(t)\in L_p$. Note that a function is in L_∞ if and only if it is bounded. For detailed treatment, refer to Lewis, Abdallah, and Dawson (1993) and Lewis, Jagannathan, and Yesilderek (1999).

In the discrete-time case, let $Z_+ = \{0, 1, 2, ...\}$ be the set of natural numbers and $f(k): Z_+ \to \Re^n$. The L_p (function) norm is given in terms of the vector $\|f(k)\|_p$ at each value of k by

$$\|f(\cdot)\|_p = \left(\sum_{k=0}^\infty \|f(k)\|_p^p\right)^{\frac{1}{p}}, \tag{2.25}$$

and, if $p = \infty$,

$$\|f(\cdot)\|_\infty = \sup_k \|f(k)\|_\infty \tag{2.26}$$

If the L_p norm is finite, we say $f(k)\in L_p$. Note that a function is in L_∞ if and only if it is bounded.

2.3 Properties of Dynamical Systems

In this section are discussed some properties of dynamical systems, including stability. For observability and controllability, please refer to Jagannathan (2006), and Goodwin and Sin (1984). If the original open-loop system is controllable and observable, then a feedback control system can be designed to meet desired performance. If the system has certain passivity properties, this design procedure is simplified, and additional closed-loop properties such as robustness can be guaranteed. On the other hand, properties such as stability may not be present in the original open-loop system but are design requirements for closed-loop performance.

Stability along with robustness (discussed in the following subsection) is a performance requirement for closed-loop systems. In other words, though the open-loop stability properties of the original system may not be satisfactory, it is desirable to design a feedback control system such that the closed-loop stability is adequate. We will discuss stability for discrete-time systems, but the same definitions also hold for continuous-time systems with obvious modifications.

Consider the dynamical system

$$x(k+1) = f(x(k),k) \qquad (2.27)$$

where $x(k) \in \Re^n$, which might represent either an uncontrolled open-loop system, or a closed-loop system after the control input $u(k)$ has been specified in terms of the state $x(k)$. Let the initial time be k_0 and the initial condition be $x(k_0) = x_0$. This system is said to be nonautonomous because the time k appears explicitly. If k does not appear explicitly in $f(\cdot)$, then the system is autonomous. A primary cause of explicit time dependence in control systems is the presence of time-dependent disturbances $d(k)$.

A state x_e is an equilibrium point of the system $f(x_e,k) = 0, k \ge k_0$. If $x_0 = x_e$, so that the system starts out in the equilibrium state, then it will remain there forever. For linear systems, the only possible equilibrium point is $x_e = 0$; for nonlinear systems x_e may be nonzero. In fact, there may even be an equilibrium set, such as a limit cycle.

2.3.1   Asymptotic Stability

An equilibrium point x_e is locally asymptotically stable (AS) at k_0, if there exists a compact set $S \subset \Re^n$ such that, for every initial condition $x_0 \in S$, one has $\|x(k) - x_e\| \to 0$ as $k \to \infty$. That is, the state $x(k)$ converges to x_e. If $S = \Re^n$, so that $x(k) \to x_e$ for all $x(k_0)$, then x_e is said to be globally asymptotically stable (GAS) at k_0. If the conditions hold for all k_0, the stability is said to be uniform (e.g., UAS, GUAS).

Asymptotic stability is a very strong property that is extremely difficult to achieve in closed-loop systems, even using advanced feedback controller design techniques. The primary reason is the presence of unknown but bounded system disturbances. A milder requirement is provided as follows.

2.3.2 Lyapunov Stability

An equilibrium point x_e is stable in the sense of Lyapunov (SISL) at k_0, if for every $\varepsilon > 0$, there exists $\delta(\varepsilon, x_0)$, such that $\|x_0 - x_e\| < \delta(\varepsilon, k_0)$ implies that $\|x(k) - x_e\| < \varepsilon$ for $k \geq k_0$. The stability is said to be uniform (e.g., uniformly SISL) if $\delta(\cdot)$ is independent of k_0; that is, the system is SISL for all k_0.

It is extremely interesting to compare these definitions to those of function continuity and uniform continuity. SISL is a notion of continuity for dynamical systems. Note that for SISL, there is a requirement that the state $x(k)$ be kept arbitrarily close to x_e by starting sufficiently close to it. This is still too strong a requirement for closed-loop control in the presence of unknown disturbances. Therefore, a practical definition of stability to be used as a performance objective for feedback controller design in this book is as follows.

2.3.3 Boundedness

This is illustrated in Figure 2.2. The equilibrium point x_e is said to be uniformly ultimately bounded (UUB), if there exists a compact set $S \subset \mathfrak{R}^n$, so that for all $x_0 \in S$, there exists a bound $\mu \geq 0$, and a number $N(\mu, x_0)$, such that $\|x(k)\| \leq \mu$ for all $k \geq k_0 + N$. The intent here is to capture the notion that for all initial states in the compact set S, the system trajectory eventually reaches, after a lapsed time of N, a bounded neighborhood of x_e.

The difference between UUB and SISL is that, in UUB, the bound μ cannot be made arbitrarily small by starting closer to x_e. In fact, the Vander Pol oscillator is UUB but not SISL. In practical closed-loop applications, μ depends on the disturbance magnitudes and other factors. If the controller is suitably designed, however, μ will be small enough for practical purposes. The term uniform indicates that N does not depend upon k_0.

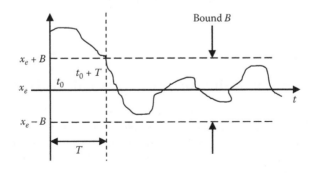

FIGURE 2.2
Illustration of uniform ultimate boundedness (UUB).

The term ultimate indicates that the boundedness property holds after a time lapse N. If $S = \mathfrak{R}^n$, the system is said to be globally UUB (GUUB).

2.3.4 A Note on Autonomous Systems and Linear Systems

If the system is autonomous so that

$$x(k+1) = f(x(k)) \qquad (2.28)$$

where $f(x(k))$ is not an explicit function of time, the state trajectory is independent of the initial time. This means that if an equilibrium point is stable by any of the three definitions, the stability is automatically uniform. Nonuniformity is only a problem with nonautonomous systems.

If the system is linear so that

$$x(k+1) = A(k)x(k) \qquad (2.29)$$

with $A(k)$ being an $n \times n$ matrix, then the only possible equilibrium point is the origin.

For linear time-invariant (LTI) systems, matrix A is time-invariant. Then the system poles are given by the roots of the characteristic equation

$$\Delta(z) = |zI - A| = 0, \qquad (2.30)$$

where $|\cdot|$ is the matrix determinant, and z is the Z transform variable. For LTI systems, AS corresponds to the requirement that all the system poles stay within the unit disc (i.e., none of them are allowed on the unit disc). SISL corresponds to marginal stability, that is, all the poles are within the unit disc, and those on the unit disc are not repeated.

2.4   Nonlinear Stability Analysis and Controls Design

For LTI systems, it is straightforward to investigate stability by examining the locations of the poles in the s-plane. However, for nonlinear or nonautonomous (e.g., time-varying) systems there are no direct techniques. The (direct) Lyapunov approach provides methods for studying the stability of nonlinear systems and shows how to design control systems for such complex nonlinear systems. For more information see Lewis, Abdallah, and Dawson (1993), which deals with robot manipulator control, as well as Jagannathan (2006), Goodwin and Sin (1984), Landau (1979), Sastry and Bodson (1989), and Slotine and Li (1991), which have proofs and many excellent examples in continuous and discrete-time.

2.4.1 Lyapunov Analysis for Autonomous Systems

The autonomous (time-invariant) dynamical system

$$x(k+1) = f(x(k)), \tag{2.31}$$

$x \in \Re^n$, could represent a closed-loop system after the controller has been designed. In Section 2.3.1, we defined several types of stability. We shall show here how to examine stability properties using a generalized energy approach. An isolated equilibrium point x_e can always be brought to the origin by redefinition of coordinates; therefore, let us assume without loss of generality that the origin is an equilibrium point. First, we give some definitions and results. Then, some examples are presented to illustrate the power of the Lyapunov approach.

Let $L(x): \Re^n \to \Re$ be a scalar function such that $L(0) = 0$, and S be a compact subset of $\Re^n$. Then $L(x)$ is said to be,

Locally positive definite, if $L(x) > 0$ when $x \neq 0$, for all $x \in S$. (Denoted by $L(x) > 0$.)

Locally positive semidefinite, if $L(x) \geq 0$ when $x \neq 0$, for all $x \in S$. (Denoted by $L(x) \geq 0$.)

Locally negative definite, if $L(x) < 0$ when $x \neq 0$, for all $x \in S$. (Denoted by $L(x) < 0$.)

Locally negative semidefinite, if $L(x) \leq 0$ when $x \neq 0$, for all $x \in S$. (Denoted by $L(x) \leq 0$.)

An example of a positive definite function is the quadratic form $L(x) = x^T P x$, where P is any matrix that is symmetric and positive definite. A definite function is allowed to be zero only when $x = 0$, a semidefinite function may vanish at points where $x \neq 0$. All these definitions are said to hold globally if $S = \Re^n$.

A function $L(x): \Re^n \to \Re$ with continuous partial differences (or derivatives) is said to be a Lyapunov function for the system described in Theorem 2.4.1, if, for some compact set $S \subset \Re^n$, one has locally:

$$L(x) \text{ is positive definite, } L(x) > 0 \tag{2.32}$$

$\Delta L(x)$ is negative semidefinite,

$$\Delta L(x) \leq 0 \tag{2.33}$$

where $\Delta L(x)$ is evaluated along the trajectories of (2.4.1) (as shown in subsequent example). That is,

$$\Delta L(x(k)) = L(x(k+1)) - L(x(k)) \tag{2.34}$$

THEOREM 2.4.1 (LYAPUNOV STABILITY)
If there exists a Lyapunov function for a system as shown in Equation 2.31, then the equilibrium point is stable in the sense of Lyapunov (SISL).
This powerful result allows one to analyze stability using a generalized notion of energy. The Lyapunov function performs the role of an energy function. If L(x) is positive definite and its derivative is negative semidefinite, then L(x) is nonincreasing, which implies that the state x(t) is bounded. The next result shows what happens if the Lyapunov derivative is negative definite — then L(x) continues to decrease until ∥x(k)∥ vanishes.

THEOREM 2.4.2 (ASYMPTOTIC STABILITY)
If there exists a Lyapunov function L(x) for system as shown in Equation 2.31 with the strengthened condition on its derivative,
$\Delta L(x)$ *is negative definite,*

$$\Delta L(x) < 0 \tag{2.35}$$

then the equilibrium point is AS.
To obtain global stability results, one needs to expand the set S to all of $\Re^n$, but also required is an additional radial unboundedness property.

THEOREM 2.4.3 (GLOBAL STABILITY)

Globally SISL: If there exists a Lyapunov function $L(x)$ for the system as shown In Equation 2.31 such that Equation 2.32 and Equation 2.33 hold globally, and

$$L(x) \to \infty \text{ as } \|x\| \to \infty \tag{2.36}$$

then the equilibrium point is globally SISL.

Globally AS: If there exists a Lyapunov function $L(x)$ for a system as shown in Equation 2.31 such that Equation 2.32 and Equation 2.35 hold globally, and also the unboundedness condition as in Equation 2.36 holds, then the equilibrium point is GAS.

The global nature of this result of course implies that the equilibrium point mentioned is the only equilibrium point.

The subsequent examples show the utility of the Lyapunov approach and make several points. Among the points of emphasis are those stating that the Lyapunov function is intimately related to the energy properties of a system.

Example 2.4.1: Local and Global Stability

LOCAL STABILITY
Consider the system

$$x_1(k+1) = x_1(k)\left(\sqrt{x_1^2(k) + x_2^2(k)} - 2 \right)$$

$$x_2(k+1) = x_2(k)\left(\sqrt{x_1^2(k) + x_2^2(k)} - 2 \right)$$

Stability for nonlinear discrete-time systems can be examined by selecting the quadratic Lyapunov function candidate

$$L(x(k)) = x_1^2(k) + x_2^2(k)$$

which is a direct realization of an energy function and has first difference

$$\Delta L(x(k)) = x_1^2(k+1) - x_1^2(k) + x_2^2(k+1) - x_2^2(k)$$

Evaluating this along the system trajectories simply involves substituting the state differences from the dynamics to obtain, in this case,

$$\Delta L(x(k)) = -\left(x_1^2(k) + x_2^2(k) \right)\left(1 - x_1^2(k) - x_2^2(k) \right)$$

which is negative as long as

$$\|x(k)\| = x_1^2(k) + x_2^2(k) < 1$$

Therefore, $L(x(k))$ serves as a (local) Lyapunov function for the system, which is locally asymptotically stable. The system is said to have a domain of attraction with a radius of one. Trajectories beginning outside $\|x(k)\| = 1$ in the phase plane cannot be guaranteed to converge.

GLOBAL STABILITY

Consider now the system

$$x_1(k+1) = x_1(k)x_2^2(k)$$

$$x_2(k+1) = x_2(k)x_1^2(k)$$

where the states satisfy $(x_1(k)x_2(k))^2 < 1$.
Selecting the Lyapunov function candidate

$$L(x(k)) = x_1^2(k) + x_2^2(k)$$

which is a direct realization of an energy function and has first difference

$$\Delta L(x(k)) = x_1^2(k+1) - x_1^2(k) + x_2^2(k+1) - x_2^2(k)$$

Evaluating this along the system trajectories simply involves substituting the state differences from the dynamics to obtain, in this case,

$$\Delta L(x(k)) = -\left(x_1^2(k) + x_2^2(k)\right)\left(1 - x_1^2(k)x_2^2(k)\right)$$

Applying the constraint, the system is globally stable because the states are restricted.

Example 2.4.2: Lyapunov Stability

Consider now the system

$$x_1(k+1) = x_1(k) - x_2(k)$$

$$x_1(k+1) = \sqrt{2x_1(k)x_2(k) - x_1^2(k)}$$

Selecting the Lyapunov function candidate

$$L(x(k)) = x_1^2(k) + x_2^2(k)$$

which is a direct realization of an energy function and has first difference

$$\Delta L(x(k)) = x_1^2(k+1) - x_1^2(k) + x_2^2(k+1) - x_2^2(k)$$

Evaluating this along the system trajectories simply involves substituting the state differences from the dynamics to obtain, in this case,

$$\Delta L(x(k)) = -x_1^2(k)$$

This is only negative semidefinite (note that $\Delta L(x(k))$ can be zero, when $x_2(k) \neq 0$). Therefore, $L(x(k))$ is a Lyapunov function, but the system is only shown by this method to be SISL — that is, $\|x_1(k)\|$ and $\|x_2(k)\|$ are both bounded.

2.4.2 Controller Design Using Lyapunov Techniques

Though we have presented Lyapunov analysis only for unforced systems in the form described in Theorem 2.4.1, which have no control input, these techniques also provide a powerful set of tools for designing feedback control systems of the form

$$x(k+1) = f(x(k)) + g(x(k))u(k) \qquad (2.37)$$

Thus, select a Lyapunov function candidate $L(x) > 0$ and differentiate along the system trajectories to obtain

$$\Delta L(x) = L(x(k+1)) - L(x(k)) = x^T(k+1)x(k+1) - x^T(k)x(k)$$
$$= (f(x(k)) + g(x(k))u(k))^T(f(x(k)) + g(x(k))u(k)) - x^T(k)x(k)$$
$$\qquad (2.38)$$

Then, it is often possible to ensure that $\Delta L \leq 0$ by appropriate selection of $u(k)$. When this is possible, it generally yields controllers in state-feedback form, that is, where $u(k)$ is a function of the states $x(k)$.

Practical systems with actuator limits and saturation often contain discontinuous functions including the signum function defined for scalars $x \in \Re$ as

$$\mathrm{sgn}(x) = \begin{cases} 1, & x \geq 0 \\ -1, & x < 0 \end{cases} \qquad (2.39)$$

shown in Figure 2.3, and for vectors $x = [x_1 \quad x_2 \quad \cdots \quad x_n]^T \in \Re^n$ as

$$\mathrm{sgn}(x) = [\mathrm{sgn}(x_i)] \qquad (2.40)$$

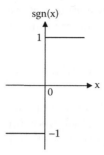

FIGURE 2.3
Signum function.

where $[z_i]$ denotes a vector z with components z_i. The discontinuous nature of such functions often makes it impossible to apply input/output feedback linearization where differentiation is required. In some cases, controller design can be carried out for systems containing discontinuities using Lyapunov techniques.

Example 2.4.3: Controller Design by Lyapunov Analysis
Consider the system

$$x_1(k+1) = x_2(k)\mathrm{sgn}(x_1(k))$$

$$x_2(k+1) = \sqrt{x_1(k)x_2(k) + u(k)}$$

having an actuator nonlinearity. A control input has to be designed using feedback linearization techniques (i.e., cancels all nonlinearities). A stabilizing controller can be easily designed using Lyapunov techniques.
Select the Lyapunov function candidate

$$L(x(k)) = x_1^2(k) + x_2^2(k),$$

and evaluate

$$\Delta L(x(k)) = x_1^2(k+1) - x_1^2(k) + x_2^2(k+1) - x_2^2(k).$$

Substituting the system dynamics in the previous equation results in

$$\Delta L(x(k)) = x_2^2(k)\mathrm{sgn}^2(x_1(k)) - x_1^2(k) + (x_1(k)x_2(k) + u(k)) - x_2^2(k)$$

Now, select the feedback control

$$u(k) = -x_2^2(k)\text{sgn}^2(x_1(k)) + x_1^2(k) - x_1(k)x_2(k)$$

This yields,

$$\Delta L(x(k)) = -x_2^2(k)$$

so that $L(x(k))$ is rendered a (closed-loop) Lyapunov function. Because $\Delta L(x(k))$ is negative semidefinite, the closed-loop system with this controller is SISL.

It is important to note that by slightly changing the controller, one can also show global asymptotic stability of the closed-loop system. Moreover, note that this controller has elements of feedback linearization (discussed in Lewis, Jagannathan, and Yesiderek 1999), in that the control input $u(k)$ is selected to cancel nonlinearities. However, no difference of the right-hand side of the state equation is needed in the Lyapunov approach, but the right-hand side becomes quadratic, which makes it hard to design controllers and show stability. This will be a problem for the discrete-time systems, and we will be presenting how to select suitable Lyapunov function candidates for complex systems when standard adaptive control and NN-based controllers are deployed. Finally, there are some issues in this example, such as the selection of the discontinuous control signal, which could cause chattering. In practice, the system dynamics act as a low-pass filter so that the controllers work well.

2.4.3 Lyapunov Analysis and Controls Design for Linear Systems

For general nonlinear systems, it is not always easy to find a Lyapunov function. Thus, failure to find a Lyapunov function may be because the system is not stable, or because the designer simply lacks insight and experience. However, in the case of LTI systems

$$x(k+1) = Ax \tag{2.41}$$

Lyapunov analysis is simplified and a Lyapunov function is easy to find, if one exists.

2.4.4 Stability Analysis

Select as a Lyapunov function candidate, the quadratic form

$$L(x(k)) = \frac{1}{2}x^T(k)Px(k), \tag{2.42}$$

where P is a constant symmetric positive definite matrix. Because $P > 0$, then $x^T P x$ is a positive function. This function is a generalized norm, which serves as a system energy function. Then,

$$\Delta L(x(k)) = L(x(k+1)) - L(x(k)) = \frac{1}{2}[x^T(k+1)Px(k+1) - x^T(k)Px(k)] \quad (2.43)$$

$$= \frac{1}{2}x^T(k)[A^T PA - P]x(k) \quad (2.44)$$

For stability, one requires negative semidefiniteness. Thus, there must exist a symmetric positive semidefinite matrix Q, such that

$$\Delta L(x) = -x^T(k)Qx(k) \quad (2.45)$$

This results in the next theorem.

THEOREM 2.4.4 (LYAPUNOV THEOREM FOR LINEAR SYSTEMS)
The system discussed in Equation 2.41 is SISL, if there exist matrices $P > 0$, $Q \geq 0$ that satisfy the Lyapunov equation

$$A^T PA - P = -Q \quad (2.46)$$

If there exists a solution such that both P and Q are positive definite, the system is AS.

It can be shown that this theorem is both necessary and sufficient. That is, for LTI systems, if there is no Lyapunov function of the quadratic form described in Equation 2.42, then there is no Lyapunov function. This result provides an alternative to examining the eigenvalues of the A matrix.

2.4.5 Lyapunov Design of LTI Feedback Controllers

These notions offer a valuable procedure for LTI control system design. Note that the closed-loop system with state feedback

$$x(k+1) = Ax(k) + Bu(k) \quad (2.47)$$

$$u = -Kx \quad (2.48)$$

is SISL if and only if there exist matrices $P > 0$, $Q \geq 0$ that satisfy the closed-loop Lyapunov equation

$$(A - BK)^T P(A - BK) - P = -Q \tag{2.49}$$

If there exists a solution such that both P and Q are positive definite, the system is AS.

Now suppose there exist $P > 0$, $Q > 0$ that satisfy the Riccati equation

$$P(k) = A^T P(k+1)(I + BR^{-1}B^T P(k+1))^{-1} A + Q \tag{2.50}$$

Now select the feedback gain as

$$K(k) = -(R + B^T P(k+1)B)^{-1} B^T P(k+1)A \tag{2.51}$$

and the control input as

$$u(k) = -K(k)x(k) \tag{2.52}$$

for some matrix $R > 0$.

These equations verify that this selection of the control input guarantees closed-loop asymptotic stability.

Note that the Riccati equation depends only on known matrices — the system (A, B) and two symmetric design matrices Q and R that need to be selected positive definite. There are many good routines that can find the solution P to this equation provided that (A, B) is controllable (e.g., MATLAB). Then, a stabilizing gain is given by Equation 2.51. If different design matrices Q and R are selected, different closed-loop poles will result. This approach goes far beyond classical frequency domain or root locus design techniques in that it allows the determination of stabilizing feedbacks for complex multivariable systems by simply solving a matrix design equation. For more details on this linear quadratic (LQ) design technique, see Lewis and Syrmos (1995).

2.4.6 Lyapunov Analysis for Nonautonomous Systems

We now consider nonautonomous (time-varying) dynamical systems of the form

$$x(k+1) = f(x(k), k), k \geq k_0 \tag{2.53}$$

$x \in \mathfrak{R}^n$. Assume again that the origin is an equilibrium point. For non-autonomous systems, the basic concepts just introduced still hold, but the explicit time dependence of the system must be taken into account. The basic issue is that the Lyapunov function may now depend on time. In this situation, the definitions of definiteness must be modified, and the notion of "decrescence" is needed.

Let $L(x(k),k): \mathfrak{R}^n \times \mathfrak{R} \to \mathfrak{R}$ be a scalar function such that $L(0,k) = 0$ and S be a compact subset of $\mathfrak{R}^n$. Then $L(x(k),k)$ is said to be:

Locally positive definite if $L(x(k),k) \geq L_0(x(k))$ for some time-invariant positive definite $L_0(x(k))$, for all $k \geq 0$ and $x \in S$. (Denoted by $L(x(k),k) > 0$.)

Locally positive semidefinite if $L(x(k),k) \geq L_0(x(k))$ for some time-invariant positive semidefinite $L_0(x(k))$, for all $k \geq 0$ and $x \in S$. (Denoted by $L(x(k),k) \geq 0$.)

Locally negative definite if $L(x(k),k) \leq L_0(x(k))$ for some time-invariant negative definite $L_0(x(k))$, for all $k \geq 0$ and $x \in S$. (Denoted by $L(x(k),k) < 0$.)

Locally negative semidefinite if $L(x(k),k) \leq L_0(x(k))$ for some time-invariant negative semidefinite $L_0(x(k))$, for all $k \geq 0$ and $x \in S$. (Denoted by $L(x(k),k) \geq 0$.)

Thus, for definiteness of time-varying functions, a time-invariant definite function must be dominated. All these definitions are said to hold *globally* if $S \in \mathfrak{R}^n$.

A time-varying function $L(x(k),k): \mathfrak{R}^n \times \mathfrak{R} \to \mathfrak{R}$ is said to be decrescent if $L(0,k) = 0$ and there exists a time-invariant positive definite function $L_1(x(k))$ such that

$$L(x(k),\ k) \leq L_1(x(k)), \quad \forall k \geq 0 \qquad (2.54)$$

The notions of decrescence and positive definiteness for time-varying functions are depicted in Figure 2.4.

Example 2.4.4: Decrescent Function

Consider the time-varying function

$$L(x(k),k) = x_1^2(k) + \frac{x_2^2(k)}{3 + \sin kT}$$

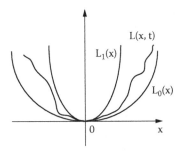

FIGURE 2.4
Time-varying function $L(x(k), k)$ that is positive definite $((L_0(x(k))) < L(x(k), k))$ and decrescent $(L(x(k), k) \leq L_1(x(k)))$.

Note that $2 \leq 3 + \sin kT \leq 4$, so that

$$L(x(k), k) \geq L_0(x(k)) \equiv x_1^2(k) + \frac{x_2^2(k)}{4},$$

and $L(x(k), k)$ is globally positive definite. Also,

$$L(x(k), k) \leq L_1(x(k)) \equiv x_1^2(k) + x_2^2(k)$$

so that it is decrescent.

THEOREM 2.4.5 (LYAPUNOV RESULTS FOR NONAUTONOMOUS SYSTEMS)

1. *Lyapunov stability:* If, for system described in Equation 2.53, there exists a function $L(x(k), k)$ with continuous partial derivatives, such that for x in a compact set $S \subset \Re^n$

$$L(x(k), k) \text{ is positive definite, } L(x(k), k) > 0 \qquad (2.55)$$

$$\Delta L(x(k), k) \text{ is negative semidefinite, } \Delta L(x(k), k) \leq 0 \qquad (2.56)$$

then the equilibrium point in SISL.
2. *Asymptotic stability:* If, furthermore, condition (2.56) is strengthened to

$$\Delta L(x(k), k) \text{ is negative definite, } \Delta L(x(k), k) < 0 \qquad (2.57)$$

then the equilibrium point is AS.

3. *Global stability:* If the equilibrium point is SISL or AS, if $S = \Re^n$ and, in addition, the radial unboundedness condition holds,

$$L(x(k), k) \to \infty \quad \text{as} \quad \|x(k)\| \to \infty, \quad \forall k \qquad (2.58)$$

then the stability is global.

4. *Uniform stability:* If the equilibrium point is SISL or AS, and in addition $L(x(k), k)$ is decrescent (e.g., Equation 2.54 holds), then the stability is uniform (e.g., independent of k_0).

The equilibrium point may be both uniformly and globally stable — e.g., if all the conditions of the theorem hold, then one has GUAS.

2.4.7 Extensions of Lyapunov Techniques and Bounded Stability

The Lyapunov results so far presented have allowed the determination of SISL. If there exists a function such that $L(x(k), k) > 0$, $\Delta L(x(k), k) \le 0$, and AS, if there exists a function such that $L(x(k), k) > 0$, $\Delta L(x(k), k) < 0$. Various extensions of these results allow one to determine more about the stability properties by further examining the deeper structure of the system dynamics.

2.4.7.1 UUB Analysis and Controls Design

We have seen how to demonstrate that a system is SISL or AS using Lyapunov techniques. However, in practical applications, there are often unknown disturbances or modeling errors that make even SISL too much to expect in closed-loop systems. Typical examples are systems of the form

$$x(k+1) = f(x(k), k) + d(k) \qquad (2.59)$$

with $d(k)$ an unknown but bounded disturbance. A more practical notion of stability is uniform ultimate boundedness (UUB). The next result shows that UUB is guaranteed if the Lyapunov derivative is negative outside some bounded region of $\Re^n$.

THEOREM 2.4.6 (UUB BY LYAPUNOV ANALYSIS)
If, for system as described in Equation 2.59, there exists a function $L(x, k)$ with continuous partial differences such that for x in a compact set $S \subset \Re^n$

$$L(x(k), k) \text{ is positive definite,} \quad L(x(k), k) > 0$$

$$\Delta L(x(k), k) < 0 \quad \text{for} \quad \|x\| > R$$

for some $R > 0$ such that the ball of radius R is contained in S, then the system is UUB, and the norm of the state is bounded to within a neighborhood of R. In this result, note that ΔL must be strictly less than zero outside the ball of radius R. If one only has $\Delta L(x(k), k) \leq 0$ for all $\|x\| > R$, then nothing may be concluded about the system stability.

For systems that satisfy the theorem, there may be some disturbance effects that push the state away from the equilibrium. However, if the state becomes too large, the dynamics tend to pull it back towards the equilibrium. Because of these two opposing effects that balance when $\|x\| \approx R$, the time histories tend to remain in the vicinity of $\|x\| = R$. In effect, the norm of the state is effectively or practically bounded by R.

The notion of the ball outside which ΔL is negative should not be confused with that of domain of attraction — in Example 2.4.1a. It was shown there that the system is AS provided $\|x\| < 1$, defining a domain of attraction of radius one.

The subsequent examples show how to use this result. They make the point that it can also be used as a control design technique where the control input is selected to guarantee that the conditions of the theorem hold.

Example 2.4.5: UUB of Linear Systems with Disturbance

It is common in practical systems to have unknown disturbances, which are often bounded by some known amount. Such disturbances result in UUB and require the UUB extension for analysis. Suppose the system

$$x(k+1) = Ax(k) + d(k)$$

has A stable, and a disturbance $d(k)$ that is unknown, but bounded so that $\|d(k)\| < d_M$, with the bound d_M known.

Select the Lyapunov function candidate

$$L(x(k)) = x^T(k)Px(k)$$

and evaluate

$$\Delta L(x(k)) = x^T(k+1)Px(k+1) - x^T(k)Px(k)$$

$$= x^T(k)(A^TPA - P)x(k) + 2x^T(k)A^TPd(k) + d^T(k)Pd(k)$$

$$= -x^T(k)Qx(k) + 2x^T(k)A^TPd(k) + d^T(k)Pd(k)$$

where (P,Q) satisfy the Lyapunov equation

$$A^T P A - P = -Q$$

One may now use the norm equalities to write

$$\Delta L(x(k)) \leq -[\sigma_{\min}(Q) \|x(k)\|^2 - 2\|x(k)\|\sigma_{\max}(A^T P)\|d(k)\| - \sigma_{\max}(P)\|d(k)\|^2]$$

which is negative as long as

$$\|x(k)\| \geq \frac{\sigma_{\max}(A^T P)d_M + \sqrt{\sigma_{\max}^2(A^T P)d_M^2 + \sigma_{\min}(Q)\sigma_{\max}(P)d_M^2}}{\sigma_{\min}(Q)}$$

Thus, if the disturbance magnitude bound increases, the norm of the state will also increase.

Example 2.4.6: UUB of Closed-Loop System

The UUB extension can be utilized to design stable closed-loop systems. The system described by

$$x(k+1) = x^2(k) - 10x(k)\sin x(k) + d(k) + u(k)$$

is excited by an unknown disturbance whose magnitude is bounded so that $\|d(k)\| < d_M$. To find a control that stabilizes the system and mitigates the effect of disturbances, select the control input as

$$u(k) = -x^2(k) + 10x(k)\sin x(k) + k_v x(k)$$

This helps cancel the sinusoidal nonlinearity and provides a stabilizing term yielding the closed-loop system

$$x(k+1) = k_v x(k) + d(k)$$

Select the Lyapunov function candidate

$$L(x(k)) = x^2(k)$$

whose first difference is given by

$$\Delta L(x(k)) = x^2(k+1) - x^2(k)$$

Evaluating the first difference along the closed-loop system trajectories yields

$$\Delta L(x(k)) \leq -x^2(k)\left(1 - k_{v\max}^2\right) - 2x(k)k_v d(k) + d^2(k)$$

which is negative, as long as

$$\|x(k)\| > \frac{k_{v\max}d_M + \sqrt{k_{v\max}^2 d_M^2 + \left(1 - k_{v\max}^2\right)d_M^2}}{\left(1 - k_{v\max}^2\right)}$$

which, after simplification, results in

$$\|x(k)\| > \frac{(1 + k_{v\max})}{\left(1 - k_{v\max}^2\right)} d_M$$

The UUB bound can be made smaller by moving the closed-loop poles near the origin. Placing the poles at the origin will result in a deadbeat controller and it should be avoided under all circumstances.

References

Goodwin, C.G. and Sin, K.S., *Adaptive Filtering, Prediction, and Control*, Prentice Hall, Englewood Cliffs, NJ, 1984.

Ioannou, P. and Kokotovic, P., *Adaptive Systems with Reduced Models*, Springer-Verlag, New York, 1983.

Jagannathan, S., *Neural Network Control of Nonlinear Discrete-Time Systems*, Taylor and Francis (CRC Press), Boca Raton, FL, 2006.

Kalkkuhl, J.C. and Hunt, K.J., Discrete-time neural model structures for continuous-time nonlinear systems, *Neural Adaptive Control Technology*, Zbikowski, R. and Hunt, K.J., Eds., World Scientific, Singapore 1996, chap. 1.

Landau, Y.D., *Adaptive Control: The Model Reference Approach*, Marcel Dekker, Basel, 1979.

Lewis, F.L. and Syrmos, V.L., *Optimal Control*, 2nd ed., John Wiley and Sons, New York, 1995.

Lewis, F.L., Abdallah, C.T., and Dawson, D.M., *Control of Robot Manipulators*, Macmillan, New York, 1993.

Lewis, F.L., Jagannathan, S., and Yesiderek, A., *Neural Network Control of Robot Manipulators and Nonlinear Systems*, Taylor and Francis, London, 1999.

Luenberger, D.G., *Introduction to Dynamic Systems*, Wiley, New York, 1979.

Narendra, K.S. and Annaswamy, A.M., *Stable Adaptive Systems*, Prentice Hall, Englewood Cliffs, NJ, 1989.

Sastry, S. and Bodson, M., *Adaptive Control*, Prentice Hall, Englewood Cliffs, NJ, 1989.
Slotine, J.-J.E. and Li, W., *Applied Nonlinear Control*, Prentice Hall, NJ, 1991.
Von Bertalanffy, L., *General System Theory*, Braziller, New York, 1998.
Whitehead, A.N., *Science and the Modern World*, Lowell Lectures (1925), Macmillan, New York, 1953.

Problems

Section 2.1

Problem 2.1.1: Simulation of compound interest system. Simulate the system of Example 2.1.3 and plot the state with respect to time.

Problem 2.1.2: Genetics. Many congenital diseases can be explained as a result of both genes at a single location being the same recessive gene (Luenberger 1979). Under some assumptions, the frequency of the recessive gene at generation k is given by the recursion

$$x(k+1) = \frac{x(k)}{1 + x(k)}$$

Simulate in MATLAB using $x(0) = 75$. Observe that $x(k)$ converges to zero, but very slowly. This explains why deadly genetic diseases can remain active for a very long time. Simulate the system starting for a small negative value of $x(0)$ and observe that it tends away from zero.

Problem 2.1.3: Discrete-time system. Simulate the system

$$x_1(k+1) = \frac{x_2(k)}{1 + x_2(k)}$$

$$x_2(k+1) = \frac{x_1(k)}{1 + x_2(k)}$$

using MATLAB. Plot the phase-plane plot.

Section 2.4

Problem 2.4.1: Lyapunov stability analysis. Using the Lyapunov stability analysis, examine stability for the following systems. Plot time histories to substantiate your claims.

$$\text{a. } x_1(k+1) = x_1(k)\sqrt{\left(x_1^2(k) + x_2^2(k)\right)}$$

$$x_2(k+1) = x_2(k)\sqrt{\left(x_1^2(k) + x_2^2(k)\right)}$$

$$\text{b. } x(k+1) = -x^2(k) - 10x(k)\sin x(k)$$

Problem 2.4.2: Lyapunov control design. Using Lyapunov techniques design controllers to stabilize the following system. Plot time histories of the states to verify your design. Verify passivity and dissipativity of the systems.

$$\text{a. } x(k+1) = -x^2(k)\cos x(k) - 10x(k)\sin x(k) + u(k)$$

$$\text{b. } \begin{aligned} x_1(k+1) &= x_1(k)x_2(k) \\ x_2(k+1) &= x_1^2(k) - \sin x_2(k) + u(k) \end{aligned}$$

Problem 2.4.3: Stability improvement using feedback. The system

$$x(k+1) = Ax(k) + Bu(k) + d(k)$$

has a disturbance $d(k)$ that is unknown but bounded so that $\|d(k)\| < d_M$, with the bounding constant known. In Example 2.4.6, the system with no control input, $B = 0$, and A stable was shown to be UUB. Show that by selecting the control input as $u(k) = -Kx(k)$ it is possible to improve the UUB stability properties of the system by making the bound on $\|x(k)\|$ smaller. In fact, if feedback is allowed, the initial system matrix A need not be stable as long as (A, B) is stabilizable.

3

Congestion Control in ATM Networks and the Internet

In the last chapter, background information on dynamic systems and Lyapunov stability was introduced. In the remainder of the book, we will apply these concepts from control theory to the design and development of protocols for wired and wireless networks. It can be observed that by using the control theory concepts, one can demonstrate not only stability but also performance of these protocols, in terms of quality of service (QoS) metrics such as throughput, packet/cell losses, end-to-end delay, jitter, energy efficiency, etc. In particular, we will focus on the development of congestion and admission control, distributed power control and rate adaptation, distributed fair scheduling, and routing protocols for wired and wireless networks. First, we will apply the control system concepts to the design of congestion control protocol for wired networks.

This chapter proposes an adaptive methodology for the available bit rate (ABR) service class of an asynchronous transfer mode network (ATM) first, and then an end-to-end congestion control scheme for the Internet, both being high-speed networks. In this methodology, the transmission rates of the sources are controlled, in response to the feedback information from the network nodes, to prevent congestion. The network is modeled as a nonlinear discrete-time system. As the traffic behavior on a network is self-similar, and its behavior is not typically known *a priori*, an adaptive scheme using a neural network (NN)-based controller is designed to prevent congestion, where the NN is used to estimate the traffic accumulation in the buffers at the given switch/destination. Tuning methods are provided for the NN, based on the delta rule, to estimate the unknown traffic. Mathematical analysis is presented to demonstrate the stability of the closed-loop error in a buffer occupancy system, so that a desired QoS can be guaranteed. The QoS for an ATM network is defined in terms of cell loss ratio (CLR), transmission or transfer delay (latency), and fairness. No learning phase is required for the NN, and initialization of the network weights is straightforward. However, by adding an initial learning phase, the QoS is shown to improve in terms of cell losses during transient conditions.

We derive design rules mathematically for selecting the parameters of the NN algorithm such that the desired performance is guaranteed during congestion, and potential trade-offs are shown. Simulation results are provided to justify the theoretical conclusions. The nonlinear-system-theory-based methodology can be readily applied to designing routing algorithms, transmission-link bandwidth estimation and allocation, and so on. An ATM network uses hop-by-hop feedback whereas the Internet uses an end-to-end network. First, we present the issue of congestion control for ATM networks and then end-to-end congestion control for the Internet.

3.1   ATM Network Congestion Control

ATM is a key technology for integrating broad-band multimedia services (B-ISDN) in heterogeneous networks, where data, video, and voice sources transmit information. ATM provides services to these sources with different traffic characteristics, by statistically multiplexing cells in terms of fixed length packets of 53 bytes long. The uncertainties of broadband traffic patterns, unpredictable statistical fluctuations of traffic flows, and self-similarity of network traffic can cause congestion in the network switches, concentrators, and communication links.

The ATM forum has specified several service categories in relation to traffic management in an ATM network. Two major classes, guaranteed service and best-effort service, resulted from the service categories (Kelly et al. 2000). The best-effort service category is further divided into two subclasses, namely, unspecified bit-rate (UBR) and ABR. The UBR source neither specifies nor receives a bandwidth, delay, or loss guarantee. In contrast, the ABR service type guarantees a zero-loss cell rate if the source obeys the dynamically varying traffic management signals from the network. The network uses resource management (RM) cells to inform the ABR source about the available bandwidth. If the source obeys these signals, it is guaranteed zero loss.

In a B-ISDN, the traffic and congestion controls describe different aspects of ATM operation. Congestion is defined as a condition of an ATM network, where the network does not meet a stated performance objective. By contrast, a traffic control, such as the connection admission control (CAC), defines a set of actions taken by the network to avoid congestion. Because of the uncertainty in the traffic flows of multimedia services, network congestion may still occur, despite the fact that an appropriate connection admission control scheme is provided. To prevent the QoS from severely degrading during short-term congestion, a suitable congestion control scheme is required.

Because the ATM forum decided to use a closed-loop rate-based congestion control scheme as the standard for the ABR service (Lakshman et al. 1999), several feedback control schemes were proposed in the literature (Jagannathan 2002, Chen et al. 1999, Fuhrman et al. 1996, Cheng and Chang 1996, Bonomi and Fendick 1995, Benmohamed and Meerkov 1993, Lakshman et al. 1999, Jain 1996, Mascola 1997, Qiu 1997, Izmailov 1995, Bae and Suda 1991, Liew and Yin 1998, Liu and Douligeris 1997, Bonomi et al. 1995, Jagannathan and Talluri 2002) with (Chen et al. 1999, Fuhrman et al. 1996, Cheng and Chang 1996, Qiu 1997, Jagannathan and Talluri 2002) and without neural and fuzzy logic (Bonomi and Fendick 1995, Benmohamed and Meerkov 1993, Jain 1996, Mascola 1997, Izmailov 1995, Bae and Suda 1991, Liu and Douligeris 1997, Bonomi et al. 1995, Jagannathan and Talluri 2000). In terms of feedback control, the congestion control is viewed as changing the source rates and regulating the traffic submitted by these sources onto the network connections while ensuring fairness. Most of these schemes are based upon a linear model of the ABR buffer. Each ABR buffer has a corresponding congestion controller, which sends congestion notification cells (CNC) back to the sources. These feedback control schemes from the literature were based on buffer length, buffer change rate, cell loss rate, buffer thresholds, and so on.

The most simple form of congestion control scheme is the binary feedback with FIFO queuing. Here, when the buffer occupancy exceeds a predefined threshold value, the switch begins to issue binary notifications to the sources and it continues to do so until the buffer occupancy falls below the threshold. This is referred to as thresholding in this chapter. Unfortunately, a FIFO queue may unfairly penalize connections that pass through a number of switches and therefore more advanced schemes, including explicit rate feedback schemes such as ERICA have appeared in the literature. Here, the feedback would provide an explicit rate to the sources via RM cells and these schemes ensure a certain degree of fairness. However, these schemes are not supported by rigorous mathematical analyses to demonstrate their performance, in terms of QoS, and rely only on some simulation studies. The author, in Benmohamed and Meerkov (1993), has reported that certain congestion control schemes even resulted in oscillatory behavior. In Benmohamed and Meerkov (1993), a stable feedback congestion controller is developed for packet switching networks based on the linear control theory notion. Although a stable operation is obtained with a linear controller, the network traffic rate is not estimated and the impact of the self-similar nature of network traffic on their controller design is not considered in the analysis.

In this chapter (Jagannathan and Tohmaz 2001, Jagannathan and Talluri 2002), the network traffic accumulation at the switch is considered as nonlinear. A novel adaptive multilayer NN congestion controller scheme is developed using a learning methodology, where an explicit rate for each connection is sent back to the source. Tuning laws are provided for the NN weights and closed-loop convergence and stability is proven using a Lyapunov-based analysis. It is shown that the controller guarantees the desired performance, even in the presence of self-similar traffic with bounded uncertainties, without any initial learning phase for the NN. However, by providing offline training to the NN, the QoS is shown to improve. Here, the objective is to obtain a finite bound on the cell losses while reducing the transmission delay, utilizing the available buffer space and simultaneously ensuring fairness. Simulation results are provided to justify the theoretical conclusions during simulated congestion. Though the simulation results and analysis are geared towards ATM, the methodology can be easily applicable to the Internet in the end-to-end scenario.

3.2 Background

Background on the approximation property is presented in the following subsection.

3.2.1 Neural Networks and Approximation Property

For the past several decades, NN-based algorithms have sprung up in computer science and engineering. This popularity is because the NN-based schemes possess function approximation and learning capabilities that can be used readily in several applications. One such application, as described in this chapter, is to predict the network traffic for congestion control. The NN-based methods can be broadly categorized based on the learning scheme they employ both offline and online. The offline learning schemes are used to train the NN. Once trained, the NN weights are not updated and the NN is inserted in the application. Unfortunately, the offline learning scheme pursued using backpropagation algorithm (Jagannathan 2006) is known to have convergence and weight-initialization problems.

The online learning NN scheme proposed in this chapter — though it requires more real-time computations, relaxes the offline training phase, avoids the weight initialization problems, and performs learning and adaptation simultaneously. This scheme also bypasses the problem of gathering data for offline training because such data is usually hard to come by for several applications. However, if similar *a priori* data is available, one can use the data to train the network offline and use the weights obtained from the training as initial weights for online training. Overall convergence and stability is proven for the proposed approach. Once the weights are trained, both offline and online, the weights can be fixed.

A general function, $f(x) \in C^{(s)}$, can be approximated using the two-layer NN as

$$f(x(k)) = W^T \varphi_2(V^T \varphi_1(x(k))) + \varepsilon(k), \qquad (3.1)$$

where W and V are constant weights and $\varphi_2(V^T(k)\varphi_1(x(k)))$, $\varphi_1(x(k))$ denote the vectors of activation functions at the instant k, with $\varepsilon(k)$ an NN functional reconstruction error vector. The net output is defined as

$$\hat{f}(x(k)) = \hat{W}^T \varphi_2(\hat{V}^T \varphi_1(x(k))). \qquad (3.2)$$

From now on, $\varphi_1(x(k))$ is denoted as $\varphi_1^{(k)}$ and $\varphi_2(\hat{V}^T \varphi_1(x(k)))$ is denoted as $\hat{\varphi}_2^{(k)}$. The main challenge now is to develop suitable weight update equations such that Equation 3.1 and Equation 3.2 are satisfied.

3.2.2 Stability of Systems

To formulate the discrete-time controller, the following stability notion is needed (Jagannathan 2006). Consider the nonlinear system given by:

$$x(k+1) = f(x(k), u(k))$$

$$y(k) = h(x(k)), \qquad (3.3)$$

where $x(k)$ is a state vector, $u(k)$ is the input vector, and $y(k)$ is the output vector. The solution is said to be uniformly ultimately bounded (UUB) if, for all $x(k_0) = x_0$, there exists an $\mu \geq 0$ and a number $N(\mu, x_0)$, such that $\| x(k) \| \leq \mu$ for all $k \geq k_0 + N$.

3.2.3 Network Modeling

Figure 3.1a shows the popular parking lot network configuration for evaluating the proposed congestion control schemes. This configuration and its name are derived from theatre parking lots, which consist of several parking areas connected via a single exit path as shown in the figure. For computer networks, an *n*-stage parking lot configuration consists of *n* switches connected in a series as *n* virtual circuits (VCs). Congestion happens whenever the input rate is more than the available link capacity. In other words,

$$\sum \text{Input rate} > \text{available link capacity} \qquad (3.4)$$

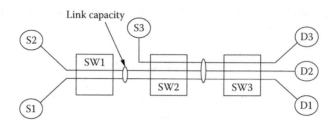

FIGURE 3.1A
Parking lot configuration.

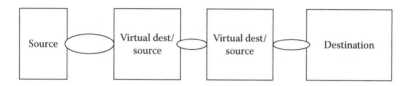

FIGURE 3.1B
Hop-by-hop rate control.

Most congestion schemes consist of adjusting the input rates to match the available link capacity (or rate) so that a desired QoS can be met. Depending upon the duration of congestion, different schemes can be applied. For congestion lasting less than the duration of connection, an end-to-end feedback control scheme can be used. For short-term congestion, providing sufficient buffers in the switches is the best solution and a design procedure is required. Though end-to-end rate control can be applicable, the round-trip delay can be very large. This problem is fixed by segmenting the network into smaller pieces and letting the switches act as "virtual source" and/or "virtual destination." Segmenting the network using virtual source/destination reduces the size of the feedback loops (Jain 1996), but the intermediate switches/routers have to be updated with software so that the entire system is transparent. To overcome this problem, an end-to-end congestion control scheme is typically proposed.

The most important step in preventing congestion is to estimate the network traffic that is being accumulated at the switch in an intelligent manner. In the worst-case scenario, every ATM switch could act as a virtual source/destination and one could apply hop-by-hop rate control, as shown in Figure 3.1b, for preventing congestion. Tracking queue length (whose exact value is determined based on QoS measurements) is desirable to avoid cell losses due to overflows and to maintain high network utilization.

A computer network can be viewed as a set of virtual source/destination pairs and they can then be modeled as multiinput multioutput

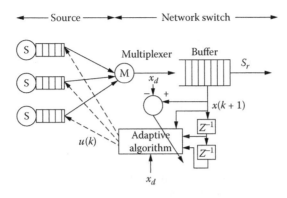

FIGURE 3.2
Network model with adaptation.

discrete-time nonlinear systems to be controlled, as shown in Figure 3.2, given in the following form:

$$x(k+1) = f(x(k)) + Tu(k) + d(k) \qquad (3.5)$$

with state $x(k) \in \Re^n$ being the buffer length (or occupancy) at time instant k, T being the sampling instant, $f(\cdot)$ being the nonlinear traffic accumulation or buffer occupancy, and control $u(k) \in \Re^n$ being the traffic rate calculated by feedback, so that the sources modify their transmission rates from their intended rates. The unknown nonlinear function, $f(\cdot) = sat[x(k) - q(t - T_{fb}) + I_{ni}^{(k)} - S_r^{(k)}]$, is defined as the actual traffic flow and it is a function of current buffer occupancy $x(k)$, buffer size, x_d, traffic arrival rate at the destination buffer, $I_{ni}^{(k)}$, bottleneck queue level $q(t - T_{fb})$ (end-to-end), service capacity Sr, at a given switch with $sat(\cdot)$ applied to the function. The unknown disturbance vector, which can be an unexpected traffic burst/load or change in available bandwidth due to the presence of a network fault, acting on the system at the instant k is $d(k) \in \Re^n$, is assumed to be bounded by a known constant, $\|d(k)\| \le d_M$. The state $x(k)$ is a scalar if a single switch/single buffer scenario is considered, whereas it becomes a vector when multiple network switches/multiple buffers are involved, as is the case of multiple switches in tandem.

Given a finite buffer size x_d, define the performance criterion in terms of buffer occupancy error as

$$e(k) = x(k) - x_d, \qquad (3.6)$$

where the cell losses are given by

$$c(k) = e(k), \quad \text{if} \quad e(k) > 0,$$
$$= 0 \quad \text{Otherwise.}$$

(3.7)

Equation 3.6 can be expressed as:

$$e(k+1) = x(k+1) - x_d,$$

(3.8)

where $e(k+1)$ and $x(k+1)$ denote the error in buffer occupancy and the state at the instant $k+1$, respectively. Using Equation 3.5 in Equation 3.8, the dynamics of the nonlinear system described in Equation 3.5 can be written in terms of the buffer-length errors as:

$$e(k+1) = f(x(k)) - x_d + Tu(k) + d(k)$$

(3.9)

In high-speed networks, each source should be able to use the available bandwidth up to its peak cell rate (PCR) or, at least, should meet the minimum-rate requirement for ensuring fairness. The goal of a traffic-rate controller then, is to fully utilize the available network bandwidth while maintaining a good QoS, where the QoS parameters are the CLR, fairness, and transmission/cell-transfer delay or latency. This objective is achieved by selecting a suitable traffic rate using feedback, $u(k)$, via minimizing the difference between the actual and desired buffer length using a predicted congestion level obtained via buffer occupancy errors, its rate of change, cell losses, and so on.

Define the traffic rate input calculated using feedback, $u(k)$, as

$$u(k) = \frac{1}{T}(x_d - \hat{f}(x(k)) + k_v e(k)),$$

(3.10)

where $\hat{f}(x(k))$ is the traffic estimate of the unknown nonlinear traffic accumulation function, $f(x(k))$, at the network switch and k_v is a diagonal gain matrix. Then, the closed-loop error in buffer occupancy system is given by Equation 3.9. One can also select $u(k)$ as

$$u(k) = \frac{1}{T}(x_d - \hat{f}(x(k)) + k_v e_1(k))$$

(3.11)

where $\hat{f}(x(k))$ is the traffic-input estimate of the unknown nonlinear traffic function $f(x(k))$, at the network switch, a diagonal gain matrix

$k_v = [k_{v1} \quad k_{v2}]^T$, the error vector $e_1^{(k)} = [e(k) \; \frac{e(k) - e(k-1)}{T}]^t$ with $e(k-1)$ being the past value of buffer occupancy. Then, the closed-loop error in buffer occupancy system becomes:

$$e(k+1) = k_v e(k) + \tilde{f}(x(k)) + d(k), \tag{3.12}$$

where the traffic-flow modeling error is given by

$$\tilde{f}(x(k)) = f(x(k)) - \hat{f}(x(k)) \tag{3.13}$$

The controller presented in Equation 3.10 without the traffic-estimation term is defined to be a proportional controller, whereas the controller presented in Equation 3.11, without the traffic-estimation term, is defined as a proportional plus derivative traffic controller. In either case, the error in buffer occupancy system expressed in terms of cell losses is driven by the network traffic flow modeling error and the unknown disturbances. In this chapter, it is envisioned that by appropriately using an adaptive NN in discrete-time to provide the traffic estimate, $\hat{f}(\cdot)$, the error in buffer length, and hence the cell losses and delay can be minimized and fairness can be ensured. Note, that for an end-to-end congestion control scheme, the nonlinear function $f(\cdot)$, includes the unknown bottleneck queue and therefore, estimating the nonlinear function would help design an efficient controller because the bottleneck queue level is always unknown. The proposed approach is readily applicable to both hop-by-hop, as well as end-to-end scenarios. A comparison in terms of QoS, obtained for the multilayered NN with and without offline training is also included.

The location of the closed-loop poles in the gain matrix k_v, dictates the trade off between the cell transfer or transmission delay and cell losses with the buffer utilization (Jagannathan and Talluri 2002). In other words, smaller cell losses imply long latency or transmission delay and *vice versa*. The buffer-length error system expressed in Equation 3.12 is used to focus on selecting discrete-time NN-tuning algorithms that guarantee the cell losses, $c(k)$, by appropriately controlling the buffer-length error $e(k)$.

For the case of a single source sending traffic onto the network switch, the feedback rate calculated using Equation 3.10 or Equation 3.11 is a scalar variable, whereas if multiple sources are sending traffic, the feedback rate, $u(k)$, has to be shared fairly among the sources. The most widely accepted notion of fairness is the max-min fairness criterion, and it is defined as

$$
\text{Fair share} = MR_p + \frac{a - \left(\sum_{i=1}^{M^\tau} \tau_i + \sum_{i=1}^{l} MR \right)}{N - M^\tau}, \tag{3.14}
$$

where MR_p is the minimum rate for a source p, $\sum_{i=1}^{l} MR$ is the sum of minimum rates of l active sources, a is the total available bandwidth on the link, N is the number of active sources, τ is the sum of bandwidth of M^τ active sources bottlenecked elsewhere. In order to achieve fairness, the feedback rate is selected for adjustable sources as

$$
u(k) = \sum_{i=1}^{m} u_i^{(k)} + \sum_{i=m+1}^{M} max(u_i(k), Fairshare(k)) \tag{3.15}
$$

where $u_i(k); i = 1, ..., M$ are source rates and for the last M sources, the feedback is used to alter their rates. Note, that in many cases, the number of controllable sources is not accurately known and hence the proposed scheme can also be used to alter some real-time traffic as the real-time video, for instance, is compressed and can tolerate bounded delays that are small.

The proposed methodology can be described as follows: the NN at the switch estimates the network traffic; this estimated value together with cell losses is used to compute the rate, $u(k)$, by which the source has to reduce or increase its rate from the original intended rate. If $u(k)$ is a positive value, it indicates that the buffer space is available and no congestion is predicted, and the sources can increase their rates by an amount given by Equation 3.14 or Equation 3.15. On the other hand, if $u(k)$ is negative, it indicates that there are cell losses at the switch, and the sources have to decrease their rates using Equation 3.14 or Equation 3.15. In any event, the NN-based controller should ensure performance and stability while learning.

3.3   Traffic Rate Controller Design for ATM Networks

In the case of two-layer NN, the tunable NN weights enter in a nonlinear fashion. Stability analysis by Lyapunov's direct method is performed for a family of weight-tuning algorithms for a two-layer NN. Assume, therefore, that the approximation error is bounded above $\|\varepsilon(k)\| \leq \varepsilon_N$ with the bounding constant ε_N known. For suitable approximation properties, it is necessary to select a large-enough number of hidden-layer neurons. It is not known how to compute this number for general multilayer NN. Typically, the number of hidden-layer neurons is selected after careful analysis.

3.3.1 Controller Structure

Defining the NN functional estimate in the controller by

$$\hat{f}(x(k)) = \hat{W}^T \varphi_2(\hat{V}^T \varphi_1(k)), \tag{3.16}$$

with $\hat{W}(k)$ and $\hat{V}(k)$ being the current NN weights, the next step is to determine the weight updates so that the performance of the closed-loop buffer error dynamics of the network is guaranteed. The structure of the controller is shown in Figure 3.3 and it is envisioned that a controller exists at every ATM switch.

Let W and V be the unknown ideal weights required for the approximation to hold in Equation 3.1 and assume they are bounded by known values so that

$$\|W\| \le W_{\max}, \quad \|V\| \le V_{\max} \tag{3.17}$$

Then, the error in the weights during estimation is given by:

$$\tilde{W}(k) = W - \hat{W}(k), \quad \text{and} \quad \tilde{V}(k) = V - \hat{V}(k), \quad \tilde{Z}(k) = Z - \hat{Z}(k), \tag{3.18}$$

where $Z = \begin{bmatrix} W & 0 \\ 0 & V \end{bmatrix}$, and $\hat{Z} = \begin{bmatrix} \hat{W} & 0 \\ 0 & \hat{V} \end{bmatrix}$.

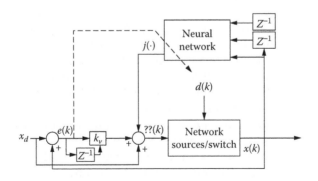

FIGURE 3.3
Neural network (NN) controller structure.

FACT 3.3.1

The activation functions are bounded by known positive values so that $\|\hat{\varphi}_1(k)\|$ $\leq \varphi_{1\max}$, $\|\hat{\varphi}_2(k)\| \leq \varphi_{2\max}$, $\|\tilde{\varphi}_1(k)\| \leq \tilde{\varphi}_{1\max}$, *and* $\|\tilde{\varphi}_2(k)\| \leq \tilde{\varphi}_{2\max}$

The traffic rate using the feedback, $u(k)$, is obtained as

$$u(k) = \frac{1}{T}(x_d(k+1) - \hat{W}^T(k)\hat{\varphi}_2(k) + k_v e(k)) \qquad (3.19)$$

and the closed-loop dynamics become

$$e(k+1) = k_v e(k) + \bar{e}_i(k) + W^T(k)\tilde{\varphi}_2(k) + \varepsilon(k) + d(k) \qquad (3.20)$$

where the traffic-flow modeling *error* is defined by

$$\bar{e}_i(k) = \tilde{W}^T(k)\tilde{\varphi}_2(k). \qquad (3.21)$$

3.3.1.1   Weight Initialization and Online Learning

For the NN control scheme that is presented in this chapter, there is no preliminary offline analysis required for tuning the weights. The weights are simply initialized at zero. Figure 3.3 shows that the controller is just a simple and conventional rate-based proportional and derivative controller because the NN outputs become zero. As a result, the unknown traffic accumulation at the switch is not approximated by the NNs initially or during the transient phase when the transmission starts. Standard results (Jagannathan and Talluri 2000) indicate that for controlling traffic congestion, a conventional rate controller may result in bounded buffer-occupancy errors if the gains are suitably selected. Then, the closed-loop system remains stable until the NN system begins to learn. Subsequently, the buffer occupancy errors become smaller and the cell losses eventually become small as well. Providing offline training will minimize the cell losses in the transient condition, and the weights from offline can serve as initial weights for online training. Therefore, the NN weights are tuned online, and the QoS improves as the NN learns the unknown traffic accumulation, $f(x(k))$. A comparison is shown in terms of QoS between multilayer NN with, and without, an initial learning phase. Here, the backpropagation algorithm is used to train the NN, and the training is described in subsection 3.3.3.

3.3.1.2 Implementation of the Proposed Scheme

Two types of ATM cells, namely, flow-data cells and RM cells, on an ABR connection are considered. A source/virtual receives a regular sequence of RM cells that provide feedback to enable it to adjust its rate of cell transmission. The bulk of the RM cells are initiated by the source, which transmit one forward RM (FRM) cell for every (Nrm –1) data cells, where Nrm is a preset parameter. As each FRM is received at the destination, it is turned around and transmitted back to the source as a backward RM (BRM) cell. Each FRM cell contains a congestion indication (CI), no increase (NI), and explicit cell rate (ER) fields, besides current cell rate (CCR) and minimum cell rate (minCR) fields. Any intermediate switch can set CI or NI fields, the source will set the CCR and MinCR field, whereas the intermediate or destination switches set the ER field. Our proposed scheme at each intermediate/destination switch will provide the feedback, $u(k)$, to be used in the ER field for its source/virtual source and, accordingly, the sources/virtual sources will alter their rate based on the ER field. Therefore, the implementation is straightforward. The next step is to determine the weight updates so that the tracking performance of the closed-loop tracking error dynamics is guaranteed.

3.3.2 Weight Updates

It is required to demonstrate that the performance criterion in terms of cell losses, $c(k)$, transmission delay, and buffer utilization monitored using the error in buffer occupancy, $e(k)$, is suitably small, source rates adjusted fairly, and that the NN weights, $W(k)$, $\hat{V}(k)$, remain bounded for the traffic rate, and $u(k)$, is bounded and finite. In the following theorem, discrete-time weight-tuning algorithms based on the error in buffer occupancy are given, which guarantee that both the error in buffer occupancy and weight estimates are bounded.

THEOREM 3.3.1 (TRAFFIC CONTROLLER DESIGN)
Let the desired buffer length, x_d, be finite and the NN traffic approximation error bound, ε_N, which is equal to the target CLR, and the disturbance bound, d_M, be known constants. Take the control input for Equation 3.5 as Equation 3.19 with weight tuning provided by

$$\hat{V}(k+1) = \hat{V}(k) - \alpha_1 \hat{\varphi}_1(k)[\hat{y}_1(k) + B_1 k_v e(k)]^T - \Gamma \left\| I - \alpha \hat{\varphi}_1(k) \hat{\varphi}_1^T(k) \right\| \hat{V}^T(k), \quad (3.22)$$

$$\hat{W}(k+1) = \hat{W}(k) - \alpha_2 \hat{\varphi}_2(k) e^T(k+1) - \Gamma \left\| I - \alpha_2 \hat{\varphi}_2(k) \hat{\varphi}_2^T(k) \right\| \hat{W}(k), \quad (3.23)$$

where $\hat{y}_1(k) = \hat{V}^T(k)\hat{\varphi}_1(k)$, and $\Gamma > 0$ is a design parameter. Then, the buffer occupancy error, $e(k)$, the NN weight estimates, $\hat{V}(k)$, $\hat{W}(k)$, are *UUB*, with the bounds specifically given by Equation 3.A.9 and Equation 3.A.10, provided the design parameters are selected as:

$$(1) \quad \alpha_1 \varphi_{1\max}^2 < 2, \tag{3.24}$$

$$(2) \quad \alpha_2 \varphi_{2\max}^2 < 1, \tag{3.25}$$

$$(3) \quad 0 < \Gamma < 1, \tag{3.26}$$

$$(4) \quad k_{v\max} < \frac{1}{\sqrt{\bar{\sigma}}}, \tag{3.27}$$

where $\bar{\sigma}$ is given by

$$\bar{\sigma} = \beta_1 + k_1^2 \beta_2, \tag{3.28}$$

$$\beta_1 = 1 + a_2 \varphi_{2\max}^2 + \frac{\left[-a_2 \varphi_{2\max}^2 + \Gamma\left(1 - a_2 \varphi_{2\max}^2\right)\right]^2}{1 - a_2 \varphi_{2\max}^2} \tag{3.29}$$

$$\beta_2 = 1 + a_1 \varphi_{1\max}^2 + \frac{\left[a_1 \varphi_{1\max}^2 + \Gamma\left(1 - a_1 \varphi_{1\max}^2\right)\right]^2}{2 - a_1 \varphi_{1\max}^2}. \tag{3.30}$$

PROOF   See Appendix 3.A.

3.3.3   Simulation Examples

This section describes the NN model, parameters, and constants used in the simulations. The traffic sources used in the simulation are also discussed and the simulation results are explained.

3.3.3.1   Neural Network (NN) Model

The NN used in this approach is shown in Figure 3.4. Previous six values of the buffer occupancy were used as inputs to the first layer NN as a trade-off between approximation and computation. The output of the NN

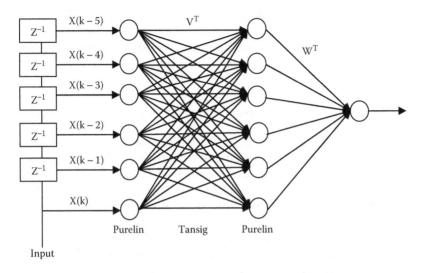

FIGURE 3.4
Two-layer neural network architecture for traffic approximation.

is a scalar value, which gives the approximated value of the traffic at each switch. The size of the weight matrix for $\hat{V}(k)$ is 6 by 6 and for $\hat{W}(k)$ it is a 6 by 1 vector at each switch. The activation function, $\phi(\cdot)$, used in the input and the output layers at each node is linear, whereas a tangent sigmoid function is used at each node in the hidden layer. Each switch will have its own NN congestion controller.

The feedback gain, k_v, is selected as $[0.1\ 0]^T$. The initial adaptation gain α, is taken as 0.9 and is updated using the projection algorithm as $\alpha(k) = \frac{0.1}{(0.1 + \phi(x(k))^T \phi(x(k)))}$. The parameter Γ is chosen to be 0.001 as per the requirement, and the buffer occupancy is considered empty initially. The initial weights are all chosen as zero for the case of NN with no offline training. Here, one can select any initial values for the weights. In this paper, a comparison is shown between NN, with and without a training phase.

Using 4000 frames of data, which is randomly chosen from a Star Wars movie, the NN is trained by the well-known backpropagation algorithm (Jagannathan 2006). We saved the weights after training and deployed them as initial weights for the online training. The weight-tuning updates derived therein were deployed for online training. The network model and the traffic sources used in this work are discussed next.

3.3.3.2 Network Model and Performance Metrics

The network model used in simulations is similar to that shown in Figure 3.2 with several ATM switches in tandem. The maximum buffer size at the switch x_d, is varied along 150, 200, 250, 300, and 350 cells. Cell losses $c(k)$ are defined as $x(k) - x_d$, when $x(k) > x_d$. The CLR is defined as the total number of cells discarded at the receiver, due to buffer overflows, divided by the total number of cells sent onto the network. The transmission delay is defined as

$$\text{Transmission delay} = T_c - T_o, \tag{3.31}$$

where T_c and T_o denote the time to complete the transmission by a source, with and without feedback, respectively. Initially, no additional delays are injected into the feedback, besides the delayed feedback from the simulator due to congestion, whereas additional delays were injected later to evaluate the performance of the controller, in the event that the feedback delays increase due to other uncertainties or hardware.

Each source is assumed to possess infinite buffer and the service rate at the buffer is altered as a result of the feedback from the ingress switch. If the virtual source is an ATM switch, then the buffer size is finite and it receives feedback from other switches. Also, if there is no feedback, the arrival and service rates at the buffer of a source/virtual are equal and this is referred to as an open-loop scenario. With feedback, the service rate at the buffer of a source or a virtual source is adjusted using the feedback, $u(k)$, in Equation 3.19. The service rate at the source will become the arrival rate at the destination. In the case of multiple sources, the feedback $u(k)$ is computed using Equation 3.19 and is divided among the sources fairly using Equation 3.15 to maintain a minimum rate. In our simulations, several traffic sources were considered. Because the proposed methodology is compared with thresholding, adaptive ARMAX and one-layer NN-based methods, they are discussed subsequently. The simulation scenario and the results presented along with the parameters are on par with that of other works (Chang and Chang 1993, Benhamohamed and Meerkov 1993, Jain 1996, Liu and Douligeris 1997, Bonomi et al. 1995).

3.3.3.3   Threshold Methodology

In the threshold-based controller (Liu and Douligeris 1997), a CNC is generated when the ratio of occupied buffer length is greater than a threshold value Q_t. Consequently, all sources reduce their transmission rates to 50% from their current rates when a CNC is received. For simplicity, a buffer threshold value of 40% is selected, similar to the work published by other researchers, which implies that we have chosen $Q_t = 0.4$.

3.3.3.4   Adaptive ARMAX and One-Layer Neural Network

In the adaptive ARMAX method (Jagannathan and Talluri 2000), the traffic accumulation is given by a linear in the parameter ARMAX approach, whereas in the case of a one-layer NN basis, functions in the hidden layer are employed (Jagannathan and Talluri 2002) to approximate the traffic. For details of these methods, please refer to (Jagannathan and Talluri 2000, 2002).

3.3.3.5 Traffic Sources

This section describes different traffic sources and the NN model used in the simulations. Both ON/OFF and MPEG data were used. Intially, to evaluate the performance of the controller, ON/OFF and MPEG sources are used separately. In the case of fairness, both ON/OFF and MPEG are used together.

3.3.3.5.1 Multiple ON/OFF Sources

To evaluate the performance of the controllers, multiple ON/OFF sources were used initially. Figure 3.5 shows the multiplexed ON/OFF traffic created through 40 ON/OFF sources. The PCR for each source is 1200 cells/sec, the MCR is 1105 cells/sec, and the PCR of the combined source is 48,000 cells/sec and the MCR is 44,200 cells/sec. A sampling interval of 1 msec was used in the simulations.

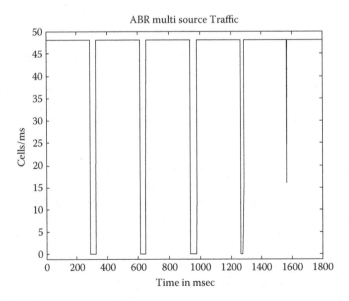

FIGURE 3.5
Traffic from ON/OFF sources.

3.3.3.5.2   VBR Sources

In high-speed networks such as ATM and Internet, real-time video applications will become major traffic sources. The moving picture expert group (MPEG) is one of the most promising interframe compression techniques for such applications (Jagannathan and Talluri 2002). Therefore, one of the critical issues is to realize the effective MPEG video transfer through networks. Our scheme takes advantage of the flexibility of MPEG video encoding and the feedback mechanisms of ABR service.

Compressed video also is often rate adaptive, that is, it is possible to modify the source rate dynamically by adjusting the compression parameters of a video coder (Liu and Yin 1998). The video source rate can be matched to the rate returned in the RM cell by modifying the quantization level used during compression. Therefore, a feedback approach can be employed even with variable bit rate (VBR) type traffic.

The MPEG data set used in the simulations was found at a Bellcore ftp site (Liu and Douligeris 1997). This data set comes from the movie "Star Wars." The movie length is approximately 2 h and contains a diverse mixture of material ranging from low-complexity scenes to scenes with high action. The data set has 174,138 patterns, each pattern representing the number of bits generated in a frame time, F. In this trace, 24 frames are coded per second, so F is equal to 1/24 sec. The peak bit rate of this trace is 185,267 b/frame, the mean bit rate is 15,611 b/frame and the standard deviation of the bit rate is about 18,157. We used 4000 frames from this trace to run our simulations. The trace used for offline training is different from the one used for testing.

The sources were driven with MPEG data, which cannot be adequately modeled using a conventional traffic model. This source behavior can be accommodated by the ABR service, because the explicit-rate scheme allows sources to request varying amounts of bandwidth over time. When the bandwidth demand cannot be met, the network provides feedback to modify the source rate.

Figure 3.10 shows the multiplexed MPEG traffic created through multiple VBR sources. The combined PCR for all the sources is given by 13,247 cells/sec with the MCR being 3444 cells/sec. Simulations using multiple VBR sources were performed with a measurement interval of 41.67 msec. Simulation Example 3.3.1 deals with results obtained when the sources are of ON/OFF type, whereas the Example 3.3.2 illustrates the results for VBR type. In Simulation Example 3.3.3, additional delays were injected in the feedback and the controller performance was compared, whereas Example 3.3.4 presents the performance of the controller in the presence of cross-traffic and fairness in the end-to end case. Example 3.3.5 deals with multiple bottlenecks with switches connected in tandem.

Example 3.3.1: Multiple ON/OFF Sources

To evaluate the performance of the control schemes in the presence of ON/OFF traffic, network congestion was created by altering the service capacity at the destination switch as follows:

$$S_r = 48{,}000 \text{ cells/sec}, \qquad 0 \leq t \leq 360m\sec,$$

$$= 45{,}000 \text{ cells/sec}, \qquad 361 \leq t \leq 720m\sec,$$

$$= 40{,}000 \text{ cells/sec}, \qquad 721 \leq t \leq 1080m\sec, \qquad (3.32)$$

$$= 45{,}000 \text{ cells/sec}, \qquad 1081 \leq t \leq 1440m\sec,$$

$$= 48{,}000 \text{ cells/sec}, \qquad 1441 \leq t \leq 1800m\sec.$$

Congestion was created by reducing the service capacity to 40,000 cells/ sec (less than the MCR). Figure 3.6 and Figure 3.7 (same as Figure 3.6, but with threshold scheme removed) show the CLR with time when threshold, adaptive, one-layer, two-layer — with and without offline training-based

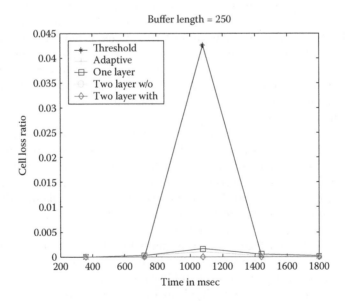

FIGURE 3.6
Cell loss ratio with congestion.

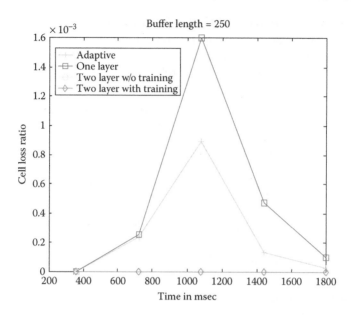

FIGURE 3.7
Cell loss ratio with congestion.

congestion control schemes — were deployed at the network switch with buffer length of 250 cells. The service capacity at the switch for different intervals of time is given by Equation 3.32. As expected, the CLR for the threshold, ARMAX, the one-layer NN method increases as the service capacity is decreased, reaches a maximum value when the service capacity is reduced to a small value during the time interval of $721 \leq t \leq 1080 m \sec$. The CLR again decreases as the service capacity is increased toward the PCR of the combined traffic.

In contrast, the CLR for the two-layer NN method, with and without *a priori* training, remains near zero throughout the simulation time, even when the service capacity was reduced to 20,000 cells/sec, which implies that the two-layer NN controller performs better than all other methods during congestion in controlling the arrival rate of the cells into the destination buffer. As all the traffic used was of ON/OFF type, the source rates were reduced fairly and quickly using the proposed scheme, resulting in a low CLR. Further, a transmission delay of approximately 25 msec (<2% of total time of transmission) was only noticed for the proposed two-layer NN during congestion and because of the feedback delay. Note that the cells are being transmitted during OFF periods from the source buffer if they are queued. The transmission delay is considered acceptable, whereas for other adaptive and one-layer NN schemes, variable delays

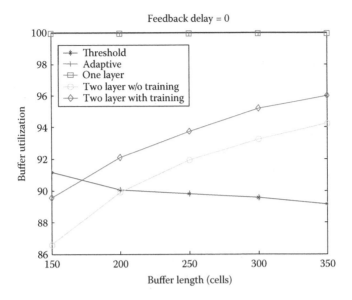

FIGURE 3.8
Buffer utilization.

of much higher than 50 msec (>2% of the total time of transmission) were observed, partly due to the unsatisfactory controller performance in appropriately reducing/increasing the rate when required. As a result, higher CLR was observed for thresholding, adaptive ARMAX, and one-layer NN whereas the two-layer NN controller was able to store cells in the buffer (as presented next in the buffer utilization) and able to transmit it during OFF periods, reducing cell delay and cell losses.

Figure 3.8 and Figure 3.9 illustrate the buffer utilization for different congestion control schemes. Higher buffer utilization resulted in slightly long cell-transfer delays because of queuing at the buffer. In addition, the buffer utilization for the case of multilayer NN with training is higher than the buffer utilization of the NN without training. In spite of larger queuing delays for the two-layer NN, the overall delay is much smaller than for the other schemes. Moreover, by suitably adjusting the gain parameter, one can alter the transmission delay and CLR and one can observe a trade-off.

Example 3.3.2: Multiple MPEG Sources

Figure 3.10 shows the multiplexed VBR input traffic created using several MPEG data sources. Figure 3.11 and Figure 3.12 (same as in Figure 3.7, but with threshold scheme removed) show the CLR with buffer size when the service capacity is kept constant at 2400 cells/sec, which is slightly

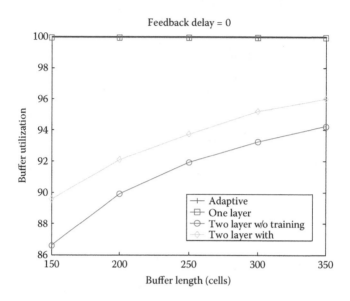

FIGURE 3.9
Buffer utilization.

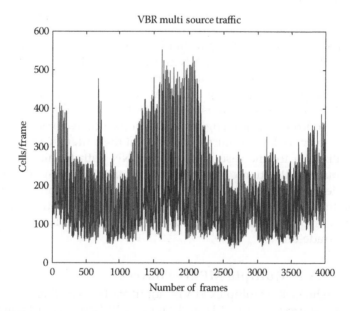

FIGURE 3.10
Traffic from MPEG sources.

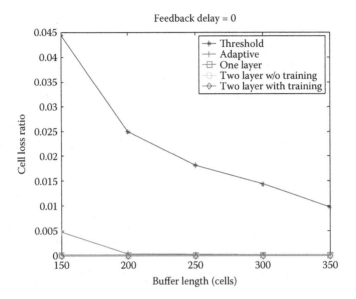

FIGURE 3.11
Cell loss ratio with buffer length.

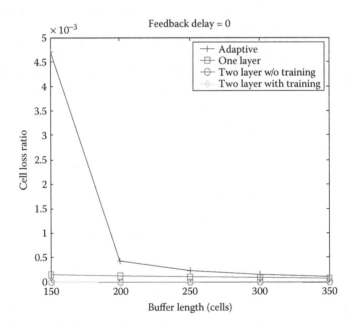

FIGURE 3.12
Cell loss ratio with buffer length.

less than the MCR of 3444 cells/sec. The cell losses decrease with an increase in buffer size for the case of one- and two-layer NNs. Due to a slight decrease in CLR with increasing buffer size, the two-layer NN with training method indicates that a buffer size of 150 cells provides a result similar to having buffer size of 350 cells. First of all, low CLR observed for all methods are due to the feedback, and large buffer size or feedback implies that longer delays occur with low CLR.

Further, the one-layer and multilayer NN methods render a low CLR compared to the threshold and adaptive ARMAX methods because the NN approximation property helps to accurately predict the traffic, based on past measurements. In fact, the two-layer NN method with *a priori* training provides a CLR near zero. The NN-based congestion controllers were able to achieve the low CLR because the NN accurately predicted traffic inflow and onset of congestion. Subsequently, the NN controller was able to generate a desirable source rate to prevent congestion and the sources were able to quickly accommodate their behavior in response to the delayed feedback. On the other hand, threshold approach generates a much higher CLR, compared to other schemes, even when the threshold value is kept at 40%, taking longer to transmit all the cells. Increasing the threshold value will increase CLR but the delay will be smaller. Similar to the case of ON/OFF traffic, transmission delays were noticed for all the controller schemes and the NN-based scheme was able to transfer the information with a delay of less than 2%, whereas others took much longer.

Figure 3.13 and Figure 3.14 (same as Figure 3.13, but with thresholding scheme removed) show the result of the simulated congestion created by throttling the outgoing transmission link at the network switch when a buffer size of 250 cells was used. Throttling was performed by reducing the service capacity at the switch to 1920 cells/sec during the time interval of 66.71 to 100 sec as follows:

$$S_r = 4680 \text{ cells/sec} \qquad 0 \leq t \leq 33.33 \sec,$$

$$= 3360 \text{ cells/sec} \qquad 33.37 \leq t \leq 66.66 \sec,$$

$$= 1920 \text{ cells/sec} \qquad 66.71 \leq t \leq 100 \sec,$$

$$= 3360 \text{ cells/sec} \qquad 100.04 \leq t \leq 133.33 \sec, \qquad (3.33)$$

The CLR for all the methods remains close to 0 when the service capacity is higher than the MCR during the time intervals of 0 to 33.33 sec and 133.37 to 167 sec. As expected, the CLR attains a maximum value when the service capacity is reduced to less than the MCR of the combined traffic.

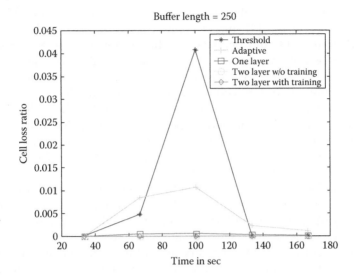

FIGURE 3.13
Cell loss ratio with congestion.

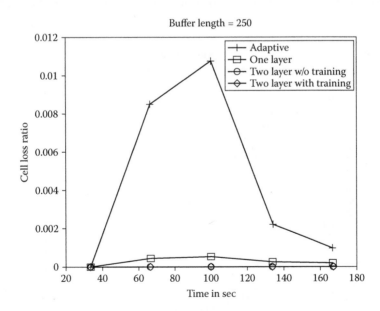

FIGURE 3.14
Cell loss ratio with congestion.

Moreover, it is observed that the two-layer NN method performed better in providing a low CLR during congestion, compared to the adaptive ARMAX, threshold, and one-layer methods. However, as expected, the multilayer NN method takes longer to transfer cells from the source to the destination compared to the open-loop scenario. In addition, providing offline training did indeed reduce the cell-transfer delay, whereas the CLR is still near zero. From the figures, it is clear that the CLR resulting from the two-layer NN method outperforms the adaptive, thresholding, and one-layer NN methods. Finally, the overall delay observed for two-layer NN was within 1% of the total transmission time of 167 sec, whereas other schemes took longer than 3%.

Figure 3.15 and Figure 3.16 present the buffer utilization using adaptive ARMAX, one-layer NN, multilayer NN with and without *a priori* training. The buffer utilization is very low for the thresholding method because of a threshold value of 40%, whereas the one-layer NN stores cells in the buffer frequently resulting in a very high utilization. The multilayer NN without *a priori* training does not utilize the buffer as much as the multilayer NN with training. This is due to the inaccurate knowledge of the traffic flow with no *a priori* training. Also, as more buffer space is provided to the multilayer NN with no *a priori* training, the utilization increases with buffer size. However, as pointed out earlier, the overall delay is smaller than other schemes even though the queuing delay is slightly higher.

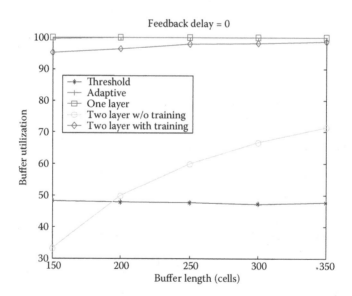

FIGURE 3.15
Cell loss ratio with congestion.

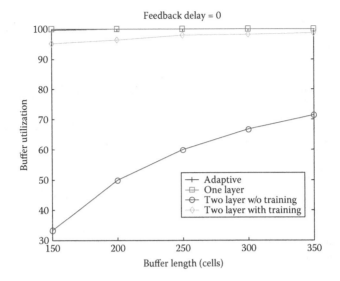

FIGURE 3.16
Cell loss ratio with congestion.

Example 3.3.3: Performance Comparison with Additional Feedback Delays

Note that the delays observed in Example 3.3.1 and Example 3.3.2 were due to feedback because the hardware delays were neglected. Example 3.3 presents the case when additional delays that are multiples of the measurement interval T were injected in the feedback. Figure 3.17

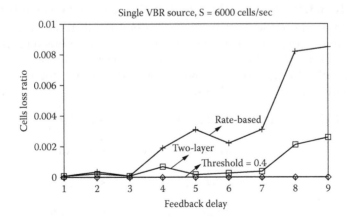

FIGURE 3.17
Cell loss ratio with feedback delays.

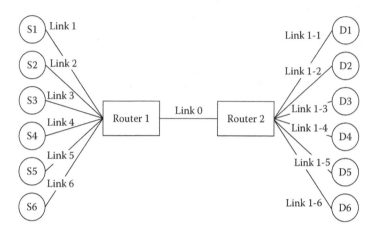

FIGURE 3.18
End-to-end topology.

shows the CLR with delay when additional delays were injected in the feedback for different congestion control methods. From the figure, it is clear that the two-layer NN-based controller is more stable and robust compared to the adaptive ARMAX (rate-based) method when feedback delays were present. Here, the thresholding approach (with a value of 0.4) appears to be more stable as the rate adjustment depends only upon the magnitude of the buffer occupancy; however, during congestion, it performs unsatisfactorily in terms of CLR and delay compared to the two-layer NN approach. Similar results were observed when more cross-traffic was used.

Example 3.3.4: Fairness Test in the Presence of Cross-Traffic

We use two data traffic sources (elastic traffic), 1 VBR traffic source, and 3 CBR traffic sources (inelastic). Elastic source rates are adjusted using the feedback $u(k)$ and fair share equation, Equation 3.15. The bottleneck buffer size is 25 cells. Congestion was created by cutting down the bandwidth of Link0 in Figure 3.18 as follows:

$$
\begin{aligned}
\text{Bandwidth of Link0} &= 23{,}585 \text{ cells/sec}, & 0 \le t &< 3\text{sec}; \\
&= 18{,}868, \text{ cells/sec} & 3 \le t &< 6\text{sec}; \\
&= 9{,}434 \text{ cells/sec}, & 6 \le t &< 24\text{sec}; \qquad (3.34) \\
&= 18{,}868 \text{ cells/sec}, & 24 \le t &< 27\text{sec}; \\
&= 23{,}585 \text{ cells/sec}, & 27 &\le t.
\end{aligned}
$$

TABLE 3.1

Comparison of Results

Example 4	NN	Threshold
Cell loss ratio	0	1.172%
System power	7.475	7.0917

The results are presented in Table 3.1. The transmission time for individual traffic sources are shown in Table 3.2. From Table 3.1 and Table 3.2, we can say that the threshold scheme cannot guarantee fair share to all the users and our proposed NN scheme can achieve fair share for the users by adjusting the elastic traffic rates (because these traffic rates can tolerate wide-range changes in delay and throughput) while assuring the QoS for the inelastic traffic. The metric *power* is defined as:

$$\text{Power} = \frac{\text{average throughput}}{\text{transmission delay}} \qquad (3.35)$$

Example 3.3.5: Extended Topology with Multiple Bottlenecks

We use the same traffic sources: 3 VBR traffic sources and 3 CBR traffic sources introduced in Example 3.4. Source rates are adjusted using the feedback $u(k)$ separately. The buffer size is taken at 50 cells at each switch. For results in Figure 3.11, congestion was created by reducing the bandwidth of different links of the topology in Figure 3.19, as follows, at the specified time interval and, held at that rate, till the completion of the total transmission:

Bandwidth of Link0 = 23,584 cells/sec->5896 cells/sec, t =3 sec;

Bandwidth of Link1 = 23,584 cells/sec->7862 cells/sec, t = 6 sec;   (3.36)

Bandwidth of Link2 = 23,584 cells/sec->11,793 cells/sec, t =24 sec;

TABLE 3.2

Transmission Times for Sources to Ensure Fairness

Source	NN (sec)	Threshold (sec)
Data1	31.58	30.05
Data2	31.85	30.05
VBR	30.07	30.05
CBR1	30.04	31.95
CBR2	30.05	30.04
CBR3	30.04	30.04

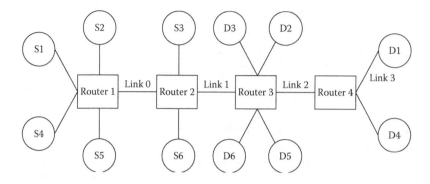

FIGURE 3.19
Extended topology with multiple bottlenecks.

From Figure 3.11, we can once again see the CLR obtained using the proposed congestion control scheme is much better than the threshold-based congestion control scheme. In fact, the CLR using the congestion control scheme is equal to zero and, as in Example 3.4, the controller reduces the source rates fairly. The results between proposed and threshold-based control are shown in Table 3.3.

3.4 End-to-End Congestion Controller Design for the Internet

In recent years, the emergence of high-speed Internet and increase in multimedia traffic has spawned several challenging problems for the smooth transfer of data. As the Internet offers best-effort service to multimedia traffic and, because, this traffic has bursty traffic characteristics and various QoS and bandwidth requirements, a high-speed Internet requires a real-time control scheme that manages the admission and control of traffic flows to avoid congestion.

Congestion and admission control protocols are too numerous to mention; only the important ones are referred to here. Several schemes

TABLE 3.3

Comparison of Results for the Extended Topology

3VBR+3CBR	NN	Threshold-Based
Total cell loss	0	73
Cell loss ratio	0	0.42%
Normalized delay	0.6587	0.6643
System power	6.9984	6.968

(Jagannathan and Talluri 2002, Benmohamed and Meerkov 1993, Chen and Chang 1996, Widmer et al. 2001, Floyd 2001, Jain 1996, Kelly et al. 2000, Floyd and Fall 1999, Low et al. 2002, Jagannathan and Talluri 2002, Chandrayana et al. 2002) have been proposed on congestion control. Some (Jagannathan 2002, Benmohamed and Meerkov 1993, Cheng and Chang 1996) are based on nonlinear system theory (Low et al. 2002, Jagannathan and Talluri 2002, Chandrayana et al. 2002), whereas others are based on heuristics and past experience (Floyd 2001). Most schemes adjust the input rates to match the available link capacity (or rate) (Jain 1996, Lakshman et al. 1999) or carefully admit new traffic (Kelly et al. 2000, Mortier et al. 2000). A comparison of the performance of the TCP-based end-to-end congestion control variants are presented in (Fall and Floyd 1996).

The basis of the congestion control scheme in the Internet (Floyd 2001) lies in additive increase and multiplicative decrease (AIMD) algorithm, halving the congestion window for every window containing packet loss, and otherwise increasing the congestion window roughly one segment per round trip time (RTT) (Floyd 2001, Jain 1996). The second component is the retransmit timer, which includes the exponential backoff when a transmitted packet itself is dropped. Another component is the slow-start mechanism for initial probing for available bandwidth instead of sending it at a high rate. The fourth component is acknowledgement (ACK) clocking, where the arrival of acknowledgments at the sender is used to clock out the transmission of new data. Within this general congestion control framework of slow start, AIMD, retransmit timers, and ACK clocking, there is a wide range of possible behaviors.

Unfortunately, the AIMD-based congestion control scheme appears to have several problems:

1. It does not guarantee fair share to all the users (elastic and inelastic) (Widmer et al. 2001).

2. It is not proven analytically to converge and the stability of the entire system is not shown.

3. The loss of packets in the Internet may be construed as congestion.

4. The initial congestion window is currently selected heuristically.

5. It is a reactive scheme; a predictive scheme is preferable due to feedback delays. Finally, the congestion control problem in the Internet is exacerbated as the Internet is increasingly transformed into a multiservice high-speed network (Widmer et al. 2001).

Random early detection (RED) scheme for controlling average queue size is proposed so that unnecessary packet drops can be avoided (Floyd 2001). Further, explicit congestion notification (ECN) bit is introduced, which, when built upon queue management, allows routers the option of marking, rather than dropping packets, as indication of congestion at end nodes (Floyd 2001). However, many of the problems (Items 1,2,4, and 5 in the previous list) still remain.

Congestion control mechanisms in today's Internet already represent one of the largest deployed artificial feedback systems; a protocol should be implemented, in a decentralized way — using sources and links, and which satisfies some basic objectives, such as high network utilization in equilibrium and local stability for arbitrary delays, capacities, and routing (Low et al. 2002). Under the assumption that queuing delays will eventually become small, relative to propagation delays, stability results for a fluid-flow model of end-to-end Internet congestion control is derived in Johari and Tan (2001). The choice of congestion control mechanism is constrained to adjust the sending rate on a timescale fixed by the propagation delay. The end users are naturally placed to implement a control scheme that acts on the timescale of the end-to-end propagation delay, and thus, one might hope for a stable, distributed end-to-end congestion control algorithm.

Another approach, outlined in Floyd and Fall (1999), is for routers to support the continued use of end-to-end congestion control as the primary mechanism for best-effort traffic to share scarce bandwidth, and to deploy incentives for its continued use. These incentives would be in the form of router mechanisms to restrict the bandwidth of best-effort flows, using a disproportionate share of the bandwidth, in times of congestion. These mechanisms would give a concrete incentive to end users, application developers, and protocol designers to use end-to-end congestion control for best-effort traffic. More recently, there is a surge in the development of novel end-to-end congestion control schemes (Jagannathan and Talluri 2002, Chandrayana et al. 2002, Bansal and Balakrishnan 2001, Borri and Merani 2004, Kuzmanovic and Knightly 2004, Sastry and Lam 2005), such as binomial congestion control (Bansal and Balakrishnan 2001), TCP friendly schemes (Borri and Merani 2004), and CYRF (Sastry and Lam 2005), to overcome the limitations of TCP-based networking protocols. However, it was shown through simulation studies in Chandrayana et al. (2002) that AIMD-based schemes still outperform TCP-compliant binomial schemes for congestion control in a wide range of networking environments. In contrast to ATM networks where hop-by-hop type congestion control is utilized (Section 3.3), end-to-end congestion control is discussed for the Internet.

This chapter presents a new rate-based end-to-end control scheme from Peng et al. 2006 to overcome some of the problems of the existing Internet congestion (Bansal and Balakrishnan 2001, Borri and Merani 2004). The novelty of the scheme is its distributed network modeling based on nonlinear system theory, using which, the performance and stability of the scheme can be mathematically analyzed and established. The proposed scheme uses NN to estimate the traffic-accumulation rate and uses this estimate to design a congestion controller, in contrast with other schemes (Floyd and Fall 1999, Sastry and Lam 2005), where the accumulation is assumed to be known beforehand. This scheme is implemented in NS-2, and the simulation results justify theoretical conclusions. The simulation uses measurements of buffer occupancy, packet arrival, and service rates at the destination, to determine the source data rates for congestion control. The proposed work appears to provide better results in a wide range of networking conditions compared to TCP friendly schemes.

3.4.1 Network Model

Figure 3.20 illustrates an end-to-end network configuration for evaluating the proposed congestion control schemes. Congestion happens whenever the input rate (*Ini*) is more than the available link capacity at the ingress node, or when the network has a congested node inside the network. In Figure 3.20, $q(k)$ denotes the network modeled as a single buffer, I_r refers to the rate at which packets arrive at the egress node, *Uk* is the feedback variable from the destination, *Xd* is the desired buffer occupancy at the egress node buffer, S_r refers to the departure rate of the packets to the destination, and *Vk* is the departure rate of the packets into the network from the ingress node.

Most congestion schemes consist of adjusting the input rates to match the available link capacity (or rate) (Jain 1996) or carefully admit new traffic (Qiu 1997, Bae and Suda 1991). Depending upon the duration of congestion, different schemes can be applied. For long-term congestion,

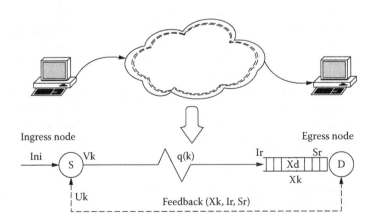

FIGURE 3.20
End-to-end network configuration.

an end-to-end feedback control scheme can be used. For short-term congestion, providing sufficient buffers in the switches is the best solution.

Motivation of the end-to-end protocol is that they place the complex functionalities in the hosts and not inside the network and, hence, the hosts typically upgrade the software to receive a better service. The second reason is that by keeping the network simple, it can scale more easily. Throughout this chapter, we define the ingress and egress nodes/routers as end hosts. These hosts possess finite buffers and they are connected to the network via physical links of finite bandwidth. Depending upon the type of source, we use the feedback accordingly to meet the fairness criteria. For instance, the feedback can be used to alter the source rate or compression ratios/quantization parameters of the codec (Lakshman et al., 1999) and/or to dynamically route the packets. The process of selecting the best option is not dealt with in this paper and will be proposed in the future.

The most important step in preventing congestion is to estimate the network traffic via network modeling. Modeling a high-speed network, in general, is quite complex as it involves different time scales (packet-level congestion control, source level admission control, error control) in the networking layers, and the control schemes involve both discrete event and discrete/continuous dynamics. In this chapter, the congestion control is accomplished via buffer dynamics modeling in discrete time.

Consider the buffer dynamics of a pair of ingress/egress node pairs connected to a network and characterized as distributed discrete time nonlinear systems to be controlled, given in the following form:

$$x(k+1) = Sat_p(f(x(k)) + Tu(k) + d(k)). \qquad (3.37)$$

$x(k) \in \mathfrak{R}^n$ is a state variable for n buffers, which is the buffer occupancy at time instant k, $u(k) \in \mathfrak{R}^n$ a control signal, which is the regulated traffic rate, and T is the measurement interval. The function, $f(\cdot)$, represents the actual packet accumulation at the network router/node, which is further defined as $f(\cdot) = [x(k) + q(k) + (I_{ni}(k) - S_r(k))T]$, where $I_{(k)}$ is the packet arrival rate at the ingress node from n sources, $q(k)$ is the bottleneck queue level, this value is estimated by $(I_{ni}(k) - I_r(k))T$, $I_r(k)$ is the arrival rate at the egress/destination node, $S_r(k)$ is the service rate at the egress node, and $Sat_p(\cdot)$ is the saturation function. The unknown disturbance vector, $d(k) \in \mathfrak{R}^n$, which can be an unexpected traffic burst/load or a network fault, is assumed to be bounded by a constant, $\|d(k)\| \le d_M$.

Given a desired buffer size at the destination x_d, define the performance in terms of buffer occupancy error

$$e(k) = x(k) - x_d \qquad (3.38)$$

where the packet losses are defined because of overflows as $c(k) = \max(0, e(k))$. The dynamics of the buffer can be written in terms of the buffer occupancy errors as:

$$e(k+1) = Sat_p[f(x(k)) - x_d + Tu(k) + d(k)].$$ (3.39)

The objective is to select a suitable traffic rate $u(k)$, such that the available bandwidth can be exploited up to its peak packet rate (PPR).

3.4.2 End-to-End Congestion Control Scheme

Define the traffic rate input, $u(k)$, as

$$u(k) = \frac{1}{T}(x_d - f(x(k)) + k_v e(k))$$ (3.40)

where k_v is a diagonal gain matrix, $f(x(k))$ is the traffic accumulation at the bottleneck link/egress buffer, and T is the measurement interval. Then, the error in buffer length system becomes

$$e(k+1) = Sat_p[k_v e(k) + d(k)]$$ (3.41)

If an adaptive estimator is employed to estimate the bottleneck queue and hence the traffic accumulation, $\hat{f}(\cdot)$, at the destination buffer, then the buffer-length error is written as:

$$e(k+1) = Sat_p[k_v e(k) + \hat{f}(\cdot) + d(k)]$$ (3.42)

Then $u(k)$ will be defined as

$$u(k) = \frac{1}{T}(x_d - \hat{f}(x(k)) + k_v e(k))$$ (3.43)

The proposed rate-based methodology uses packet-arrival rate at the ingress node and packet-arrival rate at the egress node to estimate the network traffic accumulation or flow. The estimated value and the packet accumulation at egress node are together used to compute the rate $u(k)$ by which the source has to reduce or increase its rate. If $u(k)$ is a positive value, it indicates that the buffer space is available and the sources can increase their rates, whereas if $u(k)$ is negative, it indicates that there are packet losses at the egress node and the sources have to decrease their rates. The updated traffic rate based on $u(k)$ minimizes the difference between the actual and desired buffer length, while providing fairness among all sources. The details of the stability analysis shown in the following section suggests that the error in queue length $e(k)$ is bounded and the packet losses are also bounded and finite for any initial state of the network.

3.4.2.1   Stability Analysis

LEMMA 3.4.1
The nonlinear system $y(k+1) = Sat(Ay(k))$, *for* $A \in \mathfrak{R}^{2 \times 2}$, *where sat(y) is of the form*

$$sat(y_i) = 1, > 1$$

$$= y_i, |y_i| \leq 1 \qquad (3.44)$$

$$= -1, y_i < -1$$

is asymptotically stable if the symmetric matrix $A = [a_{ij}]$ is stable, and in addition the following condition is met, that is $|a_{11} - a_{22}| \leq 2$ min $\{|a_{11}|, |a_{22}|\} + 1 - \det(A)$, where $\det(A)$ is the determinant of A. For the systems defined in Equation 3.41, Equation 3.43 and for multiple buffers, these conditions are met as long as $0 < k_{max} < 1$, where k_{vmax} is the maximum singular value of k_v.

THEOREM 3.4.1 (IDEAL CASE)
Let the desired buffer length x_d *be finite and the network disturbance bound* d_M *be equal to zero. Let the source rate for Equation 3.37 be provided by Equation 3.41, then the buffer dynamics is globally asymptotically stable, provided*

$$0 < k_v^T k_v < I \qquad (3.45)$$

PROOF   Let us consider the following Lyapunov function candidate

$$J = e(k)^T e(k). \qquad (3.46)$$

The first difference is

$$\Delta J = e(k+1)^T e(k+1) - e^T(k)e(k). \qquad (3.47)$$

Substituting the buffer error dynamics Equation 3.42 in Equation 3.47 yields

$$\Delta J \leq -\left(1 - k_{v\max}^2\right) \|e(k)\|^2 . \tag{3.48}$$

The closed-loop system is globally asymptotically stable.

COROLLARY 3.4.1
Let the desired buffer length x_d be finite and the network disturbance bound, d_M be equal to zero. Let the source rate for Equation 3.37 be provided by Equation 3.41, then the packet losses $e(k)$ approach zero asymptotically.

REMARK 1
This theorem shows that in the absence of bounded disturbing traffic patterns, the error in buffer length $e(k)$ and packet losses converges to zero asymptotically, for any initial condition. The rate of convergence and transmission delays and network utilization depends upon the gain matrix k_v.

THEOREM 3.4.2 (GENERAL CASE)
Let the desired buffer length x_d be finite and the disturbance bound d_M be a known constant. Take the control input for Equation 3.37 as Equation 3.41 with an estimate of network traffic such that the approximation error, $\tilde{f}(\cdot)$, is bounded above by f_M, then the error in buffer length $e(k)$ is GUUB, provided

$$0 < k_v < I \tag{3.49}$$

PROOF Let us consider the following Lyapunov function candidate, Equation 3.46, and substitute Equation 3.42 in the first difference Equation 3.47 to get

$$\Delta J = (k_v e + \tilde{f} + d)^T (k_v e + \tilde{f} + d) - e^T(k)e(k). \tag{3.50}$$

$\Delta J \leq 0$ if and only if

$$(k_{v\max} \|e\| + \tilde{f} + d_M) < \|e\| \tag{3.51}$$

or

$$\|e\| > \frac{f_M + d_M}{(1 - k_{v\max})} \tag{3.52}$$

Here, any rate-based scheme can be used, including the robust control approach as long as the bound on the approximate error in traffic accumulation is known.

COROLLARY 3.4.2 (GENERAL CASE)
Let the desired buffer length x_d be finite and the disturbance bound, d_M, be a known constant. Take the control input for Equation 3.37 as Equation 3.43, then the packet losses are finite.

REMARK 2
This theorem shows that the error in queue length, e(k), is bounded, and the packet losses are also bounded and finite for any initial state of the network. The rate of convergence depends upon the gain matrix.

CASE II (UNKNOWN CONGESTION LEVEL/ARRIVAL AND SERVICE RATES)
In this case, a neural network (NN)-based adaptive scheme is proposed but any type of function approximator can be applicable. Assume, therefore, that there exist some constant ideal weights W and V for a two-layer NN, so that the nonlinear traffic accumulation function in Equation 3.48 can be written as:

$$f(x) = W^T \varphi(V^T x(k)) + \varepsilon(k) \qquad (3.53)$$

where V is the input to hidden-layer weights kept constant throughout the tuning process after selecting it randomly, $\varphi(V^T x(k))$ is the hidden-layer sigmoidal activation function vector, and the approximation error $\|\varepsilon(k)\| \le \varepsilon_N$ with the bounding constant ε_N is known. Note, that by not tuning V and randomly selecting initially, the hidden-layer sigmoidal functions form a basis (Jagannathan 2006).

3.4.2.2   Congestion Controller Structure

Defining the NN traffic estimate in the controller by

$$\hat{f}(x(k)) = \hat{W}^T(k)\varphi(x(k)) \qquad (3.54)$$

with $\hat{W}(k)$ being the current value of the weights, let W be the unknown ideal weights required for the approximation to hold in Equation 3.53 and assume they are bounded so that

$$\|W\| \le W_{max} \qquad (3.55)$$

where W_{max} is the maximum bound on the unknown weights. Then, the error in the weights during estimation is given by

$$\tilde{W}(k) = W - \hat{W}(k) \tag{3.56}$$

FACT 3.4.1
Because the buffer size is finite, the activation functions are bounded by known positive values so that $\|\varphi(x(k))\| \leq \varphi_{max}$ *and* $\|\tilde{\varphi}(x(k))\| \leq \tilde{\varphi}_{max}$.
The traffic rate $u(k)$ is

$$u(k) = \frac{1}{T}(x_d - \hat{W}^T(k)\varphi(x(k)) + k_v e(k)) - \sum u(k - RTT) \tag{3.57}$$

and the closed-loop buffer occupancy dynamics become

$$e(k+1) = k_v e(k) + \overline{e}_i(k) + \varepsilon(k) + d(k) \tag{3.58}$$

where the traffic flow modeling *error* is defined by

$$\overline{e}_i(k) = \tilde{W}^T(k)\varphi(x(k)) \tag{3.59}$$

3.4.2.3 Weight Updates for Guaranteed QoS

It is required to demonstrate that the performance criterion in terms of packet losses $c(k)$, transmission delay, and network utilization monitored through error in buffer occupancy $e(k)$, is suitably small and that the NN weights $\hat{W}(k)$, remain bounded for, the traffic rate $u(k)$ is bounded and finite.

THEOREM 3.4.3 (TRAFFIC RATE CONTROLLER DESIGN)
Let the desired buffer length x_d be finite and the NN traffic reconstruction error bound ε_N and the disturbance bound d_M be known constants. Take the traffic rate input for Equation 3.37 as Equation 3.47 with weight tuning provided by

$$\hat{W}(k+1) = \hat{W}(k) + \alpha\varphi(x(k))e^T(k+1) - \Gamma \|I - \alpha\varphi(x(k))\varphi^T(x(k))\|\hat{W}(k) \tag{3.60}$$

with $\Gamma > 0$ a design parameter. Then the error in buffer length $e(k)$ and the NN weight estimates $\hat{W}(k)$ are UUB, with the bounds specifically given by Equation 3.67 and Equation 3.68 provided the following conditions hold:

$$(1) \quad \alpha \, \|\varphi(x(k))\|^2 < 1 \tag{3.61}$$

$$(2) \quad 0 < \Gamma < 1 \tag{3.62}$$

$$(3) \quad k_{v\max} < \frac{1}{\sqrt{\bar{\sigma}}}, \tag{3.63}$$

where $k_{v\max}$ is the maximum singular value of k_v and $\bar{\sigma}$ is given by

$$\bar{\sigma} = \eta + \frac{[\Gamma^2(1 - \alpha \, \|\varphi(x(k))\|^2)^2 + 2\alpha\Gamma \, \|\varphi(x(k))\|^2 \, (1 - \alpha \, \|\varphi(x(k))\|^2)]}{(1 - \alpha \, \|\varphi(x(k))\|^2)} \tag{3.64}$$

PROOF Define the Lyapunov function candidate

$$J = e^T(k)e(k) + \frac{1}{a} tr(\tilde{W}^T(k)\tilde{W}(k)). \tag{3.65}$$

The first difference is given by

$$\Delta J = e^T(k+1)e(k+1) - e^T(k)e(k) + \frac{1}{\alpha} tr(\tilde{W}^T(k+1)\tilde{W}(k+1) - \tilde{W}^T(k)\tilde{W}(k)). \tag{3.66}$$

Use the buffer length error dynamics Equation 3.39 and tuning mechanism Equation 3.60 to obtain

$$\|e(k)\| > \frac{1}{\left(1 - \bar{\sigma}k_{v\max}^2\right)}\left[\gamma k_{v\max} + \sqrt{\rho_1\left(1 - \bar{\sigma}k_{v\max}^2\right)}\right] \tag{3.67}$$

or

$$\|\tilde{W}(k)\| > \frac{\Gamma(1-\Gamma)W_{\max}}{\Gamma(2-\Gamma)} + \frac{\sqrt{\Gamma^2(1-\Gamma)^2 W_{\max}^2 + \Gamma(2-\Gamma)\theta}}{\Gamma(2-\Gamma)} \tag{3.68}$$

where ρ_1 and θ are constants. In general, $\Delta J \le 0$ in a compact set as long as Equation 3.61 through Equation 3.68 are satisfied and either Equation 3.67

or Equation 3.68 holds. According to a standard Lyapunov extension theorem (Jagannathan 2006), this demonstrates that the buffer occupancy error and the error in weight estimates are UUB.

COROLLARY 3.4.3
Let the desired buffer length x_d be finite and the NN traffic reconstruction error bound ε_N and the disturbance bound d_M be known constants. Take the traffic rate input for Equation 3.37 as Equation 3.43 with weight tuning provided by Equation 3.60. Then the packet losses $c(k)$ and the NN weight estimates $\hat{W}(k)$ are globally UUB, provided the conditions hold in Equation 3.61 through Equation 3.63.

The next result describes the behavior of the closed-loop queue length error system, in the ideal case of no traffic modeling error in the network dynamics, with no disturbances present.

THEOREM 3.4.4
Let the desired buffer length x_d be finite and the network traffic estimation error bound ε_N and the disturbance bound d_M be equal to zero. Let the source rate for Equation 3.37 be provided by Equation 3.43 with weight tuning given by

$$\hat{W}(k+1) = \hat{W}(k) + \alpha\varphi(x(k))e^T(k+1), \tag{3.69}$$

where $\alpha > 0$ is a constant learning rate parameter or adaptation gain. Then the buffer occupancy error $e(k)$ approaches asymptotically to zero, and the weight estimates are bounded provided the condition in Equation 3.61 holds and

$$k_{v\,\text{max}} < \frac{1}{\sqrt{\eta}} \tag{3.70}$$

where η is given by

$$\eta = \frac{1}{(1 - \alpha\,\|\varphi(x(k))\|^2)} \tag{3.71}$$

PROOF Follow the steps as in the previous theorem.

COROLLARY 3.4.5
Let the desired buffer length x_d be finite, and the network traffic estimation error bound ε_N and the disturbance bound d_M be equal to zero. Let the source rate for

Equation 3.37 be given by Equation 3.43 with weight tuning provided by Equation 3.69. Then the error in buffer occupancy, e(k), or packet losses, c(k), approaches asymptotically to zero and the weight estimates are bounded provided Equation 3.70 and Equation 3.71 hold.

3.5   Implementation

Network simulator 2 (NS-2) is a discrete event simulator targeted at networking research. We have selected NS-2 to implement our scheme. For convenience, we denoted our scheme as TQ (transmission-based queue size).

3.5.1   NS-2 Implementation

In NS-2, agents represent endpoints where network-layer packets are constructed or consumed, and are used in the implementation of protocols at various layers. We implement our TQ scheme at transport level. The Tq agent represents a TQ sender. It sends data to a Tq sink agent, processes its acknowledgments, and performs congestion control. It has a separate object associated with it which represents an application's demand. TQ assumes that a packet has been lost (due to congestion) when it observes 3 duplicate ACKs and it retransmits the lost packet immediately. The feedback, which is generated at the egress node, provides the RTT. The following is a UML state diagram of the implementation shown in Figure 3.21.

The pseudo code for this algorithm is presented below and the implementation is explained after the pseudo code.

```
//pseudo code for TQ algorithm
//notation:
// Ik:packets arrive rate at the ingress node
// Vk:packets departure rate at the ingress node
// Uk:feedback variable from the destination
// Qlevel:bottleneck queue level
// Ir:packets arrive rate at the egress node
// Sr:packets departure rate at the egress node
// Xk:buffer occupancy at the egress node buffer
// Xd:desired buffer occupancy at the egress node
buffer
// kv:diagonal gain matrix
// T:measurement interval
100 begin
```

```
101 while (not finishing sending all packets) do
102  begin
103   TqAgent send packets using current rate
104
105 meanwhile if receive ACK packets from TqSink then
106 begin
107   get Sr,Ir,Xk from ACK packet
108       Qlevel=(Ik-Ir)*T;
109       fxk=Xk+Qlevel+(Ik-Sr)*T;
110       Uk=(Xdfxk+kv*(XkXd))/T;
111       Vk=Ik+Uk;
112        update current sending rate using Vk
113 end
114  end
115 end
```

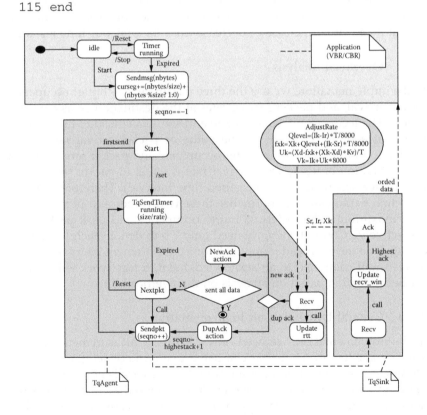

FIGURE 3.21
UML state diagrams for TQ implementation.

The above algorithm being end-to-end can be implemented at the ingress node as a congestion control software agent. No internal information from the network is necessary for the implementation. Using the acknowledgments packets from the destination, the congestion control agent estimates the number of packets in the network, including those at the destination, by treating the network as a buffer attached to the egress node. Here, packets entering the ingress and egress nodes can provide an indication of the bottleneck in the network, whereas the packets in the buffer at the egress node along with the desired buffer occupancy can provide information about the available buffer space. This information will be communicated back as part of the acknowledgment. Using this feedback information, the rate at which the packets are allowed to enter the network is determined and used in the subsequent time interval.

If packets entering the network at the ingress node are higher than the ones reaching the egress node, then a bottleneck is expected soon. Using queue length at the destination and number of packets in the network, the onset of congestion is estimated and a suitable rate to mitigate congestion is calculated. Subsequently, the rate of transmission of packets is adjusted so that it can either increase or decrease from the previous time instant.

3.5.2   Overhead Analysis

In the implementation, we use the three parameters — buffer occupancy, packet arrival, and packet service rates at the destination — as feedback inputs. The feedback information will add some overhead to the network flow, but actually, we can get the value of packet arrival rate at the destination buffer, given the buffer occupancy and packet service rate at the destination buffer (packet arrival rate is equal to packet service rate plus buffer occupancy over measured time interval). Therefore, we only need two parameters. We normalize these two values as a percentage of their maximum values (desired buffer occupancy and peak packet rate, respectively). Consequently, we need only 7 b to denote each value (percentage). So the overhead is only 14 b for a measurement interval. Though this is small compared to an acknowledgment, future work will involve further reduction based on analysis on the overhead bits.

3.5.3   General Discussion on Implementation

In the implementation, we need only 14 b for a measurement interval. This is considerably smaller as compared to an acknowledgment packet. Moreover, the internal information from the network is not needed making it an end-to-end implementation. The proposed algorithm does not depend upon number of nodes in the network because the network is treated as a buffer and the number of packets in the network is estimated as the buffer occupancy. Therefore, the algorithm is scalable.

3.6   Simulation Results

In this section, we present the simulation studies that are used to study the behavior of the algorithm. The section discusses the network topology, traffic sources, parameters, and constants used in the simulations and results. NS-2 is used in the simulations. We compare the proposed scheme with the New Reno-TCP protocol (Fall and Floyd 1996), as the latter reflects the current packet switched networks (e.g., Internet) quite well.

3.6.1   Network Topology and Traffic Sources

In our simulations, we use a typical "dumb-bell" topology as shown in Figure 3.18. All links, except the bottleneck link, are sufficiently provisioned to ensure that any drops/delays that occur are only due to congestion at the bottleneck link. All links are drop-tail links.

The default buffer length at each link is set to 50 packets. A packet is of 1 kB in size. The bandwidth of each link is set to 10 Mbps. The bottleneck link delay is 10 msec, other link delays are at 5 msec. The simulation length is taken to be 30 sec. In the simulation, totally six traffic sources S1 to S6 are used. Because the proposed methodology is compared with New-Reno TCP (which in turn is based on AIMD), they are discussed next.

3.6.2   New-Reno TCP Methodology

Reno TCP refers to TCP (Fall and Floyd 1996) and prevents the communication path from going empty after fast retransmit, thereby avoiding the need to slow-start to refill it after a single packet loss. Fast recovery operates by assuming each duplicate ACK received represents a single packet having left the pipe. During fast recovery the sender inflates its window by the number of duplicate ACKs it has received. Thus, during fast recovery, the TCP sender is able to make intelligent estimates of the amount of outstanding data. After entering fast recovery and retransmitting a single packet, the sender effectively waits until half a window of duplicate ACKs have been received, and then sends a new packet for each additional duplicate ACK that is received. Reno TCP has performance problems when multiple packets are dropped from one window of data because it has to wait for the expiration of the retransmission timer before reinitiating data flow.

The New-Reno TCP (Fall and Floyd 1996) includes a small change to the Reno algorithm at the sender. Partial ACKs received during fast recovery are treated as an indication that the packet immediately following the acknowledged packet in the sequence space has been lost, and should be retransmitted. Thus, when multiple packets are lost from a single window of data, New-Reno can recover without a retransmission timeout, retransmitting one lost packet per round-trip time until all of the lost packets from that window have been retransmitted.

3.6.3   Performance Metrics

Transmission delay, packet loss ratio, system power, and fairness were used as the major measures of performance. The transmission delay is defined as:

$$\text{Transmission delay} = \frac{T_c - T_o}{T_o}. \tag{3.72}$$

where T_c and T_o denote the time to complete the transmission by a source, with and without feedback, respectively.

The network utilization is defined as:

$$\text{Network utilization} = \frac{\text{current throughput}}{\text{bandwidth}} \times 100\%. \tag{3.73}$$

Packet losses $c(k)$ are defined as $x(k) - x_d$, when $x(k) > x_d$. The packet loss ratio (PLR) is defined as the total number of packets discarded at the receiver because of buffer overflows divided by the total number of packets sent onto the network.

$$\text{PLR} = \frac{\text{total number of packets loss}}{\text{total number of packets sent}} \times 100\% \tag{3.74}$$

The system power is defined as,

$$\text{System power} = \frac{\text{average throughput}}{\text{transmission delay}} \tag{3.75}$$

System power metric encodes the throughput over delay as a single metric. Power appears to be a concise metric that can be used to compare different congestion control schemes because, during congestion, higher throughput results in higher delay.

The most widely accepted notion of fairness is the max-min fairness criterion, and it is defined as

$$\text{Fair share} = MR_p + \frac{a - \left(\sum_{i=1}^{M^\tau} \tau_i + \sum_{i=1}^{l} MR \right)}{N - M^\tau} \tag{3.76}$$

where MR_p is the minimum rate for a source p, $\sum_{i=1}^{l} MR$ is the sum of minimum rates of l active sources at the ingress node, a is the total available bandwidth on the link, N is the number of active sources, and τ is the sum of bandwidth of M^τ active sources bottlenecked elsewhere. Note that in many cases, the number of controllable sources is not accurately known and, hence, the proposed scheme also can be used to alter some real-time traffic as the real-time video is compressed and can tolerate bounded delays.

3.6.4 Simulation Scenarios

To show the performance of the proposed rate-based end-to-end congestion control algorithm, we test it in the following scenarios.

Example 3.6.1: Single Source

Figure 3.22a to Figure 3.22d show the plot of CBR traffic with time for a single source. The bandwidth of each link is 2 Mbps. The bottleneck buffer size is taken at 10 packets. The bottleneck link delay is 2 msec, other link delays are at 10 msec. In the simulations, the bandwidth of bottleneck Link0 between the routers was reduced from 2 to 0.5 Mbps and various performance measures were observed in the Link0. The associated cumulative packets loss is shown in Figure 3.22a, the buffer utilization is shown in Figure 3.22b, the PLR is shown in Figure 3.22c, and network utilization is shown in Figure 3.22d. From this figure, it is clear that during congestion, the packets are stored in the buffer to prevent losses. When the bandwidth of the links between the nodes is reduced to 25%, the delay is significant as the control scheme appears to minimize losses. Further, the losses happen only during transient conditions whereas no losses are observed during steady state. Finally, when there is no congestion, the time to transmit all the packets took approximately 60 msec more than without feedback because of packet

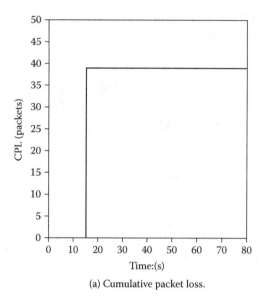

(a) Cumulative packet loss.

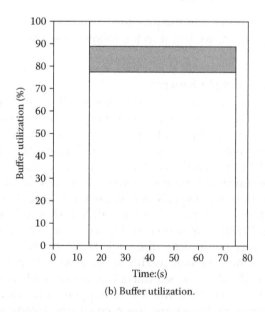

(b) Buffer utilization.

FIGURE 3.22
Performance in TQ scheme with single source: (a) cumulative packet loss, (b) buffer utilization, (c) PLR, and (d) network utilization.

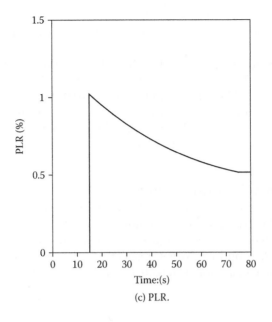

(c) PLR.

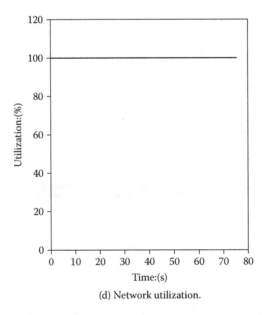

(d) Network utilization.

FIGURE 3.22 (Continued).

retransmission. In comparison, New-Reno TCP takes about 210 msec over its open-loop case. This is about 3.5 times longer than the proposed TQ scheme.

The corresponding results for TCP New-Reno are shown in Figure 3.23a to Figure 3.23d. The congestion window in New Reno-TCP scheme is shown in Figure 3.23a, which exhibits a sawtooth-like behavior. We also tested with some other single traffic sources and the results are similar to those shown in earlier text. The compared results between TQ and New-Reno TCP are shown in the Table 3.4.

Example 3.6.2: Multiple Sources

Here S1 to S3 were MPEG data obtained from a Web site (MPEG trace files). These data sets are from the movies "Star Wars IV," "Jurassic Park I," and sport event "Soccer" (Figure 3.24). The CBR sources S4 to S6 had data rates of 2, 1, and 0.5 Mbps, respectively. Source rates are adjusted using the feedback $u(k)$ for each source. The bottleneck buffer size is taken

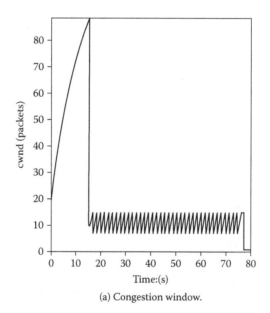

(a) Congestion window.

FIGURE 3.23
Performance in TCP scheme with single source: (a) congestion window, (b) cumulative packet loss, (c) buffer utilization, and (d) PLR.

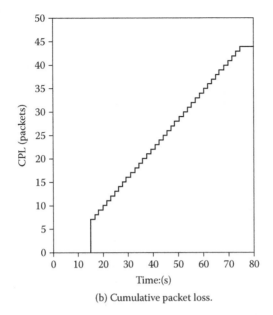

(b) Cumulative packet loss.

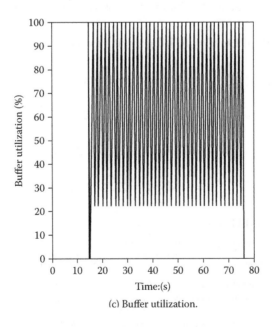

(c) Buffer utilization.

FIGURE 3.23 (Continued).

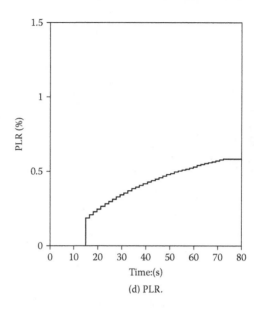

(d) PLR.

FIGURE 3.23 (Continued).

at 60 packets. For results in Figure 3.25 through Figure 3.28, congestion was created by reducing the bandwidth of Link0 as follows:

$$\text{Bandwidth of Link0} = 10 \text{ Mbps,} \quad 0 \le t < 3 \text{sec;}$$

$$= 2 \text{ Mbps,} \quad 3 \le t < 6 \text{sec;}$$

$$= 1 \text{ Mbps,} \quad 6 \le t < 24 \text{sec;}$$

$$= 2 \text{ Mbps,} \quad 24 \le t < 27 \text{ sec;}$$

$$= 10 \text{Mbps,} \quad 27 \le t.$$

TABLE 3.4

Compared Results for Single Source

Case I	TQ	New-Reno TCP
Packet loss ratio	0.517%	0.583%
Transmission delay	1.5033	1.5360
Power	1.3304	1.3021
Transmission time (no congestion)	30.06sec	30.21sec

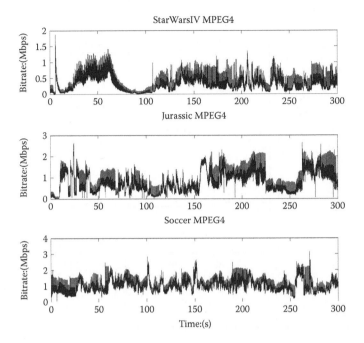

FIGURE 3.24
MPEG source traffic.

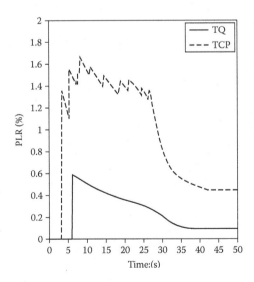

FIGURE 3.25
PLR with time for multiple traffic sources.

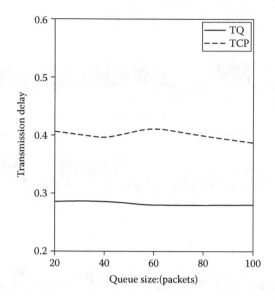

FIGURE 3.26
Transmission delay with queue size for multiple traffic sources.

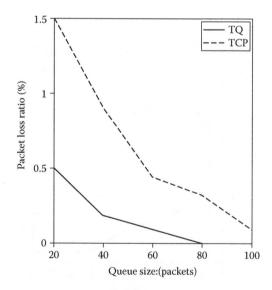

FIGURE 3.27
PLR with queue size for multiple traffic sources.

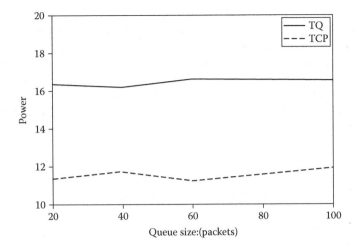

FIGURE 3.28
System power with queue size for multiple traffic sources.

Figure 3.25 shows the PLR with time when TQ and New-Reno TCP-based congestion control schemes were deployed. It can be observed that the PLR for both schemes increases as the service capacity is decreased, reaches a maximum value when the service capacity is a small value during the time interval of $6 \leq t < 24\,\mathrm{sec}$. The PLR again decreases as the service capacity is increased. From Figure 3.26, the PLR obtained using TQ congestion control scheme is much better than the New-Reno TCP congestion control scheme. The complete analysis results are shown in Table 3.5.

Figure 3.26 through Figure 3.28 display transmission delay (TD), PLR, and system power (SP), with different queue size for TQ and New-Reno TCP. The packet losses decrease with an increase in queue size for both schemes. From the three figures, we can see that the TD, PLR, and SP obtained using TQ congestion control scheme is much better than the New-Reno TCP congestion control scheme.

TABLE 3.5

TQ vs. New-Reno TCP Comparison

Case II	TQ	New-Reno TCP
Packet loss ratio	0.087%	0.443%
Transmission delay	0.2783	0.4110
System power	16.5701	11.2162

Example 3.6.3: Extended Topology

We also test our scheme's performance with multiple bottlenecks, using the extended topology of Figure 3.19. We use the following traffic sources: 3 VBR traffic sources and 3 CBR traffic sources introduced in Case II. Source rates are adjusted using the feedback, $u(k)$, separately. The bottleneck buffer size is taken at 50 packets. For results that are included in Figure 3.29, congestion was created by reducing the bandwidth of different Links as follows:

Bandwidth of Link0 = 10 to 4 Mbps,   t = 3 sec;

Bandwidth of Link1 = 10 to 3 Mbps,   t = 6 sec;

Bandwidth of Link2 = 10 to 2 Mbps,   t = 24 sec;

Bandwidth of Link3 = 10 to 1 Mbps,   t = 27 sec.

From Figure 3.29, we can see once again that the PLR obtained using TQ congestion control scheme is much better than the New-Reno TCP congestion control scheme because, in the proposed scheme, mathematical analysis ensures the convergence and the performance of the scheme. In fact, the PLR using TQ congestion control scheme is equal to zero while the transmission delay time is smaller. This is due to the fact that our

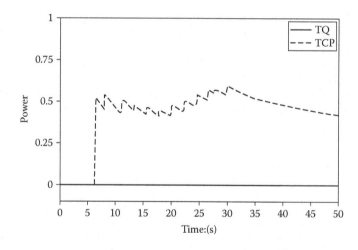

FIGURE 3.29
PLR with time for extended topology.

TABLE 3.6

Comparison of TQ with New-Reno TCP

Case III	TQ	New-Reno TCP
Packet loss ratio	0	0.42%
Transmission delay	0.6587	0.6643
System power	6.9984	6.968

scheme actively controls the packets by using packet losses and end-to-end delay while ensuring minimal retransmissions, which in turn, could reduce delay and the system power. On the contrary, the New-Reno TCP does not limit losses and causes higher retransmission of packets. Results comparing TQ and New-Reno TCP are shown in Table 3.6.

Example 3.6.4: Fairness

Here we use 2 Data traffic sources (elastic traffic), 1 VBR traffic source, and 3 CBR traffic sources (inelastic). Elastic source rates are adjusted using the feedback, $u(k)$, and fair share equation 3.20. The bottleneck buffer size is taken as 10 packets. For results presented in Figure 3.27 through Figure 3.29, congestion was created by reducing the bandwidth of Link0 as follows:

$$\text{Bandwidth of Link0} = 10 \text{ Mbps}, \quad 0 \leq t < 3 \sec;$$

$$= 8 \text{ Mbps}, \quad 3 \leq t < 6 \sec;$$

$$= 4 \text{ Mbps}, \quad 6 \leq t < 24 \sec;$$

$$= 8 \text{ Mbps}, \quad 24 \leq t < 27 \sec;$$

$$= 10 \text{ Mbps}, \quad 27 \leq t.$$

From Figure 3.31 and Figure 3.32, we can see that the PLR obtained using the TQ congestion control scheme is much better than the New-Reno TCP. In particular, from Figure 3.30, we can see that the throughput obtained using the TQ scheme is smoother than the New-Reno TCP. The comparison is given in Table 3.7. The transmission time for individual traffic sources are shown in Table 3.8. From this table, we can say that TCP cannot guarantee fair share to all the users, whereas TQ scheme can achieve fair share for the users by adjusting the elastic traffics rate (because these traffics can tolerate a wide range in delay and throughput) while

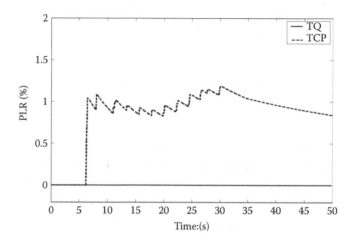

FIGURE 3.30
PLR with time for fairness policy.

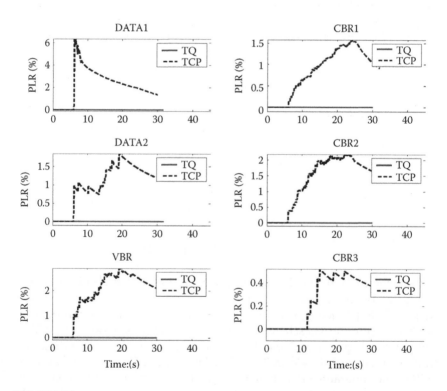

FIGURE 3.31
Individual PLR with time for fairness policy.

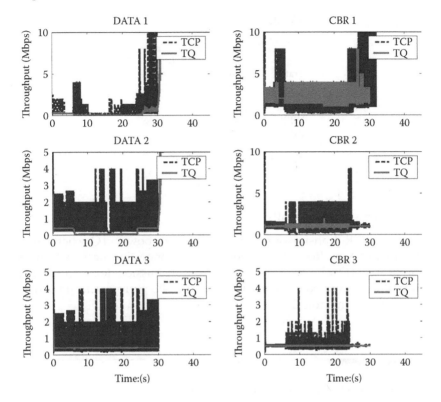

FIGURE 3.32
Individual throughput with time for fairness policy.

guaranteeing the QoS for the inelastic users. The rates are adjusted from the theoretical results presented in the previous sections, which guarantee performance. Consequently, the TQ scheme displays superior performance in comparison with the TCP. These results are significant as the Internet congestion control is not fair, and therefore a best-effort service results. On the contrary, the proposed end-to-end congestion scheme developed using rigorous mathematics guarantees the performance of the scheme, both theoretically and via extensive simulation studies.

TABLE 3.7

Fairness Comparison

Case IV	TQ	New-Reno TCP
Packet loss ratio	0	1.172%
Transmission delay	0.0617	0.0650
System power	7.475	7.0917

TABLE 3.8

Transmission Time

Traffic	TQ (sec)	New-Reno TCP (sec)
Data1	31.58	30.05
Data2	31.85	30.05
VBR	30.07	30.05
CBR1	30.04	31.95
CBR2	30.05	30.04
CBR3	30.04	30.04

3.6.5 Discussion of Results

In terms of PLR, transmission delay, and system power, TQ scheme outperforms the New-Reno TCP scheme when congestion occurred, in all our test cases. TQ scheme has less PLR, less transmission delay, and higher system power over New-Reno TCP. The specific performance improvement is shown in Table 3.9.

3.7 Summary and Conclusions

This chapter detailed a multilayer NN-based adaptive traffic-rate controller for ATM networks. The ATM switch/buffer dynamics along with the traffic flow is modeled as a nonlinear dynamical system and an NN controller is designed to prevent congestion. This learning-based approach does not require accurate information about the network system dynamics or traffic rate, it estimates the traffic rate at the switch and uses this estimate to arrive at the controller. No initial offline learning phase is required for the NN. By providing the NN with offline training using the backpropagation algorithm, the QoS improved slightly. However, in

TABLE 3.9

Performance Improvement for TQ over New-Reno TCP

Case	PLR (%)	Transmission Delay (%)	System Power (%)
Single source	11	2	2
Multiple source	80	32	45
Extended topology	100	1	0.4
Fairness test	100	0.8	0.4

general, it is very difficult to use offline training as the bursty MPEG data is statistically different from one source to the other.

Novel weight-tuning methods are derived for the NN. A rigorous mathematical analysis is provided for the closed-loop system, defined in terms of error in buffer occupancy, unlike in other available congestion control techniques. In fact, the proposed NN controller guaranteed performance as shown through the Lyapunov analysis. In addition, the performance in terms of CLR can be reduced to arbitrarily small values by choosing the gains appropriately.

Results are presented by evaluating the performance of the proposed controller during congestion, when both ON/OFF and bursty MPEG sources were transmitting data onto the network. Based on the results, it was concluded that the proposed methodology generates a low CLR and overall delay compared to the adaptive ARMAX, one layer, and the thresholding methods, and the NN controller ensured fairness during congestion. In addition, in most cases, one can design a switch with a small buffer due to the weight adaptation. In other words, simulation results show that the performance of the proposed two-layer NN-based congestion controller varies insignificantly with an increase in buffer size. This is an important result, which can be used to design buffers for network switches. Finally, the learning-based methodology provides an efficient queue management for ATM networks.

Next, this chapter discussed a rate-based end-to-end congestion control scheme for packet switched networks. The buffer dynamics were modeled as a nonlinear dynamical system, which is driven by the outgoing bandwidth, bottleneck queue level, and inflow rate. The bottleneck queue level is estimated by packet-arrival rate at ingress/egress nodes. A rigorous mathematical analysis is provided for the closed-loop system, defined in terms of error in queue length unlike in other available congestion control techniques. In fact, the proposed controller guaranteed performance, as shown through the Lyapunov analysis. Simulation examples were given to justify the theoretical conclusions. Based on the results, it was concluded that the proposed methodology performs better compared to the conventional New-Reno TCP method.

References

Bae, J.J. and Suda, T., Survey of traffic control schemes and protocols in ATM networks, *Proceedings of the IEEE*, Vol. 79, No. 2, February 1991, pp. 170–189.

Bansal, D. and Balakrishnan, H., Binomial congestion control algorithms, *Proceedings of the INFOCOM*, Vol. 2, 2001, pp. 631–640.

Benmohamed, L. and Meerkov, S.M., Feedback control of congestion in packet switching networks: the case of a single congested node, *IEEE/ACM Transaction on Networking*, Vol. 1, No. 6, 693–707, 1993.

Bonomi, F. and Fendick, K., The Rate-Based Flow Control Framework for the ABR ATM Service, *IEEE Network*, Vol. 9, No. 2, 1995, pp. 25–39.

Bonomi, F., Mitra, D. and Seery, J.B., Adaptive algorithms for feedback-based flow control in high-speed, wide-area ATM networks, *IEEE Journal on Selected Areas in Communications*, Vol. 13, No. 7, 1267–1283, September 1995.

Borri, M. and Merani, M.L., Performance and TCP-friendliness of the SQRT congestion control protocols, *IEEE Communication Letters*, Vol. 8, No. 8, 541–543, August 2004.

Chandrayana, K., Sikdar, B., and Kalyanaraman, S., Comparative study of TCP compatible binomial congestion control schemes, *Proceedings of the High Performance Switching and Routing*, 2002, pp. 224–228.

Chen, B., Zhang, Y., Yen, J., and Zhao, W., Fuzzy adaptive connection admission control for real-time applications in ATM-based heterogeneous networks, *Journal of Intelligent and Fuzzy Systems*, Vol. 7, No. 2, 1999.

Cheng, R.-G. and Chang, C.-J., Design of a fuzzy traffic controller for ATM networks, *IEEE/ACM Transactions on Networking*, Vol. 4, No. 3, June 1996.

Fall, K. and Floyd, S., Simulation-Based Comparisons of Tahoe, Reno and SACK TCP, *Computer Communication Review*, Vol. 26, No. 3, July 1996, pp. 5–21.

Floyd, S. and Fall, K., Promoting the use of end-to-end congestion control in the Internet, *IEEE/ACM Transactions on Networking*, Vol. 7, No. 4, 458–472, August 1999.

Floyd, S., A Report on Recent Developments in TCP Congestion Control, *IEEE Communications Magazine*, April 2001, pp. 84–90.

Fuhrman, S., Kogan, Y., and Milito, R.A., An adaptive autonomous network congestion controller, *Proceedings of the IEEE conference of Decision and Control*, 1996, pp. 301–306.

Izmailov, R., Adaptive feedback control algorithms for large data transfers in high-speed networks, *IEEE Transactions on Automatic Control*, Vol. 40, No. 8, 1469–1471, August 1995.

Jagannathan, S. and Talluri, J., Adaptive predictive congestion control of high-speed networks, *IEEE Transaction on Broadcasting*, Vol. 48, No. 2, 129–139, June 2002.

Jagannathan, S. and Talluri, J., Predictive congestion control of ATM networks: multiple sources/single buffer scenario, *Proceedings of IEEE Conference on Decision and Control*, Vol. 1, December 2000, pp. 47–52.

Jagannathan, S. and Tohmaz, A., Congestion control of ATM networks using a learning methodology, *Proceedings of the IEEE Conference on Control Applications*, September 2001, pp. 135–140.

Jagannathan, S., End to end congestion control of packet switched networks, *Proceedings of the IEEE Conference on Control Applications*, September 2002, pp. 519–524.

Jagannathan, S., *Neural Network Control of Nonlinear Discrete-Time Systems*, Taylor and Francis, Boca Raton, FL, 2006.

Jain, R., Congestion control and traffic management in ATM networks: recent advances and a survey, *Computer Networks and ISDN Systems*, Vol. 28, 1723–1738, 1996.

Johari, R. and Tan, D.K.H, End-to-end congestion control for the Internet: delays and stability, *IEEE/ACM Transactions on Networking*, Vol. 9, No. 6, 818–832, December 2001.

Kelly, F.P., Key, P.B., and Zachary, S., Distributed admission control, *IEEE Journal on Selected Areas in Communications*, Vol. 18, No. 12, 2617–2628, December 2000.

Kuzmanovic, A. and Knightly, E., A performance vs. trust perspective in the design of end-point congestion control protocols, *Proceedings of the 12th IEEE Conference on Network Protocols*, 2004, pp. 96–107.

Lakshman, T.V., Mishra, P.P., and Ramakrishan, K.K., Transporting compressed video over ATM networks with explicit-rate feedback control, *IEEE/ACM Transactions on Networking*, Vol. 7, No. 5, October 1999.

Liew, S. and Yin, D.C., A control-theoretic approach to adopting VBR compressed video for transport over a CBR communications channel, *IEEE/ACM Transactions on Networking*, Vol. 6, No. 1, 42–55, February 1998.

Liu, Y.-C. and Douligeris, C., Rate regulation with feedback controller in ATM networks—a neural network approach, *IEEE/ACM Transactions on Networking*, Vol. 15, No. 2, 200–208, February 1997.

Low, S.H., Paganini, F., and Doyle, J.C., Internet Congestion Control, *IEEE Control Systems Magazine* , Vol. 22, No.1, February 2002, pp. 28–43.

Mascolo, S., Smith's principle for congestion control in high speed ATM networks, *Proceedings of the IEEE Conference on Decision and Control*, 1997, pp. 4595–4600.

Mortier, R., Pratt, I., Clark, C., and Crosby, S., Implicit admission control, *IEEE Journal on Selected Areas in Communications*, Vol. 18, No. 12, 2629–2639, December 2000.

MPEG-4 trace files, available: http://www-tkn.ee.tu-berlin.de/~fitzek/TRACE/ltvt.html.

NS-2 manual, available at http://www.isi.edu/nsnam/ns/ns-documentation.html.

Peng, M., Jagannathan, S., and Subramanya, S., End to end congestion control of multimedia high speed Internet, *Journal of High Speed Networks*, accepted for publication, 2006.

Pitsillides, A., Ioannou, P., and Rossides, L., Congestion control for differentiated services using nonlinear control theory, *Proceedings of the IEEE INFOCOM*, 2001, pp. 726–733.

Qiu, B., A predictive fuzzy logic congestion avoidance scheme, *Proceedings of the IEEE Conference on Decision and Control*, 1997, pp. 967–971.

Sastry, N.R. and Lam, S.S., CYRF: a theory of window-based unicast congestion control, *IEEE Transactions on Networking*, Vol. 13, No. 2, 330–342, April 2005.

Widmer, J., Denda, R., and Mauve, M., A Survey on TCP-Friendly Congestion Control, *IEEE Network*, May/June 2001, pp. 28–37.

Problems

Section 3.2

Problem 3.2.1: Network modeling. Search the literature and identify network modeling techniques. Compare queuing theory-based modeling scheme with that of the state-space-based network modeling.

Problem 3.2.2: Networking modeling with delays. Model the ATM network with delays included.

Section 3.3

Problem 3.3.1: Traffic rate controller. Compare the conventional traffic rate controller with that of the NN-based method presented in the chapter.

Problem 3.3.2: Overhead analysis. Evaluate the overhead in terms of multiplications for the NN-based traffic-rate controller design.

Problem 3.3.3: End-to-end scheme evaluation. Use the end-to-end topology from Figure 3.18 with number of sources transmitting data from 6 to 15. Evaluate the performance of the congestion control scheme.

Problem 3.3.4: Extended topology. Use Figure 3.19 with the number of sources increased from 6 to 10, redo the performance evaluation of the congestion control scheme.

Section 3.4

Problem 3.4.1: Network modeling. Search the literature and identify network modeling techniques for the end-to-end congestion-control development for the network illustrated in Figure 3.20. Compare the queuing theory-based modeling scheme with that of the state-space-based network modeling.

Problem 3.4.2: Networking modeling with delays. Model the Internet with delays included.

Section 3.6

Problem 3.6.1: Traffic rate controller. Compare a conventional controller with that of the NN-based method presented in the chapter.

Problem 3.6.2: Overhead analysis. Evaluate the overhead in terms of multiplications for the TQ-based controller design.

Problem 3.6.3: End-to-end scheme. Use the end-to-end topology from Figure 3.18 with number of sources transmitting data from 6 to 15. Evaluate the performance of the TQ control scheme.

Problem 3.6.4: Extended topology. Use Figure 3.19 with number of sources increased from 6 to 10, study the performance evaluation of the TQ control scheme.

Appendix 3.A

PROOF FOR THEOREM 3.3.1 Define the Lyapunov function candidate

$$J = e^T(k)e(k) + \frac{1}{\alpha_1}tr[\tilde{V}^T(k)\tilde{V}(k)] + \frac{1}{\alpha_2}tr[\tilde{W}^T(k)\tilde{W}(k)], \qquad (3.A.1)$$

The first difference is given by

$$\Delta J = \Delta J_1 + \Delta J_2 \qquad (3.A.2)$$

$$\Delta J_1 = e^T(k+1)e(k+1) - e^T(k)e(k) \qquad (3.A.3)$$

$$\Delta J_2 = \frac{1}{\alpha_2}tr[\tilde{W}^T(k+1)\tilde{W}(k+1) - \tilde{W}^T(k)\tilde{W}(k)]$$
$$+ \frac{1}{\alpha_1}tr[\tilde{V}^T(k+1)\tilde{V}(k+1) - \tilde{V}^T(k)\tilde{V}(k)] \qquad (3.A.4)$$

Use the buffer length error dynamics of Equation 3.20 to obtain

$$\Delta J_1 = -e^T(k)\left[1 - k_v^T k_v\right] + 2(k_v e(k))^T(\bar{e}_i(k) + W^T\tilde{\varphi}_2(k) + \varepsilon(k) + d(k))$$

$$\times \bar{e}_i^T(k)(\bar{e}_i(k) + 2W^T\tilde{\varphi}_2(k) + 2(\varepsilon(k) + d(k))) + (W^T\tilde{\varphi}_2(k))^T \qquad (3.A.5)$$

$$\times (W^T\tilde{\varphi}_2(k) + 2(\varepsilon(k) + d(k)))(\varepsilon(k) + d(k))^T(\varepsilon(k) + d(k))$$

Considering the input and hidden-layer weight updates, and using these in Equation 3.A.1 and combining with Equation 3.A.5, one obtains

$$\Delta J \leq -\left(1 - \bar{\sigma}k_{v\max}^2\right)\|e(k)\|^2 + 2\gamma k_{v\max}\|e(k)\|$$

$$+\rho - \left\|\tilde{V}(k)\hat{\varphi}_1(k)\frac{\left(1 - \alpha_1\hat{\varphi}_1^T(k)\hat{\varphi}_1(k)\right) - \Gamma\left\|1 - \alpha_1\hat{\varphi}_1^T(k)\hat{\varphi}_1(k)\right\| \times (V^T\hat{\varphi}_1(k) + B_1 k_v e(k))}{2 - \alpha_1\hat{\varphi}_1^T(k)\hat{\varphi}_1(k)}\right\|$$

$$\times\left[2 - \alpha_1\hat{\varphi}_1^T(k)\hat{\varphi}_1(k)\right]$$

$$-\left\|\bar{e}_i(k) - \frac{\alpha_2\hat{\varphi}_2^T(k)\hat{\varphi}_2(k) + \Gamma\left\|I - \alpha_2\hat{\varphi}_2^T(k)\hat{\varphi}_2(k)\right\| \times (k_v e(k) + W^T\tilde{\varphi}_2(k) + \varepsilon(k) + d(k))}{1 - \alpha_2\hat{\varphi}_2^T(k)\hat{\varphi}_2(k)}\right\|^2$$

$$\times\left[I - \alpha_2\hat{\varphi}_2^T(k)\hat{\varphi}_2(k)\right] + \frac{1}{\alpha_1}\left\|I - \alpha_1\hat{\varphi}_1^T(k)\hat{\varphi}_1(k)\right\|^2 tr[\Gamma^2\hat{V}^T(k)\hat{V}(k) + 2\Gamma\hat{V}^T(k)\hat{V}(k)]$$

$$+\frac{1}{\alpha_2}\|I - \alpha_2\hat{\varphi}_2^T(k)\hat{\varphi}_2(k)\|^2 tr[\Gamma^2\hat{W}^T(k)\hat{W}(k) + 2\Gamma\hat{W}^T(k)\hat{W}(k)] \qquad (3.A.6)$$

where

$$\gamma = \beta_1\left(W_{\max}\tilde{\varphi}_{2\max} + \varepsilon_N + d_M + \Gamma(1 - \alpha_2)\varphi_{2\max}^2\right)\varphi_{2\max}W_{\max}$$

$$+k_1\left(\beta_2 + \Gamma\left(1 - \alpha_1\varphi_{1\max}^2\right)\right)\varphi_{1\max}V_{\max} \qquad (3.A.7)$$

and

$$\rho = \left[\beta_1(W_{\max}\tilde{\varphi}_{\max} + \varepsilon_N + d_M) + 2\Gamma\left(1 - \alpha_2\varphi_{2\max}^2\right)\varphi_{2\max}W_{\max}\right]$$

$$\times (W_{\max}\tilde{\varphi}_{2\max} + \varepsilon_N + d_M)\left[\beta_2 + \Gamma\left(1 - \alpha_1\varphi_{1\max}^2\right)\varphi_{1\max}^2 V_{\max}^2\right] \qquad (3.A.8)$$

Completing the squares for $\|\tilde{Z}(k)\|$ in Equation 3.A.6 results in $\Delta J \le 0$ as long as the conditions in Equation 3.24 through Equation 3.27 are satisfied and with the upper bound on the buffer-length error by

$$\|e(k)\| > \frac{1}{\left(1 - \bar{\sigma}k_{v\max}^2\right)} \left(\gamma k_{v\max} + \sqrt{\gamma^2 k_{v\max}^2 + \left(\rho + \frac{\Gamma}{(2-\Gamma)}Z_M^2\right)\left(1 - \bar{\sigma}k_{v\max}^2\right)} \right)$$

$$(3.A.9)$$

On the other hand, completing the squares for $\|e(k)\|$ in Equation 3.A.6 results in $\Delta J \le 0$ as long as the conditions Equation 3.24 through Equation 3.27 are satisfied and

$$\|\tilde{Z}(k)\| < \frac{1}{\Gamma(2-\Gamma)} \left(\Gamma(1-\Gamma)Z_M + \sqrt{\Gamma^2(1-\Gamma)^2 Z_M^2 + \Gamma(2-\Gamma)\theta} \right) \quad (3.A.10)$$

where

$$\theta = \Gamma^2 Z_M^2 + \frac{\Gamma^2 k_{v\max}^2}{1 - \bar{\sigma}k_{v\max}^2} + \rho \quad (3.A.11)$$

In general, $\Delta J \le 0$ in a compact set as long as Equation 3.24 through Equation 3.27 are satisfied and either Equation 3.A.9 or Equation 3.A.10 holds. According to a standard Lyapunov extension theorem (Jagannathan 2006), this demonstrates that the cell losses and the error in weight estimates are UUB.

4

Admission Controller Design for High-Speed Networks: A Hybrid System Approach

In the previous chapter, congestion control for ATM and the Internet was discussed. In many applications, congestion control alone is not sufficient unless some sort of admission control is employed. In this chapter, we use the congestion control scheme for the development of admission control.

4.1 Introduction

High-speed networks supporting multimedia services are expected to be capable of handling bursty traffic and satisfying their various quality of service (QoS) and bandwidth requirements. Asynchronous transfer mode (ATM), a high-speed network pipe, is one of the key technologies for integrating broadband multimedia services (B-ISDN) in heterogeneous networks, where multimedia applications consisting of data, video, and voice sources transmit information (Lakshman et al. 1999, Jain 1996, Liew and Yin 1998). Owing to broadband traffic pattern uncertainties and unpredictable statistical fluctuations in network traffic, bandwidth management and traffic control in high-speed networks pose new challenges and create difficulties. Therefore, a high-speed network must have an appropriate admission control (AC) scheme, not only to maintain QoS for existing sources but also to achieve high network utilization by properly admitting new traffic. This chapter discusses an admission controller scheme based on a hybrid system theory. The performance of the proposed scheme, though it is evaluated using cells, is applicable even for packet-based networks. For instance, an ATM provides services to sources with different traffic characteristics by statistically multiplexing cells of fixed-length packets of 53 B length. As a result, we have used packets/cells interchangeably.

Conventional AC schemes (Bae and Suda 1991, Gurin et al. 1991, Jamin et al. 1997) that utilize either capacity estimation or buffer thresholds suffer from fundamental limitations, such as the requirement of input traffic characteristics. An AC scheme must be dynamic in terms of regulating the traffic flows according to changing network conditions This, however, requires an understanding of network dynamics. As a result, it is not possible to determine the equivalent capacity of buffer thresholds for multimedia high-speed networks in bursty flow conditions or under dynamic conditions (Dziong et al. 1997). On the other hand, conventional AC schemes, based on mathematical analysis, provide robust solutions for different kinds of traffic environments, but suffer from estimation (modeling error) and approximation errors (real-time), so they are not suitable for dynamic environments (Dziong et al. 1997). Networks are forced to make decisions based on incomplete information, which, if not done properly, degrades performance of the network. Therefore, a neuro-fuzzy approach for AC is proposed (Chen et al. 1999, Cheng et al. 1999). Although this approach appears to make decisions in the presence of incomplete information, it is not clear whether the approach can be implemented in real-time, and further, no mathematical analysis is provided in Chen et al. (1999) and Cheng et al. (1999) to show its performance.

Information on available bandwidth is required by the network to decide whether to accept a new source or not. Routes are typically selected and sources accepted so as to minimize a certain measure of resources while providing adequate QoS to the carried traffic. This requires an accurate estimate of traffic conditions and the impact of adding a new source. Subsequently, this information is provided for calculating the amount of bandwidth currently allocated to accommodate existing sources, and by identifying how much additional bandwidth needs to be reserved on links over which new traffic is to be routed.

Because of the statistical multiplexing of traffic and shared buffers, both accounting and reservation are based on some aggregate statistical measures matching the overall traffic demand, rather than physically dedicated bandwidth or buffer space per source. In order to obtain an accurate accounting of bandwidth currently used, the current traffic onto these links has to be determined. The available bandwidth estimation schemes are based on simulation curves (Bae and Suda 1991). The major drawback of using simulation curves is that in addition to requiring storage of the precomputed curves at the bandwidth control points, the static nature of the information may not accurately reflect the dynamic and nonlinear behavior of the network traffic conditions and connection characteristics.

Although certain bandwidth estimation and allocation methods are heuristic (Jamin et al. 1997) in nature, others use neural and fuzzy logic (Chen et al. 1999, Cheng et al. 1999). Some of the schemes that utilize an estimate of the network traffic use the backpropagation algorithm or a linear least-squares method. Because the backpropagation algorithm is not proved to converge in all circumstances and requires a learning phase initially, other online schemes were proposed. For instance, the linear least-squares algorithm, though computationally simple, may not provide an accurate estimate of the highly nonlinear network traffic. Our proposed bandwidth estimation scheme takes advantage of a simplified network model in discrete time, calculates the current bandwidth usage online, estimates the future bandwidth requirement, and assigns the available capacity fairly to new sources while meeting the QoS. Note that all available schemes reserve bandwidth equal to the peak cell rate (PCR) of the source for real-time voice and video traffic. Unfortunately, reserving bandwidth equal to PCR wastes bandwidth and other network resources, lowering the network utilization. The proposed online bandwidth scheme is shown to address this limitation.

Only estimating the bandwidth online will not suffice to admit new traffic. In fact, the issue of potential congestion also has to be considered if a new source is being admitted into the network. Therefore, for an admission control scheme to perform reasonably, a congestion indicator has to be used to decide the impact of adding a new traffic source onto a network. In the literature, a number of congestion control schemes have been proposed (Jain 1996, Bae and Suda 1991). Unfortunately, many are reactive in nature and do not prevent congestion. Therefore, the novel predictive congestion control scheme in (Peng et al. 2006) is utilized to generate a congestion indicator, and it is subsequently used in the admission controller. Finally, the past values of buffer occupancy are used to predict the availability of network resources for the new sources. The admission controller uses the resource and the bandwidth availabilities along with the congestion indicator flag to decide whether to admit or reject the new source.

In this chapter, the buffer dynamics of the ingress node, where the traffic is being admitted into the network, is modeled as a nonlinear discrete-time system (Jagannathan 2005). A novel adaptive bandwidth scheme is developed to estimate the network traffic onto the links by minimizing the buffer occupancy error. Any function approximator can be used to estimate the traffic, but for simplicity a multilayer neural-network-based adaptive estimator is proposed. Tuning laws are provided for the estimator parameters, and closed-loop convergence with network stability is proved using a Lyapunov-based analysis when an online bandwidth estimation scheme is deployed. Using the estimated and buffered traffic along with the required QoS, the bandwidth required for all sources to meet the QoS is calculated. It is shown that the proposed online bandwidth estimation scheme guarantees the desired QoS by accurately estimating the bandwidth required by the existing sources, even in the presence of bounded network traffic uncertainties. A novel adaptive admission controller architecture is presented in this chapter from Jagannathan (2005) that unifies the congestion controller, a bandwidth estimation scheme, and the proposed rule-based discrete-event controller. The performance of the adaptive admission controller is compared using service delay, cell/packet losses, and network utilization. Simulation results are provided to justify the theoretical conclusions.

4.2   Network Model

Figure 4.1a shows the popular parking lot network configuration for evaluating the proposed admission control, bandwidth estimation, and congestion control schemes (Jain 1996). It can also be viewed as an end-to-end loop, as shown in Figure 4.1b, if the traffic enters at the ingress node/switch and leaves at the egress node passing through various networks or service providers and with very little information known about the internal network state. The egress node, which also contains a limited buffer, sends a feedback signal at every measurement interval to the ingress node via the network, with regard to the congestion. It is envisioned that the proposed admission controller reside at each network switch fabric where the link bandwidth usage at the ingress node is estimated at every measurement interval. The most important step in determining bandwidth usage and allocation is to estimate the network traffic that is being accumulated at the ingress switch/node buffer in an intelligent manner, using its occupancy.

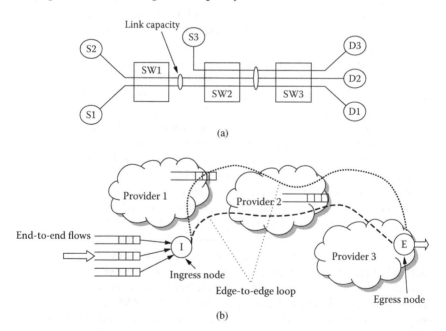

FIGURE 4.1
(a) Parking lot configuration, (b) end-to-end network.

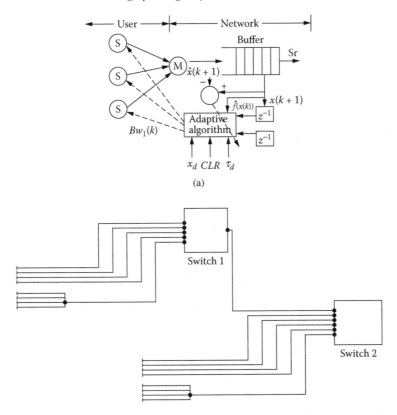

FIGURE 4.2
(a) Bandwidth allocation scheme at the ingress switch/node, (b) network schematic for admission control scenario.

Consider the buffer dynamics at an ingress node/switch fabric, shown in Figure 4.2a, given in the following form

$$x(k+1) = Sat_p(f(x(k)) + Tu(k) + d(k)) \quad (4.1)$$

with state $x(k) \in \Re^n$ being the buffer length (or occupancy) at time instant k, T being the measurement interval, and $u(k) \in \Re^n$ being the source rate that is determined via feedback. The nonlinear function, $f(x(k))$, is a function of buffer occupancy, source rate, and service capacity, S_r, at the ingress switch/node, which is given as $f(.) = [x(k) - q(t - T_{fb}) + I_{ni}(k) - S_r k)]$; $I_{ni}(k)$ is the packet arrival rate, $q(t - T_{fb})$ is the bottleneck queue level, T_{fb} is the propagation time from the bottleneck to the destination and back to the source, S_r is the service rate at the outgoing link at the ingress node, and $Sat_p(.)$ is a saturation function that satisfies the following:

$$Sat_p(z) = 0, \quad \text{if } z \le 0,$$

$$= p, \quad \text{if } z \ge p,$$

$$= z, \quad \text{otherwise}$$

This value of T_{fb} is obtained from the round-trip delay time (RTT–RTT_{min}), where RTT is the round-trip delay and RTT_{min} is the minimum propagation time obtained from the link bandwidth. The unknown disturbance vector acting on the system at the instant k is $d(k) \in \Re^n$, which is assumed to be bounded by a known constant $\|d(k)\| \le d_M$. Here, the disturbance vector $d(k)$ can be an unexpected traffic burst/load or change in available bandwidth owing to the presence of a network fault. The state $x(k)$ is a positive scalar if a single-switch or single-buffer scenario is considered, whereas it becomes a vector when multiple network switches or multiple buffers are involved. The first step in the proposed approach is to estimate the buffer occupancy, using an estimate of the network traffic at the switch. For the sake of convenience, we will eliminate saturation by not allowing $x(k)$ to saturate.

The objective here is to construct a model to identify the traffic accumulation at the switch as:

$$\hat{x}(k+1) = \hat{f}(x(k)) + Tu(k), \tag{4.2}$$

where the state of the model, $\hat{x}(k) \in \Re^n$, is the buffer occupancy estimate at time instant k, with the nonlinear function $\hat{f}(x(k))$ being the traffic accumulation estimate. Define the performance criterion in terms of buffer occupancy estimation error as

$$e(k) = x(k) - \hat{x}(k), \tag{4.3}$$

where the packet/cell losses, for a buffer size of x_d, are given by

$$c(k) = \max(0, e(k)) \tag{4.4}$$

Equation 4.3 can be expressed using Equation 4.4 and Equation 4.5 as:

$$e(k+1) = \tilde{f}(x(k)) + d(k), \tag{4.5}$$

where $e(k+1)$ and $x(k+1)$ denote the error in buffer occupancy and the buffer occupancy at the instant $k+1$, respectively, and the traffic flow modeling error is given by $\tilde{f}(x(k)) = f(x(k)) - \hat{f}(x(k))$. Equation 4.5 relates the buffer occupancy estimation error with the traffic flow modeling error.

If the traffic is estimated accurately, then the traffic flow modeling error can be made very small, which results in a small estimation error in buffer occupancy. By minimizing the buffer occupancy estimation error, it is shown that the accumulated traffic at the switch is estimated accurately.

In high-speed networks, each source should be able to use the available bandwidth up to its peak bit/cell rate (PBR/PCR). The goal of a bandwidth estimation and allocation scheme, then, is to enable the available bandwidth to be fully utilized while maintaining a good QoS, where the QoS parameters are the packet or cell loss ratio (PLR/CLR), and transfer delay. In this chapter, the objective of selecting a suitable traffic estimation scheme is to estimate the current bandwidth usage, predict the future bandwidth requirement, and to identify the available capacity for new sources in the next measurement interval.

Equation 4.5 calculates the packet/cell losses encountered at the switch fabric because of the traffic estimation scheme. In this chapter, it is envisioned that by appropriately using an adaptive estimator in discrete time to provide the traffic estimate, $\hat{f}(x(k))$, the error in buffer length, and hence the packet/cell losses, can be minimized. The actual packet/cell losses are related to the bandwidth requirement for the next measurement interval. By appropriately combining the bandwidth estimated using the predicted traffic conditions, packet/cell losses, along with the queued data at the switch, one can estimate the bandwidth required to satisfy the target QoS in the next measurement interval.

The buffer occupancy estimation error system expressed in Equation 4.5 is used to focus on selecting discrete-time parameter-tuning algorithms that guarantee the QoS by appropriately assigning the adequate bandwidth to the existing sources, provided no congestion exists. The proposed adaptive methodology can be best described as follows: The adaptive scheme at the ingress node/switch fabric estimates the network traffic; this estimated traffic, packet/cell losses, and target transfer delay are used to compute the equivalent bandwidth required to meet the target QoS.

4.3   Adaptive Traffic Estimator Design

In the two-layer NN case, the tunable weights enter in a nonlinear fashion. Stability analysis by Lyapunov's direct method is performed using novel weight-tuning algorithms that are developed in this chapter. Assume, therefore, that there exist some constant ideal weights W and V for a two-layer NN so that the nonlinear traffic accumulation function can be written as

$$f(x(k)) = W^T \varphi_2(V^T \varphi_1(x(k))) + \varepsilon(k) \qquad (4.6)$$

where $\varphi_2(k)$ is the vector of hidden-layer activation functions, $\varphi_1(k)$ is a vector linear function, and $\|\varepsilon(k)\| \le \varepsilon_N$ with the bounding constant, ε_N, known. For suitable approximation properties, it is necessary to select a large enough number of hidden-layer neurons. It is not known how to compute this number for general multilayer NN. Typically, the number of hidden-layer neurons is selected by a trial-and-error procedure.

4.3.1   Estimator Structure

Defining the NN traffic estimate in the buffer occupancy estimator by

$$\hat{f}(x(k)) = \hat{W}^T(k)\varphi_2(V^T\varphi_1(x(k))) \qquad (4.7)$$

with $\hat{W}(k)$ and $\hat{V}(k)$ being the current NN weights, the next step is to determine the weight updates so that the performance of the closed-loop buffer occupancy estimation error dynamics at the switch is guaranteed. The structure of the estimator is shown in Figure 4.3, where the current and past values of buffer occupancy are used as inputs to the NN model so that an accurate estimate of the network traffic accumulation is obtained. The current and past traffic estimates are used to derive the bandwidth equivalent.

Let W and V be the unknown target NN weights required for the approximation to hold in Equation 4.5 and assume that they are bounded so that

$$\|W\| \le W_{max}, \quad \|V\| \le V_{max}. \qquad (4.8)$$

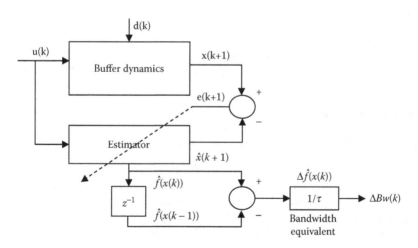

FIGURE 4.3
Neural network bandwidth estimator structure.

where W_{max} and V_{max} are the maximum bounds on the unknown weights. Then the error in weights during estimation is given by

$$\tilde{W}(k) = W - \hat{W}(k), \quad \tilde{V}(k) = V - \hat{V}(k), \quad \tilde{Z}(k) = Z - \hat{Z}(k), \quad (4.9)$$

where $Z = \begin{bmatrix} W & 0 \\ 0 & V \end{bmatrix}$, and $\hat{Z} = \begin{bmatrix} \hat{W} & 0 \\ 0 & \hat{V} \end{bmatrix}$.

FACT 4.3.1
The activation functions are bounded by known positive values so that

$$\left\| \varphi_2(x(k)) \right\| \le \varphi_{2max} \quad \text{and} \quad \left\| \tilde{\varphi}_2(x(k)) \right\| \le \tilde{\varphi}_{2max}.$$

Using the NN traffic rate estimate, the closed-loop buffer occupancy dynamics become

$$e(k+1) = k_v e(k) + \bar{e}_i(k) + W^T(k)\tilde{\varphi}_2(k) + \varepsilon(k) + d(k), \quad (4.10)$$

where the traffic flow modeling error is defined by

$$\bar{e}_i(k) = \tilde{W}^T(k)\tilde{\varphi}_2(x(k)) \quad (4.11)$$

4.3.2 Weight Updates for Guaranteed Estimation

It is required to demonstrate that the performance criterion in terms of cell losses, c(k), and transmission delay monitored through buffer occupancy estimation error, $e(k)$, is suitably small and that the NN weights $\hat{W}(k)$ and $\hat{V}(k)$ remain bounded. In the following theorem, a discrete-time weight-tuning algorithm based on the error in buffer occupancy is given, which guarantees that both the error in buffer occupancy and weight estimates are bounded.

THEOREM 4.3.1 (GENERAL CASE IN PRESENCE OF NETWORK DISTURBANCES)
Let the desired buffer length, x_d, be finite. Also, let the traffic modeling error bound, ε_N, obtained from target CLR and the disturbance bound, d_M, be known constants. Let the NN weight tuning be provided by

$$\hat{V}(k+1) = \hat{V}(k) - \alpha_1 \hat{\varphi}_1(k)[\hat{y}_1(k) + B_1 k_v e(k)]^T, \quad (4.12)$$

$$\hat{W}(k+1) = \hat{W}(k) + \alpha_2 \hat{\varphi}_2(k)e^T(k+1), \quad (4.13)$$

where $\hat{y}_1(k) = \hat{V}^T(k)\hat{\varphi}_1(k)$. Then the buffer occupancy error, $e(k)$, the NN weight estimates, $\hat{V}(k)$, $\hat{W}(k)$, are uniformly ultimately bounded (UUB), with the bounds specifically given by Equation 4.A.10 provided the following conditions hold:

$$(1) \quad \alpha_1 \varphi_{1\,\text{max}}^2 < 2, \tag{4.14}$$

$$(2) \quad \alpha_2 \varphi_{2\,\text{max}}^2 < 1, \tag{4.15}$$

$$(3) \quad c_0 < 1, \tag{4.16}$$

where c_0 is given by

$$c_0 = \frac{k_1^2}{2 - \alpha_1 \phi_{1\,\text{max}}^2} \tag{4.17}$$

REMARK 1

Note that it is very easy to verify that the conditions Equation 4.14 through Equation 4.16 are realizable, and hence the proof is avoided.

PROOF   See Appendix 4.A.

REMARK 2

As the buffer is of finite size and sigmoid is used in the hidden layer, conditions Equation 4.14 through Equation 4.16 easily follow.

REMARK 3

This theorem shows that using the parameter-tuning updates presented in Equation 4.12 and Equation 4.13, the buffer occupancy estimation error and error in weight estimates converge to a small subset provided that the design parameters are selected as Equation 4.14 through Equation 4.16. As the boundedness of buffer occupancy estimation error implies the boundedness of traffic-accumulation error, the traffic accumulated at the switch is estimated accurately.

REMARK 4

The network traffic modeling error bound, ε_N, and the disturbances, d_M, increase bounds on buffer occupancy estimation error in an interesting way.

REMARK 5

Large values of adaptation gain, α, forces smaller buffer occupancy estimation errors but large weight estimation errors, which, in turn, result in poor estimation of traffic accumulation. In contrast, a small value of α forces larger buffer occupancy and small weight estimation errors.

In the next section, the estimated traffic accumulation is used to obtain the current bandwidth usage and the future bandwidth requirement for all classes of traffic. Finally, the aggregate bandwidth required to meet the estimated traffic and target QoS will be derived for all classes of traffic for the next measurement interval. In the next section, using the aggregate bandwidth along with the physically available one, the available bandwidth is obtained and allocated fairly among sources.

4.4 Bandwidth Estimation, Allocation, and Available Capacity Determination

Assume that all the packets are transmitted at regularly spaced intervals to the buffer and consider that at the moment the new measurement interval starts, a new aggregate bandwidth value is being assigned. If the traffic that is being accumulated at the next measurement interval is known, then the additional bandwidth required to meet the packet/cell loss requirement can be determined as:

$$\Delta Bw(k) = \frac{(\hat{f}(x(k)) - \hat{f}(x(k-1)))}{T} \qquad (4.18)$$

where $\Delta Bw(k)$ is the additional bandwidth required to process the new data, $\hat{f}(x(k))$ and $\hat{f}(x(k-1))$ are the traffic estimates at the instants k and $k-1$, respectively. Further, the following equation relates the additional bandwidth required to satisfy the target CLR as:

$$\Delta Bw(k) = \frac{\text{Actual cell losses}}{TCLR * T} \qquad (4.19)$$

where TCLR represents the target or desired packet/cell loss ratio. In order to ensure that sufficient bandwidth is assigned to meet the delay requirement in the next measurement interval, the bandwidth required to meet the delay is given by

$$\Delta Bw(k) = \frac{x(k)}{\tau} \qquad (4.20)$$

where $x(k)$ is the current buffer occupancy and τ the target call transfer delay.

Equation 4.18 shows that additional bandwidth needs to be assigned to transmit the data queued in the ingress switch buffer to its destination.

Usually, the target delay is specified as a percentage of the total transfer time. Using Equation 4.18 through Equation 4.20, the additional bandwidth required in the next measurement interval to satisfy the accumulated traffic, CLR, and delay constraints is given by

$$\Delta Bw(k) = \max\left(\frac{\Delta \hat{f}(x(k))}{T}, \frac{\text{Actual Cell losses}}{TCLR * T} \right) + \frac{x(k)}{\tau} \qquad (4.21)$$

where $\Delta \hat{f}(x(k)) = \hat{f}(x(k)) - \hat{f}(x(k-1))$ is the traffic accumulated in the measurement interval. The bandwidth required at the next measurement interval for all the existing sources is given by

$$Bw(k+1) = Bw(k) + \Delta Bw(k) \qquad (4.22)$$

where $Bw(k+1), Bw(k)$ represents the bandwidth at the time instant $k + 1$ and k, respectively. Equation 4.22 dictates that the bandwidth be assigned to the sources that are sending traffic to the ingress node/switch provided the link bandwidth at the ingress switch meets the maximum available capacity requirement. If $Bw(k + 1)$ exceeds S_{max}, the maximum capacity of the outgoing link, the available bandwidth is set to S_{max}. Otherwise, the bandwidth is calculated by using Equation 4.22.

In the literature, in order to minimize short-term losses, the bandwidths of all the sources are adjusted using an ad hoc factor chosen *a priori* and the algorithm is referred to as the *overallocation algorithm*. Here, the factor $(0.8 < \rho < 1)$ is chosen by trial and error and the bandwidth required is calculated as $\frac{BW(k)}{\rho}$ instead of $BW(k)$ at the time instant k. If the overallocation factor is not chosen carefully, bandwidth will be wasted. In the literature, the PBR/PCR is used to reserve bandwidth for voice and video traffic to minimize delays and packet losses as they are categorized as real-time sources. However, assigning the bandwidth at PBR/PCR for a given source results in wastage of bandwidth and available resources, lowering the network utilization because PBR may not provide a good indicator of the arrival rate of the sources. In order to show the superiority of our scheme over the PCR and over allocation schemes, our proposed online bandwidth estimation scheme is compared with others.

For the case of single source sending traffic onto a network switch, the bandwidth required for the next measurement interval is calculated by using Equation 4.22. On the other hand, if multiple sources belonging to a particular class of traffic are sending traffic at the ingress node/switch, the bandwidth at the outgoing link has to be shared fairly among the sources. This problem is termed in the literature as *bandwidth assignment,* or *bandwidth allocation* (Fahmy et al. 1998). The bandwidth at any time instant k can be shared fairly among n sources as

$$Bw_i(k) = \frac{PCR_i}{\sum\limits_{i=1}^{n} PCR_i} Bw(k), \qquad (4.23)$$

or

$$Bw_i(k) = \frac{MCR_i}{\sum\limits_{i=1}^{n} MCR_i} Bw(k), \qquad (4.24)$$

or

$$Bw_i(k) = \frac{In_i(k)}{\sum\limits_{i=1}^{n} In_i(k)} Bw(k), \qquad (4.25)$$

where PCR_i, MCR_i, and $In_i(k)$ denote the PCR, mean cell rate, and the current intended rate of the source "i," respectively. By contrast, fairness criteria can be used to allocate the bandwidth.

4.5   Admission Control

Here the available capacity (bandwidth), congestion indicator, and buffer availability are utilized to admit new sources while maintaining QoS. Figure 4.4 shows the AC with its peripheral schemes for multimedia high-speed networks. The AC adopts several inputs: available capacity, available resource estimator, a congestion indicator, and a target packet/cell loss ratio and outputs a decision signal to indicate acceptance or rejection of the new source. The available capacity is calculated by subtracting the current bandwidth usage from all existing sources from the maximum available bandwidth of the physical link.

$$\text{Available\_capacity} = S_{\max} - \sum_{i=1}^{n} Bw_i(k) \qquad (4.26)$$

where n is the number of existing sources. This available capacity is assumed to be utilized for admitting new sources. Other call admission control methods use equivalent-capacity-based algorithm. The

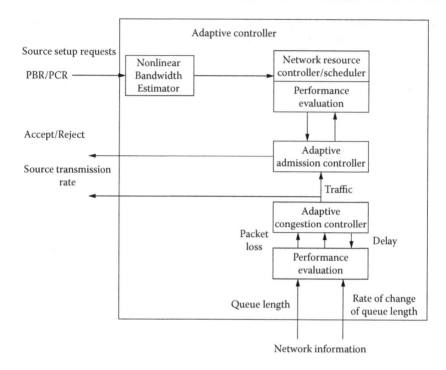

FIGURE 4.4
Adaptive admission controller.

equivalent-capacity-based algorithm transforms the traffic characteristics (usually described by three parameters: PBR/PCR, MBR/MCR, and peak packet/cell duration) of a new source into a unified metric, referred to as the *equivalent bandwidth*, to reduce the dependence of the proposed control mechanism on the traffic type. However, the transformation is based on either a static relationship or simulation curves.

The congestion indicator provides an insight into the congestion in the transmission links based on the burstiness of the existing traffic, so that future congestion can be avoided when a new source is admitted. CLR is the feedback provided by the system about its performance, which can be used to provide a closed-loop control system capable of adjusting itself to provide a stable and robust operation. Also, the available buffer space is used as an input to the controller and is calculated as:

$$\text{Available\_network\_resources} = \varphi_{\max} - \sum_{i=1}^{n} x_i(k) \qquad (4.27)$$

The network resource estimator keeps track of the available buffer space at each node/switch fabric. When a new source with a bandwidth specification is to be admitted, the available buffer space is updated by subtracting the assigned buffer space for the new source from the available network resource. Conversely, when an existing source ceases to transmit, the resource used by this source is now available to all other sources, and hence the available network resource is updated.

The peripheral schemes for the AC are a congestion controller (Jagannathan 2006, Jagannathan and Talluri 2002), a bandwidth estimator as presented in Section 4.2 through Section 4.4, and a network resource estimator. The congestion controller generates a congestion indicator according to the measured system statistics, such as the queue length, $q(k)$, and its past values, the CLR, and round-trip delays at the egress node. Here, the predictive congestion controller developed in Jagannathan and Talluri (2002) is preferred, but any congestion controller including buffer threshold can be used. The congestion indicator flag is set when the past several buffer occupancy values are about 90% full, the rate of change of the queue length is positive and high, the round-trip delays are large, and CLR occurs consistently. These values are obtained from careful analysis. In other words,

$$\text{If } ((q(k)) > 90\%) \text{ and } (q(k-1) > 90\%) \text{ and } (RTT > 2RTT_{min}) \quad \text{then}$$
$$\text{congestion\_flag} = \text{true, else false} \quad (4.28)$$

Bandwidth estimator obtains an accurate estimate of the current bandwidth that has to be assigned for the next measurement interval. Initially, the new source has to provide only its intended PBR/PCR, whereas other schemes request mean bit/cell rate (MBR/MCR), burstiness, delay, and number of packets/cells, in addition to PBR/PCR. All admitted new sources are assigned an initial bandwidth value equal to the PBR/PCR. From the next measurement interval onwards, the bandwidth to be assigned for this new source after its admission into the network is calculated using Equation 4.22.

Therefore, the approach presented in this chapter is more adaptive and does guarantee the performance in terms of QoS and network utilization. Based on this information, and to make the proposed admission controller simple and easy to implement, rules are generated as follows:

If (congestion flag is true) and (available_capacity > PCR) and (Available_network_resources > 10%), then admit source "i"; else reject.
$$(4.29)$$

The sources are expected to return to wait state, which is typically a very small value, and they are allowed to send a request again for admission into the network after the wait state. At each measurement interval, the congestion controller generates the transmission rates for all the sources to meet the QoS.

Here, the congestion controller operates at the packet level, whereas the admission controller works at the source level. These two controllers operate at two time scales and their interaction has to be carefully studied to avoid any instability and performance deterioration. Here, we propose a hybrid system theoretic approach to arrive at the overall network stability.

Because the packets arrive at discrete intervals, their transmission over the network is modeled as a continuous fluid flow behavior, and the communication among the end systems is via messaging, which is a discrete-event system; a hybrid system theory is required to analyze the overall stability of the congestion and admission control schemes. In the proposal, the admission controller design, expressed as a matrix-based discrete-event controller (DE), provides a framework for rigorous analysis of the DE system including its structure, protocols, and overall stability. It is described by the following equations, where k stands for the kth event and $z(k)$ is the state of the admission controller, ($\bar{z}(k)$ is the negation of $z(k)$), which is written as:

$$\bar{z}(k+1) = J_s\bar{v}_s(k) + J_r\bar{v}_r(k) + J_c\bar{v}_c(k) + J_u\bar{u}_s(k), \qquad (4.30)$$

where,

$$v_s = S_s z(k),$$

$$v_r = S_r z(k), \qquad (4.31)$$

$$v_c = S_c z(k),$$

and the task complete equation is

$$y(k) = S_y z(k) . \qquad (4.32)$$

Equation 4.30 through Equation 4.32 are logical equations. Input $\bar{u}_s$ (k) represents the sources waiting to get admitted and y represents the sources that ended transmission. The controller state equation (Equation 4.30) is a set of rules, so that it is formally a rule base. The coefficient matrices, J_s, J_r, J_c, and J_u are referred to as bandwidth availability, resource availability, congestion prediction, and admission request matrices, which are sparse so that real-time computations are easy for such a large interconnected DE system. The rules can be fired using an efficient Rete algorithm. The overbar in Equation 4.30 denotes logical negation (e.g., sources that have completed transmissions are denoted by '0'). The other matrices, S_s, S_r, S_c, are bandwidth, resource release, and congestion prediction matrices, whereas S_y represents the source completion matrix. Finally, all the matrix operations are defined in the 'max-min,' 'or,' and 'and' algebra. These matrices are created based on the initial state of the network and updated with time and are not discussed in detail because of space limitations.

Once the admission controller is expressed as a state equation (Equation 4.30), its performance and overall stability can be analyzed in conjunction with the congestion controller using hybrid system theory. Here, a non-smooth Lyapunov function will be used along the system trajectory defined by Equation 4.30. The continuous state space, $X \subseteq \Re^n$, is partitioned into a finite number of connected open regions, Ω_q, where $X = \cup_q \Omega_q$, $q \in Q = N$. Note that the regions, Ω_q, are not required to be disjoint. In each region a controller, u is designed by any technique. It is assumed that the controller gives a satisfactory performance only in region Ω_q for all $q \in Q$. As each closed-loop system is stable, $x(k+1) = f(x(k), u(k))$ is stable for all $x \in \Omega_q$, a standard Lyapunov function $V_q(x,t): X \times \Re \to \Re$ exists in Ω_q. These Lyapunov functions can be combined to produce a nonsmooth Lyapunov function, which globally represents the dynamics of the hybrid system. Here, the congestion controller along with Equation 4.27 becomes a local control law and the rules used for admission control switch the controller.

DEFINITION 4.5.1 (ANTSAKLIS ET AL. 1995, RAMADGE AND WONHAM 1989)
A nonsmooth Lyapunov function, $V(x(k),k)$, is assumed to be left continuous on $k \in [k_0, \infty)$ and differentiable almost everywhere on (t_O, ∞) except on the set $T_s = \{k_0, k_1, ...,\}$ when switches occur. The constraint $\Delta V < 0$ is replaced by a stronger condition: the nonincreasing condition.

$$V(x(k_i), k_i) < V(x(k_j), k_j) \quad \text{if} \quad k_i > k_j \quad (4.33)$$

THEOREM 4.5.1 (ADMISSION CONTROL SCHEME STABILITY)
Given the hybrid system H whose dynamics are governed by Equation 4.1 and Equation 4.30, if its controller is selected as Equation 4.6 (Jagannathan 2002) to the buffer dynamics and (4.31) for congestion control, and the Lyapunov functions in each region have the same values on the boundaries, Ω_q, then the origin of the discrete-state space of the hybrid system is asymptotically stable.

The next section details the simulation results.

4.6   Simulation Results

This section discusses the network model, parameters, and constants used in the simulations to test the effectiveness of the proposed algorithm. The traffic sources used in the simulation are also discussed and the simulation results are explained. In our admission control example, we drive the sources independently with both voice (ON/OFF) and video (MPEG) data. We use fixed packets of length 53 bytes referred to here as *cells*.

4.6.1   Adaptive Estimator Model

Previous six values of buffer occupancy were used as inputs to the NN as a trade-off between approximation and computation. The greater the number of past values of buffer occupancy used, the better is the approximation, but large numbers of past values increase the computation, which in turn not only causes delays at the network switches but also wastes resources. As the number of inputs to the ARMAX model at each switch is 6 and the output is a single value, the parameter vector, $\hat{W}(k)$, is a 6 by 1 vector for each switch. The regression vector, $\phi(.)$, will then consist of current and past values of buffer occupancy. The initial adaptation gain, α, is taken as 0.1 and is updated using the projection algorithm as $\alpha(k) = \frac{0.1}{(0.1+\phi(x(k))^T\phi(x(k)))}$. The buffer occupancy for both the switches is considered empty, initially. The initial parameter estimates are chosen as $\hat{\theta}(1) = [0\ 0\ 0\ 0\ 0\ 0]^T$ for each switch. However, one can select any initial values for the parameters. The parameter-tuning updates derived herein were deployed for the simulation. The network model and the traffic sources used in the simulations are discussed next.

4.6.2   Network Model

The network model used in simulations is similar to that shown in Figure 4.1 and Figure 4.2. For each simulation, the buffer length at the ingress node/switch, x_d, is selected to be as small as 10 cells and it is altered to go as high as 350 cells. Cell losses, $c(k)$, are defined as $x(k) - x_d$ when $x(k) > x_d$. The CLR is defined as the total number of cells discarded at the destination owing to buffer flows divided by the total number of cells transmitted. Target cell transfer or transmission delay is the total time taken by a source to complete its transmission from its starting point. The transmission delay is defined as $T_a - T_i$, where T_i is the intended time to transmit the data, and T_a is the total actual time needed to transmit the data. If the source sends its data before its intended time, then a negative value is obtained, indicating that the source has completed its transmission well before its target time. On the other hand, if a delay is encountered, then it is typically represented by a positive value, implying that the sources took longer than their intended time.

In the simulations, a two-switch scenario is considered. Four MPEG sources are assumed to be transmitting at Switch 1. The output of Switch 1 is connected to Switch 2. In addition, Switch 2 also allows 4 MPEG sources. Initially, it is assumed that the physically available bandwidth at both switches is equal to the PCR of all the sources. Four sources are to be admitted at Switch 1 and four at Switch 2. In addition, at 3000 units of time, the output of Switch 1 is admitted at Switch 2.

The PCR of the four sources combined at Switch 1 is 13,200 cells/sec, whereas that of the three admitted MPEG sources are 9150, 9100, and 9040 cells/sec, respectively, whereas that of the two ON/OFF sources are 4800 and 4800, respectively. The PCR of the data from Switch 1 that is admitted at Switch 2 is 137,000 cells/sec. Both the switch fabrics in this case are considered as ingress nodes in Figure 4.2b, but Switch 2 is considered as an egress node for several sources sending traffic through Switch 1. The admission condition will be checked every 500 frame times and if the conditions are met, a source will be admitted at that instant; otherwise it will be rejected.

4.6.3 Traffic Sources

Multiple ON/OFF sources: Each source was driven with 1800 msec of ON/OFF data to run our simulations. The target transfer delay is defined as 1% of the total time to transmit the data if the bandwidth is available at the PCR. This value is given as 18 msec (1% of 1800), and it is used as the target cell transfer delay requirement. The PCR of the ON/OFF data sources are given by 12,000, 11,000, 10,000, and 9,000 cells/sec, respectively, and the combined PCR for all the sources is 42,000 cells/sec, with MCR being 38,000 cells/sec. A measurement interval of 1 msec was used in the simulation run.

MPEG sources: In high-speed networks such as ATM and the Internet, real-time video applications will become major traffic sources. The Moving Picture Expert Group (MPEG) is one of the most promising interframe compression techniques for such applications (Lakshman et al. 1999). Hence, it is critical to realize effective MPEG video transfer through networks, and our scheme takes advantage of the flexibility of MPEG video.

The MPEG data set used in the simulations was found at a Bellcore ftp site (Jagannathan and Talluri 2002). This data set comes from the movie *Star Wars*. The movie length is approximately 2 h and contains a diverse mixture of material ranging from low-complexity scenes to those with high action. The data set has 174,138 patterns, each pattern representing the number of bits generated in a frame time F. In this trace, 24 frames are coded per second, so F is equal to 1/24 sec. The PBR of this trace is 185,267 bits/frame, the mean bit rate is 15,611 bits/frame, and the standard deviation of the bit rate is about 18,157. We used 4,000 frames to run our simulations.

The target transfer delay is defined as 0.1% of the total time to transmit the data if the bandwidth is available at the PCR. This value is given as

167 msec (0.1% of 167 sec), and it is used as the target cell transfer delay. The PCR of the MPEG data sources are given by 9,154, 6,106, 6,237, and 4,384 cells/sec, respectively, and the combined PCR for all the sources is 13,248 cells/sec, with MCR being 3,444 cells/sec. For VBR sources, a measurement interval of 1/24 sec, which is equal to one frame time, is used.

4.6.4   Simulation Examples

Note that our objective is to demonstrate the performance of bandwidth estimation, allocation scheme, and call admission control scheme for audio (ON/OFF) and highly bursty video (MPEG data) streaming applications, while meeting the QoS and peak bandwidth constraint of the outgoing transmission link. Physically available bandwidth at the outgoing link is taken to be the PCR of all sources and PCR of a single source for demonstrating the need for an accurate bandwidth estimation scheme.

Three methods will be compared to estimate the bandwidth, available capacity, and admission control. The first method assigns bandwidth using the PCR of the source. The second method presents an adaptive two-layer NN algorithm for estimating the traffic and then calculates the available capacity. However, in the case of NN with overallocation algorithm, the overallocation factor, ρ, is taken as 0.95 and the bandwidth is allocated as $\frac{BW(k)}{\rho}$, instead of $BW(k)$ as suggested in the second method.

In the examples, the exponential weight moving average (EWMA) filter was applied on the bandwidth error for convenience and readability. The EWMA filter can be mathematically described as,

$$EWMA(k+1) = \tau EWMA(k) + (1-\tau)^* b(k), \qquad (4.34)$$

where $EWMA(k)$ represents the filtered value of $b(k)$ at the instant k, τ is the filter constant $0 < \tau < 1$ and $b(k)$ is the bandwidth estimation or bandwidth assignment error. Here, τ is selected as 0.99. Note that a positive error value implies that not enough bandwidth is assigned (bandwidth is not estimated accurately), whereas a negative error value implies that the assigned bandwidth is not fully utilized.

Figure 4.5 shows the number of admitted sources at Switch 1 when the physically available bandwidth is equal to the PCR of all the sources. Here, all sources are admitted with no service and transmission delays or losses. Figure 4.6 shows the corresponding plot for Switch 2. The bandwidth estimation algorithm and the NN-based AC are used to admit traffic. The bandwidth prediction error at the switches is shown in Figure 4.7 and Figure 4.8. When the physically available bandwidth is equal to the

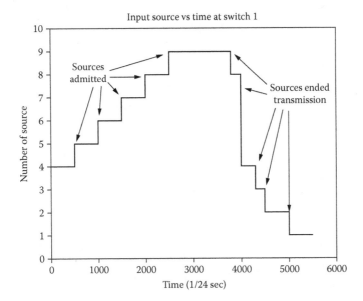

FIGURE 4.5
Sources admitted at Switch 1 (when the physically available bandwidth is equal to the peak cell rate of all the sources).

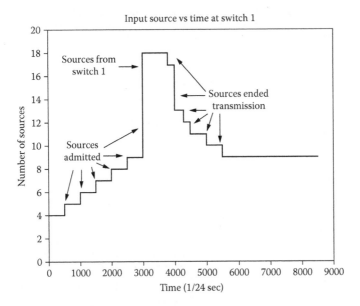

FIGURE 4.6
Sources admitted at Switch 2 (when the physically available bandwidth is equal to the peak cell rate of all the sources).

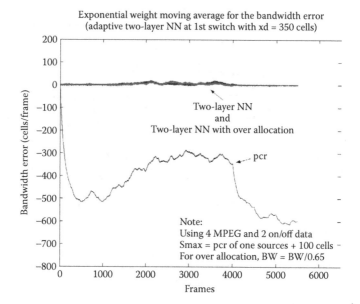

FIGURE 4.7
Prediction error at Switch 1.

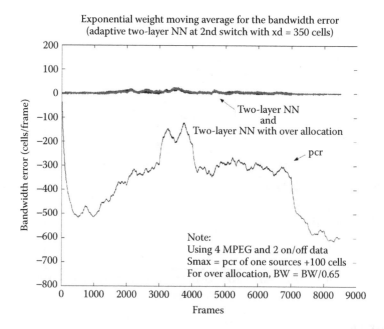

FIGURE 4.8
Prediction error at Switch 2.

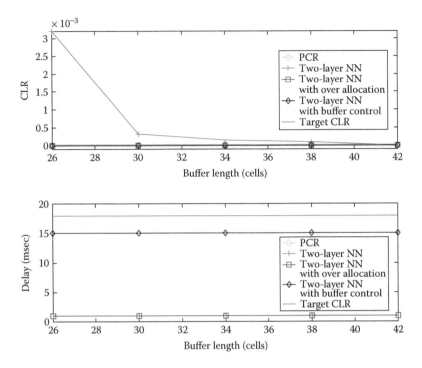

FIGURE 4.9
Prediction error at Switch 1 for several schemes in which the cell loss ratio and delay are small when peak cell rate is used.

PCR of all the sources, then all sources are admitted irrespective of the method used. However, a large amount of bandwidth is wasted when the PCR-based scheme is used. Although a small prediction error similarly is observed (blue), the selection of overallocation factor is done by trial and error whereas NN scheme in "red" generated a small error.

In Figure 4.9 and Figure 4.10, the CLR and transmission delay are given for several schemes in which the CLR and delay are small when PCR is used, compared to all other schemes. Although the two-layer NN method produces slightly higher CLR and delay than PCR, it meets the target QoS requirement.

In the second case, when the available bandwidth is not equal to the PCR of all sources, several new sources are rejected by the other methods, whereas our adaptive method admitted more sources, resulting in high network utilization. The improved performance is the result of accurate bandwidth estimation. Here, the sources are queued until the sufficient bandwidth is available. Figure 4.11 shows the comparison between the available bandwidth's being PCR of all the sources and its being PCR of one ON/OFF source at Switch 2. As expected, fewer sources are admitted.

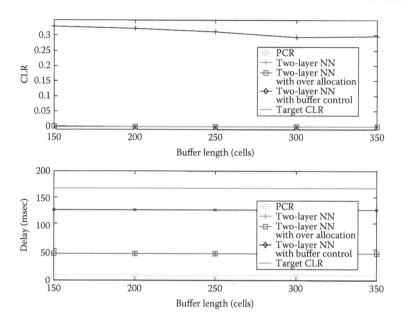

FIGURE 4.10
Prediction error at Switch 2 for several schemes in which the cell loss ratio and delay are small when peak cell rate is used.

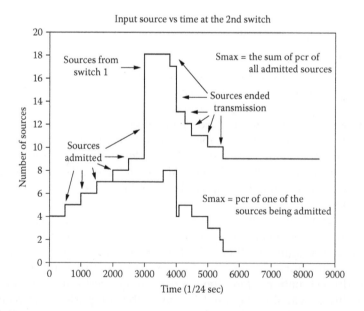

FIGURE 4.11
Admission controller performance.

The CLR for the adaptive scheme is slightly higher than for adaptive with overallocation and the PCR because the adaptive scheme is conservative; however, the CLR meets the target specification. The CLR resulting from PCR is near zero as enough resources are *a priori* reserved. These results clearly show that the bandwidth assignment has to be accurate to admit new sources. This chapter has demonstrated a novel scheme of assigning bandwidth, preventing congestion and a scheme for admitting sources based on available capacity.

4.7 Conclusions

This chapter proposes a system theoretic approach to admission controller design for high-speed networks. This admission controller explicitly computes the equivalent bandwidth required to support each class of traffic, based on online observation of aggregate traffic flow obtained via NN-based online adaptive bandwidth estimation. The network buffer dynamics are modeled as a nonlinear dynamical system and an adaptive NN algorithm is designed to estimate traffic flow. This adaptive NN approach does not require accurate information about network system dynamics or traffic rate. A novel weight tuning method was derived using rigorous mathematical analysis. In fact, the proposed adaptive estimator guarantees performance, as shown through the Lyapunov analysis.

The bandwidth required to meet the QoS is then estimated and the available capacity is derived given the actual cell losses, latency, and estimated traffic flow. This information, along with the intended PCR of the source to be admitted, predictive congestion indicator, and desired QoS metrics, is used to construct the adaptive admission controller. The admission controller is expressed as a discrete-event state equation in "min"-"max" algebra and the dynamics of the buffer along with the admission controller can be described as a hybrid system. The stability of the hybrid system is stated using nonsmooth Lyapunov function because the admission controller switches the operation of the buffer dynamics. The congestion and admission controllers operate at two different time scales.

Results are presented to evaluate the performance of the proposed admission controller with ON/OFF and bursty MPEG data. Based on these results, it was concluded that the adaptive methodology generates an accurate estimation of the bandwidth required and that needs to be assigned to the sources to meet the QoS. Accurately estimating the bandwidth online along with the admission control will result in high network utilization.

References

Antsaklis, P.J., Kohn, W., Nerode, A., and Sastry, S., Lecture Notes in Computer Science: Hybrid Systems II, New York, Springer-Verlag, Vol. 99, 1995.

Bae, J.J. and Suda, T., Survey of traffic control schemes and protocols in ATM networks, *Proceedings of the IEEE*, Vol. 79, No. 2, February 1991, pp. 170–189.

Chen, B., Zhang, Y., Yen, J., and Zhao, W., Fuzzy adaptive connection admission control for real-time applications in ATM-based heterogeneous networks, *Journal of Intelligent and Fuzzy Systems*, Vol. 7, No. 2, 1999.

Cheng, R.-G., Chang, C.-J., and Lin, L.-F., A QoS-provisioning neural fuzzy connection admission controller for multimedia high-speed networks, *IEEE/ACM Transaction on Networking*, Vol. 7, No. 1, 111–121, 1999.

Dziong, Z., Juda, M., and Mason, L.G., A framework for bandwidth management in ATM networks—aggregate equivalent bandwidth estimation approach, *IEEE/ACM Transactions on Networking*, Vol. 5, No. 1, 134–147, February 1997.

Fahmy, S., Jain, R., Kalyanaraman, S., Goyal, R., Vandalore, B., On determining the fair bandwidth share for ABR connections in ATM networks, *Proceedings of the IEEE on Communications*, Vol. 3, 1998, pp. 1485–1491.

Gurin, R., Ahmadi, H., and Naghshineh, M., Equivalent capacity and its application to bandwidth allocation in high-speed networks, *IEEE Journal on Selected Areas in Communications*, Vol. 9, No. 7, 968–981, September 1991.

Jagannathan, S. and Talluri, J., Adaptive predictive congestion control of high-speed networks, *IEEE Transactions on Broadcasting*, Vol. 48, No. 2, 129–139, June 2002.

Jagannathan, S., Admission control design for high speed networks: a hybrid system approach, *Journal of High Speed Networks*, Vol. 14, 263–281, 2005.

Jagannathan, S., *Neural Network Control of Nonlinear Discrete-Time Systems*, Taylor and Francis (CRC), Boca Raton, FL, 2006.

Jain, R., Congestion control and traffic management in ATM networks: recent advances and a survey, *Computer Networks and ISDN Systems*, Vol. 28, 1723–1738, 1996.

Jamin, S., Danzig, P.B., Shenker, S.J., Zhang, L., A measurement-based admission control algorithm for integrated service packet networks, *IEEE/ACM Transactions on Networking*, Vol. 5, No. 1, February 1997.

Lakshman, T.V., Mishra, P.P., and Ramakrishan, K.K., Transporting compressed video over ATM networks with explicit-rate feedback control, *IEEE/ACM Transactions on Networking*, Vol. 7, No. 5, October 1999.

Liew, S. and Yin, D.C., A control-theoretic approach to adopting VBR compressed video for transport over a CBR communications channel, *IEEE/ACM Transactions on Networking*, Vol. 6, No. 1, 42–55, February 1998.

Peng, M., Jagannathan, S., and Subramanya, S., End to end congestion control of multimedia high speed Internet, *Journal of High Speed Networks*, to appear in 2006.

Ramadge, P.J. and Wonham, W.M., The control of discrete event systems, *Proceedings of the IEEE*, Vol. 77, No. 1, 1989, pp. 81–89.

4.8 Problems

Section 4.5

Problem 4.5.1: Convergence analysis. Prove the admission control scheme analytically by using Lyapunov stability analysis.

Section 4.6

Problem 4.6.1: Performance evaluation. Compare the performance of the admission controller developed using Lyapunov stability with that of the conventional rule based system.

Problem 4.6.2: Performance evaluation. Evaluate the performance of the admission controller developed using Lyapunov stability for the end to end network presented in Chapter 3 by using MPEG traffic.

APPENDIX 4.A

PROOF Define the Lyapunov function candidate

$$J = e^T(k)e(k) + \frac{1}{\alpha_1} tr[\tilde{V}^T(k)\tilde{V}(k)] + \frac{1}{\alpha_2} tr[\tilde{W}^T(k)\tilde{W}(k)], \qquad (4.A.1)$$

The first difference is given by

$$\Delta J = \Delta J_1 + \Delta J_2, \qquad (4.A.2)$$

$$\Delta J_1 = e^T(k+1)e(k+1) - e^T(k)e(k), \qquad (4.A.3)$$

$$\Delta J_2 = \frac{1}{\alpha_2} tr[\tilde{W}^T(k+1)\tilde{W}(k+1) - \tilde{W}^T(k)\tilde{W}(k)]$$
$$+ \frac{1}{\alpha_1} tr[\tilde{V}^T(k+1)\tilde{V}(k+1) - \tilde{V}^T(k)\tilde{V}(k)], \qquad (4.A.4)$$

Considering the input and hidden-layer weight updates (Equation 4.13), and using these in Equation 4.A.4 and combining with Equation 4.A.3, one obtains:

$$\Delta J \le -\left(1 - \alpha_2 \hat{\varphi}_2^T(k)\hat{\varphi}_2(k)\right) \times \left[e(k) - \frac{\alpha_2 \hat{\varphi}_2^T(k)\hat{\varphi}_2(k)\delta(k)}{1 - \alpha_2 \hat{\varphi}_2^T(k)\hat{\varphi}_2(k)} \right]^T$$

$$\times \left[e(k) - \frac{\alpha_2 \hat{\varphi}_2^T(k) \hat{\varphi}_2(k) \delta(k)}{1 - \alpha_2 \hat{\varphi}_2^T(k) \hat{\varphi}_2(k)} \right] - \|e(k)\|^2 - \left(2 - \alpha_1 \hat{\varphi}_1^T(k) \hat{\varphi}_1(k) \right)$$

$$\times \left\| \hat{W}^T(k) \hat{W}(k) - \frac{\left(1 - \alpha_1 \hat{\varphi}_1^T(k) \hat{\varphi}_1(k) \right)}{2 - \alpha_1 \hat{\varphi}_1^T(k) \hat{\varphi}_1(k)} \left(W^T \hat{\varphi}_1(k) + B_1 e(k) \right) \right\|^2$$

$$+ 2 \|e(k)\| \times \frac{k_1 \phi_{1\max} W_{1\max}}{\left(2 - \alpha_1 \phi_{1\max}^2 \right)} + \frac{\delta_{\max}^2}{\left(1 - \alpha_2 \phi_{2\max}^2 \right)} + \frac{W_{1\max}^2 \phi_{1\max}^2}{\left(2 - \alpha_1 \phi_{1\max}^2 \right)}, \quad (4.A.5)$$

where

$$\delta_{\max} = W_{2\max} \tilde{\varphi}_{2\max} + \varepsilon_{N\max} + d_M \quad (4.A.6)$$

$$\Delta J \le -(1 - C_0) \times \left[\|e(k)\|^2 - 2 \frac{C_1}{(1 - C_0)} \|e(k)\| - \frac{C_2}{(1 - C_0)} \right]$$

$$- \left[1 - \alpha_2 \hat{\varphi}_2^T(k) \hat{\varphi}_2(k) \right] \left\| e(k) - \frac{\alpha_2 \hat{\varphi}_2^T(k) \hat{\varphi}_2(k) \delta(k)}{1 - \alpha_2 \hat{\varphi}_2^T(k) \hat{\varphi}_2(k)} \right\|^2$$

$$- \left(2 - \alpha_1 \hat{\varphi}_1^T(k) \hat{\varphi}_1(k) \right) \left\| \hat{W}^T(k) \hat{W}(k) - \frac{\left(1 - \alpha_1 \hat{\varphi}_1^T(k) \hat{\varphi}_1(k) \right)}{2 - \alpha_1 \hat{\varphi}_1^T(k) \hat{\varphi}_1(k)} \left(W^T \hat{\varphi}_1(k) + B_1 e(k) \right) \right\|^2$$

$$(4.A.7)$$

where

$$C_1 = \frac{k_1 \phi_{1\max} W_{1\max}}{\left(2 - \alpha_1 \phi_{1\max}^2 \right)} \quad (4.A.8)$$

and

$$C_2 = \left[\frac{\delta_{\max}^2}{\left(1 - \alpha_2 \phi_{2\max}^2 \right)} + \frac{W_{1\max}^2 \phi_{1\max}^2}{\left(2 - \alpha_1 \phi_{1\max}^2 \right)} \right] \quad (4.A.9)$$

because C_0, C_1, and C_2 are positive constants, $\Delta J \le 0$ as long as

$$\|e(k)\| > \frac{1}{(1 - C_0)} \left(C_1 + \sqrt{C_1^2 + C_2(1 - C_0)} \right). \quad (4.A.10)$$

In general $\Delta J \leq 0$ in a compact set as long as Equation 4.14 through Equation 4.16 are satisfied and Equation 4.A.10 holds. According to a standard Lyapunov extension theorem (Jagannathan 2006), this demonstrates that the cell losses and the error in weight estimates are UUB. To show the weights are bounded, consider the dynamics relative to the errors in the weight estimates are given by:

$$\tilde{W}(k+1) = (1 - \alpha_2 \varphi_2(x(k)) \varphi_2(x(k))^T) \tilde{W}(k) + e^T(k) k_v^T B_1^T \quad \text{(4.A.11)}$$

$$\tilde{V}(k+1) = (1 - \alpha_1 \varphi_1(x(k)) \varphi_1(x(k))^T) \tilde{V}(k) + e^T(k) k_v^T B_1^T \quad \text{(4.A.12)}$$

where the buffer occupancy estimation error is considered bounded. Applying the persistence of excitation condition (Jagannathan 2006) and noting that the traffic-modeling error and disturbance are bounded, the boundedness of $\tilde{W}(k)$ and $\tilde{V}(k)$ implies that the weight estimates $\hat{W}(k)$ and $\hat{V}(k)$ are bounded.

5

Distributed Power Control of Wireless Cellular and Peer-to-Peer Networks

In many cases, data from ad hoc wireless and sensor networks are routed through a cellular infrastructure or a wired network such as an Internet. To meet the quality of service (QoS), one has to ensure performance guarantees not only from the wireless network side but also from the wired network. In the previous chapter, admission-controller design was covered for wired networks so as to meet a desired level of QoS. In this chapter, we treat QoS performance for wireless cellular networks via transmitter power, active link protection, and admission-controller designs. For wireless networks, besides throughput, packet loss, and end-to-end delay guarantee, energy efficiency is another QoS performance metric. Energy efficiency, to a certain extent, can be attained by controlling transmitter power, which is the central theme of this chapter.

The huge commercial success of wireless communications along with the emerging popularity of IP-based multimedia applications are the major driving forces behind the next generation (3G/4G) evolution wherein data, voice, and video are brought into wireless networks that require diverse QoS. In such modern wireless networks, distributed power control (DPC) of transmitters allows interfering communications sharing the same channel to achieve the required QoS levels. Moreover unlike in wired networks, the channel condition affects the network capacity and QoS; therefore, any power control scheme for wireless networks must incorporate the time-varying nature of the channel.

Available power control schemes (Bambos 1998, Bambos et al. 2000, Zander 1992, Grandhi 1994, Foschini and Miljanic 1993), are efficient in combating disturbances such as path loss and user interferences that are of small magnitude, whereas they are ineffective during the process of rejecting channel uncertainty (additive white Gaussian vs. Rayleigh fading channels) and combating large disturbances because they render poor bit error rate (BER) although consuming excessive power. Assuring QoS requires novel schemes that use accurate channel state information, to reject uncertainties (Rayleigh fading channels), to combat disturbances (path loss, interference), to route data and control packets while optimizing the transmitter powers, and trading power and delay during deep channel fluctuations. In this chapter, first, a linear system theory is employed to analytically represent the interference problem among users in a cellular network. Note that state space modeling and controller design is a preferred method in other application areas as well. State space and optimal schemes (Jagannathan et al. 2002, Dontula and Jagannathan 2004) are first developed to control the power of each transmitter in the presence of path loss uncertainty. Subsequently, a channel state estimator embedded in the power control scheme (Jagannathan et al. 2006) is presented so that the receiver can suggest suitable transmitter powers required to overcome slowly varying channel conditions.

5.1   Introduction

Control of transmitter power allows communication links to be established in a channel, using minimum power to achieve required signal-to-interference ratios (SIR) reflecting QoS levels. Interference mitigation properly and quickly increases network capacity through channel reuse. Past literature on power control (Aein 1973, Alavi and Nettleton 1982) focused on balancing the SIRs on all radio links using centralized control. Later, distributed SIR-balancing schemes (Zander 1992, Grandhi et al. 1994) were developed to maintain QoS requirements of each link. In a dynamic environment, maintaining and guaranteeing the QoS is extremely difficult, because maintaining QoS of all existing links may require removal of some active links, selectively admitting new links, and/or providing differentiated QoS services. Power control protocols were addressed in the literature for cellular networks initially and later extended to ad hoc networks.

Mitra (1993) proposed a distributed asynchronous online power control scheme, which can incorporate user-specific SIR requirements and yield minimal transmitter powers while converging geometrically. However, as new users try to access the channel, the SIRs of all existing ones may drop below the threshold, causing an inadvertent drop of ongoing calls (Hanly 1996). A second-order power control scheme that uses both current and past values to determine the transmitter power for the next packet, was proposed in Jantti and Kim (2000). This scheme, developed by applying the successive overrelaxation method, converges faster, but the performance of the power control scheme can be affected negatively because of measurement errors and loop delays. In the seminal work of Bambos et al. (2000), a DPC scheme is presented with admission control. An optimum transmitter power control scheme similar to Bambos et al. (2000) with heterogeneous SIR thresholds was proposed in Wu (2000). The DPC scheme was able to incorporate inactive links with active link-protection mechanisms. The convergence of the power update scheme (Bambos et al. 2000, Wu 2000) and the process of admitting links are an issue in a highly dynamic wireless environment as illustrated in this chapter. Therefore, it is demonstrated that the state space and optimal schemes (Jagannathan et al. 2002) can converge and can admit significantly more links (Dontula and Jagannathan 2004) while guaranteeing that the links consume minimal power.

In this chapter, a networking approach to the design of DPC schemes is discussed (Dontula and Jagannathan 2004), in which the issue of admission control is made central for the peer-to-peer scenario initially. The state space and an optimal power control schemes, which sustain the SIR of active links above a certain threshold referred to as active link-protection (ALP), are introduced when new links access the channel. Intuitively, the new links gradually power up while the active ones are assured QoS with a protection margin. The DPC schemes can efficiently control the power being used by the established links in a communication channel. These schemes can achieve the required SIR, which reflects a given level of QoS. Lower interference levels in the communication channel resulting from the use of this scheme can increase the channel efficiency, capacity, and reusability for the uplink scenario.

It is proven that the DPC schemes converge and admit a greater number of links when compared to the existing schemes (Bambos et al. 2000, Jantti and Kim 2000, Wu 2000). For the proposed schemes, it is analytically proven that a trade-off exists between convergence speed and power consumed by a mobile user. This trade-off is easily observed by using the feedback gains. The power control scheme being highly distributive in nature doesn't require interlink communication, centralized computation, and reciprocity assumption as required in a centrally controlled wireless environment. As the necessity of interlink communication is eliminated, network capacity increases and easy, controlled recovery from error events is possible. The DPC/ALP scheme can be used to work in a cellular network environment. Instead of receiver–transmitter pairs as in the case with peer-to-peer networks, users in a cell communicate with the base station, which services the cell. To compare the DPC schemes discussed in the literature, users are assumed to request channel access at the same time, at the beginning of the simulation. From the results, it is indicated that the proposed schemes provide lower outage probability, consume less power, and converge in a reasonable amount of time.

The DPC schemes have to be modified in the presence of fading channels. In fact, a novel channel estimator is introduced, which is embedded in the DPC (Jagannathan et al. 2006), so that the suitable power values can be selected in the presence of fading channels. Results indicate that the modified DPC with the embedded channel estimator provides a superior performance. First, the state-space-based controls design (SSCD) and optimal DPC schemes are presented; and subsequently, the modified DPC scheme is covered.

5.2   Distributed Power Control in the Presence of Path Loss

The wireless network can be considered a collection of radio links. Each link in the network is connected between a pair of nodes, which correspond to a single-hop radio transmission. The transmission is intended from the transmitter via the link to its corresponding receiver. A collection of consecutive links corresponds to a multihop communication path. Many communication channels may be present simultaneously in the network, but it is considered that these channels are orthogonal. However, links operating in the same channel experience interference from other channels. We can, therefore, model the wireless network as a collection of interfering links for each channel. "Distributed" implies per individual link. Each receiver of the link measures the interference it is faced with and communicates this information to its transmitter. Each link decides autonomously how to adjust its power, based on exclusive information that is collected. Therefore, the decision-making is fully distributed at the link level. The overhead due to the feedback control is minimal in DPC compared with its counterpart when centralized operations are used.

Whereas a link in the peer-to-peer networking scenario corresponds to single-hop transmissions, in cellular communication network, a link corresponds to up- and downstream transmission between the users and base stations. In spread-spectrum systems, the whole spectrum can be viewed as a single channel, and interference can be attributed to the cross-correlation effects between codes in code division multiple access (CDMA) transmission.

Our goal in both peer-to-peer and cellular network paradigms is to maintain a required SIR threshold for each network link while the transmitter power is adjusted so that the least possible power is consumed. Suppose there are $N \in Z_+$ links in the network. Let G_{ij} be the power loss (or gain) from the transmitter of the jth link to the receiver of the ith link. It involves the free space loss, multipath fading, shadowing, and other radio wave propagation effects, as well as the spreading or processing gain of CDMA transmissions (Grandhi et al. 1994). The power attenuation is taken to follow the inverse fourth power law

$$G_{ij} = \frac{g}{r_{ij}^{\alpha}}, \tag{5.1}$$

where g is a constant usually equal to 1, r_{ij} is the distance between transmitter and receiver, and α is a constant that varies with the environment. Typically, $\alpha = 2$ to 4.

Calculation of SIR, R_i, at sender of ith link, (Rappaport 1999) yields

$$R_i = \frac{G_{ii} * p_i}{\left(\sum_{j \neq i} G_{ij}\, p_j + \kappa_i \right)} \tag{5.2}$$

where i , $j \in \{1, 2, 3, \dots, n\}$, p_i is the link's transmitter power, and $\kappa_i > 0$ is the thermal noise at its receiver node. For each link i, there is a lower SIR threshold γ_i. Therefore, we require

$$R_i \geq \gamma_i \tag{5.3}$$

for every $i = 1, 2, 3, \dots n$. The threshold values for all links can be taken equal to γ for convenience, reflecting a certain QoS the link has to maintain to operate properly. An upper SIR limit is also set, so that the transmitter power of a link is minimized, which, in turn, will decrease the interference due to its transmitter power at other receiver nodes. Therefore, we have

$$R_i(l) \leq \gamma_i^* \tag{5.4}$$

Let ς be the set of all links, and call a link $i \in \varsigma$ active or operational during the lth step if

$$R_i(l) \geq \gamma_i$$

where $R_i(l)$ is the measured SIR, and A_l is the set of all active links during the lth step. Alternatively, call a link $i \in \varsigma$ inactive or new link during the lth step if

$$R_i(l) < \gamma_i$$

where B_l is the set of all in active links during the lth step.

FACT 5.2.1 (EXISTENCE OF A FEASIBLE POWER VECTOR)
There exists a power vector $p^* > 0$, which is optimal, such that $\gamma_i < SIR < \gamma_i^*$. This statement is equivalent to maintaining the SIR $R_i \geq \gamma_i$ $i \in A_l$, the set of active or operational links, and $R_i \leq \gamma_i$ $i \in B_l$, the set of inactive or new links trying to get added during the lth step.

Here, B_l is the set of all inactive links during the lth step. We need a parameter η_i to provide the protection margin, which is used to guarantee that a link is active during link admissions, such that

$$R_i(l) \geq \gamma_i + \eta_i \tag{5.5}$$

Any power control scheme would have to set the individual powers at least to p^* to satisfy SIR requirements in Equation 5.3 and Equation 5.4. This solution for power p^* is optimal. A good power control scheme would set the individual powers of the links to p^* so as to minimize the power consumption.

5.2.1 Power Control Scheme by Bambos

If Equation 5.3 or Equation 5.4 fail, the transmitter power is updated. Therefore, each link independently increases its power when its current SIR is below its target γ_i, and decreases it otherwise. The associated power update can be obtained from (Bambos et al. 2000) as

$$p_i(l+1) = \frac{\gamma_i p_i(l)}{R_i(l)} \tag{5.6}$$

where $l = (1, 2, 3 \ldots)$ (see Bambos 1998, Bambos et al. 2000). If $p_i(l+1) > p_{\max}$, a new link is not added. If power slips under a minimum threshold power $p_i(l+1) < p_{\min}$ (the minimum power needed to form a link), then the power is assigned $p_i(l+1) = p_{\min}$. The DPC scheme updates the transmitter powers in steps (time slots) indexed by $l = 1, 2, 3, \ldots$.

5.2.2 Constrained Second-Order Power Control

In Jantti and Kim (2000), the SIR in Equation 5.2 is defined as a set of linear equations

$$AP = \mu \tag{5.7}$$

where $A = I - H$ and $P = (p_i)$, in turn $H = [h_{ij}]$ is defined as a Q x Q matrix, such that $h_{ij} = \frac{\gamma_i g_{ij}}{g_{ii}}$ for $i \neq j$ and $h_{ij} = 0$ for $i = j$. In addition, $\mu = (\gamma_i v_i / g_{ii})$ is a vector of length Q. Because the maximum transmission power of a node is limited, the following constraint is placed on the power vector:

$$0 \leq P \leq \overline{P} \tag{5.8}$$

Here $\bar{P} = (p_{max})$ denotes the maximum transmission power level of each transmitter. The scheme assumes that there exists a unique power vector P^*, which would solve Equation 5.7. Thus, by a feasible system, the matrix A is nonsingular, and $0 \leq P^* = A^{-1}\mu \leq \bar{P}$. Iterative methods can be executed with local measurement to find the power vector P^*. Through some manipulations, the following second-order iterative form of the algorithm is obtained as

$$p_i^{(l+1)} = \min\left\{\bar{p}_i, \max\left\{0, w^l \gamma_i p_i^l / R_i^l + (1 - w^l) p_i^{l-1}\right\}\right\} \quad (5.9)$$

where $w^l = 1 + 1 / 1.5^n$. The min and max operators in the Equation 5.9 are to guarantee that the transmitter power will be within the allowable range.

The DPC schemes proposed in Bambos et al. (2000) and Jantti and Kim (2000) appear to result in unsatisfactory performance in terms of convergence and during admission control as demonstrated in our simulations. The proposed work (Jagannathan et al. 2002, Dontula and Jagannathan 2004) is aimed at addressing these limitations. A suite of closed-loop DPC schemes is presented next. Both state space (SSCD) and optimal schemes for updating the transmitter power are described in detail. The convergence proofs of these control schemes are also given, and they are demonstrated in simulation. Later, the DPC scheme is modified so that suitable power can be selected during fading channels.

5.2.3   State-Space-Based Controls Design

The calculation of SIR can be expressed as a linear system as

$$R_i(l+1) = R_i(l) + v_i(l) \quad (5.10)$$

where $R_i(l+1)$, $R_i(l)$ are the SIR values at the time instant l and $l + 1$, respectively, with $v_i(l)$ being the power value. Equation 5.10 is obtained as follows: i.e., each user-to-user or user-to-base-station connection can be considered as a separate subsystem as described by the equation

$$R_i(l+1) = \frac{p_i(l) + u_i(l)}{I_i(l)} \quad (5.11)$$

where by definition $R_i(l) = \frac{p_i(l)}{I_i(l)}$, and interference $I_i(l) = \left(\sum_{j\neq i}^{n} p_j(l) * \frac{G_{ij}}{G_{ii}} + \frac{x_i}{G_{ii}}\right)$, with n as the number of active links. The $v_i(l)$ in each system is defined as $v_i(l) = \frac{u_i(l)}{I_i(l)}$, where $u_i(l)$ is the input to each subsystem, which depends only on the total interference produced by the other users. To maintain the SIR of each link above a desired target and to eliminate any steady-state errors, *SIR error* $e_i(l) = R_i(l) - \gamma_i$ is defined as the difference between actual SIR and its target value γ_i. Therefore, defining $e_i(l) = x_i(l)$, we have

$$x_i(l+1) = J_i x_i(l) + H_i v_i(l) \tag{5.12}$$

where $J_i = 1$, and $H_i = 1$. According to the state space theory (Lewis 1999), the equation in the preceding text represents a first-order linear state space system. Our goal is to maintain a target SIR for each network link while the transmitter power is adjusted so that the least possible power is consumed.

THEOREM 5.2.1

Given the SIR system as Equation 5.12, and if the feedback for the ith trans-mitter is chosen as $v_i(l) = -k_i x_i(l) + \eta_i$ with k_i representing the feedback gains, η_i representing the protection margin and the associated power update, $P_i(l+1) = (v_i(l)I_i(l) + p_i(l)$, given in Table 5.1 is used, then the closed-loop system is stable for each link, and the actual SIRs will converge to their corresponding target values.

PROOF Applying the feedback into Equation 5.12 the closed-loop error system in SIR Is got as

$$x_i(l+1) = (J_i - H_i k_i) x_i(l) + H_i \eta_i \tag{5.13}$$

TABLE 5.1

State-Space-Based Distributed Power Controller

SIR system state equation	$R_i(l+1) = R_i(l) + v_i(l)$
SIR error	$x_i(l) = R_i(l) - \gamma_i$
Feedback controller	$v_i(l) = -k_i x_i(l) + \eta_i$
Power update	$p_i(l+1) = (v_i(l)I_i(l) + p_i(l))$

where k_i, γ_i, and η_i are design parameters

This is a stable linear system driven by a constant bounded input η_i. Applying the well-known theory of linear systems (Lewis 1999), it is easy to show that $x_i(\infty) = \frac{\eta_i}{k_i}$. In the absence of protection margin, the SIR error x_i approaches to zero as $t \to \infty$.

REMARK 1
The transmission power is subject to the constraint $p_{min} \leq p_i \leq p_{max}$ where p_{min} is the minimum value needed to transmit, p_{max} is the maximum allowed power, and p_i is the transmission power of the user i. Hence, from Equation 5.11, $p_i(l+1)$ can be written as $p_i(l+1) = \min(p_{max}, (v_i(l)I_i(l) + p_i(l)))$.

The power update presented in Theorem 5.2.1 does not use any optimization function. Hence, it may not render optimal transmitter power values though it guarantees convergence of actual SIR of each link to its target. Therefore, in Table 5.2, an optimal DPC is proposed.

THEOREM 5.2.2 (OPTIMAL CONTROL)
Given the hypothesis presented in the previous Theorem 5.2.1, for DPC, with the feedback selected as $v_i(l) = -k_i x_i(l) + \eta_i$, where the feedback gains are taken as

$$k_i = \left(H_i^T S_\infty H_i + T_i\right)^{-1} H_i^T S_\infty J_i \qquad (5.14)$$

TABLE 5.2

Optimal Distributed Power Controller

SIR system state equation	$R_i(l+1) = R_i(l) + v_i(l)$
Performance index	$\sum_{i=1}^{\infty} x_i^T T_i x_i + v_i^T Q_i v_i$
SIR error	$x_i(l) = R_i(l) - \gamma_i$
Assumptions	$T_i \geq 0,\ Q_i > 0$ all are symmetric
Feedback controller	$S_i = J_i^T [S_i - S_i H_i (H_i^T S_i H_i + T_i)^{-1} H_i^T S_i] J + Q_i$
	$k_i = (H_i^T S_\infty H_i + T_i)^{-1} H_i^T S_\infty J_i$
	$v_i(l) = -k_i x_i(l) + \eta_i$
Power update	$p_i(l+1) = (v_i(l)I_i(l) + p_i(l))$

where $k_i, \gamma_i,$ and η_i are design parameters

where S_i is the unique positive definite solution of the algebraic Ricatti equation (ARE)

$$S_i = J_i^T \left[S_i - S_i H_i \left(H_i^T S_i H_i + T_i \right)^{-1} H_i^T S_i \right] J + Q_i \qquad (5.15)$$

Then, the resulting time-invariant closed-loop system described by

$$x_i(l+1) = (J_i - H_i K_i) x_i(l) + H_i \eta_i \qquad (5.16)$$

is asymptotically stable, if $\eta_i = 0$.

PROOF Follow the steps as in Lewis (1999).

REMARK 2
The proposed scheme minimizes the performance index $\sum_{i=1}^{\infty} x_i^T T_i x_i + v_i^T Q_i v_i$ where T_i and Q_i are weighting matrices.

The block diagram in Figure 5.1 explains the process of power control using SSCD/optimal schemes. The receiver as shown in the block diagram, after receiving the signal from the transmitter, measures the SIR value and compares it against the target SIR threshold. The difference between the desired SIR to the received signal SIR is sent to the power update block, which then calculates the optimal power level with which the transmitter has to send the next packet to maintain the required SIR. This power level is sent as feedback to the transmitter, which then uses the power level to transmit the packet in the next time slot.

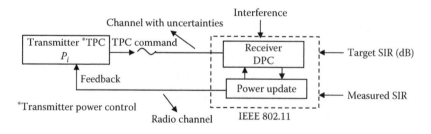

FIGURE 5.1
Block diagram representation of DPC.

Given the DPC, the issue now is whether the proposed DPC can ensure link protection of active links during inactive link admissions. As seen before, a link i belonging to the set A_l is considered active or an operational link. On the other hand, link i belonging to set B_l is termed as inactive. Certain links in this set of inactive set, B_l, update their powers according to Theorem 5.2.1 and Theorem 5.2.2 and, eventually, gain admission into the network, becoming part of active links A_l. The set of inactive links that will eventually be admitted into the channel are termed as *fully admissible set of inactive* links. There may be a set of links in B_l that can never gain admission into the network because of the following reasons: (1) the set of new links, while powering to gain admission, causes extreme interference to the existing set of active links; (2) the SIRs, $R_i(l)$, of these inactive links saturate below the target value γ_i with time after not getting active. This set of links that cannot gain admission into the network due to channel saturation, is termed as *totally inadmissible set of inactive* links.

5.2.3.1 SIR Protection of an Active Link

For any active link i, we have

$$R_i(l) \geq \gamma_i \implies R_i(l+1) \geq \gamma_i \tag{5.17}$$

This implies that a new link is added if and only if the new state of the system is stable, i.e., none of the existing links are broken. Now, using the above SSCD/optimal DPC schemes, we prove that the active links in the network continue to remain active throughout their transmission time.

THEOREM 5.2.3

For any fixed $\eta_i \in (1, \infty)$, for every $l \in \{0, 1, 2, ...\}$ and every $i \in A_l$ we have $R_i(l) \geq \gamma_i \implies R_i(l+1) \geq \gamma_i$, under the DPC/SSCD or DPC/optimal updating scheme. Therefore $i \subset A_l \implies i \in A_{l+1}$ or $A_l \subseteq A_{l+1}$ and $B_l \subseteq B_{l+1}$ for every $l \in \{0, 1, 2, ...\}$.

PROOF If $v_i(l) = -k_i x_i(l) + \eta_i$ where η_i is the protection margin, then using Equation 5.13, the system error equation can be written as $x_i(l+1) = x_i(l) - k_i x_i(l) + \eta_i$, or $x_i(l+1) = (1-k_i)x_i(l) + \eta_i$ where $0 < k_i < 1$. The system discussed in the preceding text is a linear time-invariant system with stable transition matrix driven by a constant, small, and bounded input η_i, which is the protection margin. This further implies that $x_i(l)$, which is equal to $(R_i(l) - \gamma_i)$ tends to $\frac{\eta_i}{k_i}$ at steady state. This further implies that $R_i(l) \geq \gamma_i + \frac{\eta_i}{k_i}$, and it follows that at steady state.

5.2.3.2   Bounded Power Overshoot

The following theorem proves that the transmitter power of active links is finite and can only increase in small increments while accommodating the new links that are trying to gain admission into the network.

THEOREM 5.2.4

For any fixed $\eta_i \in (1, \infty)$, we have $P_i(l+1) \le \beta(l)\,P_i(l)$, where $\beta(l)$ is a positive number for every $l \in \{0, 1, 2, \ldots\}$ and every $i \in A_l$ under the DPC/SSCD or DPC/optimal updating scheme.

PROOF   By definition, $i \in A_l$ implies that $R_i(l) \ge \gamma_i$, using $P_i(l+1) = R_i(l+1)$ $I_i(l)$, where $I_i(l) = \frac{P_i(l)}{R_i(l)}$; substituting in $P_i(l+1) = \frac{R_i(l+1)}{R_i(l)} \cdot P_i(l)$, we get $p_i(l+1) = \frac{[(1-k_i)R_i(l) + k_i\gamma_i + \eta_i]}{R_i(l)} \cdot P_i(l)$. This further implies that $P_i(l+1) \le \beta(l)\,P_i(l)$, where $\beta(l)$ is a positive number given by $\beta(l) = (1-k_i) + k_i \frac{\gamma_i}{R_i(l)} + \frac{\eta_i}{R_i(l)}$ where $\gamma_{\min} >$ $R_i(l) < (\gamma_i + \eta_i)$. This clearly shows that the overshoots of the DPC/ALP scheme are bounded by β. The value of β lies between $\beta_{\min} < \beta < \beta_{\max}$ where $\beta_{\min}$ is slightly larger than 1. Therefore, the powers of active links can only increase smoothly to accommodate the new links that are powering in the channel.

THEOREM 5.2.5 (NONACTIVE LINK SIR INCREASES)

For any fixed $\eta_i \in (1, \infty)$, we have $R_i(l) \le R_i(l+1)$ for every $l \in \{0, 1, 2, \ldots\}$ and every $i \in B_l$, in the set of fully admissible inactive links under the DPC/SSCD or DPC/optimal scheme.

PROOF   Applying the proposed DPC $v_i(l) = -k_i x_i(l) + \eta_i$ into SIR Equation 5.10 to get $R_i(l+1) = R_i(l) - k_i x_i(l) + \eta_i = R_i(l) + k_i(\gamma_i - R_i(l)) + \eta_i$. The value $(\gamma_i - R_i(l))$ is a positive number for a nonactive link in the set of fully admissible inactive links. Therefore, this implies that $R_i(l+1) \ge R_i(l)$.

THEOREM 5.2.6 (NONACTIVE LINK INTERFERENCE DECREASES)

For any fixed $\eta \in (1, \infty)$, we have $I_i(l+1) \le I_i(l)$ for every $l \in \{0, 1, 2, \ldots\}$ and every $i \in B_l$, the set of inactive links which are fully admissible, under the DPC/ SSCD or DPC/optimal scheme.

PROOF   From Theorem 5.2.5, for nonactive links we have $R_i(l+1) \ge R_i(l)$. This can equivalently be represented as $\frac{P_i(l+1)}{I_i(l+1)} \ge \frac{P_i(l)}{I_i(l)}$. From Theorem 5.2.4, we have $P_i(l+1) \le \beta(l)\,P_i(l)$ for all inactive links. Using this condition in the previous theorem results in $R_i(l+1) \ge R_i(l)$ because $\beta(l)\,P_i(l)\,I_i(l) \ge P_i(l)\,I_i(l+1)$ $\Rightarrow I_i(l+1) \le \beta(l)\,I_i(l)$.

THEOREM 5.2.7 (FINITE ADMISSION TIME)

For $i \in B_l$, each inactive link becomes active if at all admitted in a finite time.

PROOF The error in SIR of all links can be expressed as $x_i(l+1)=$ $(1-k_i)x_i(l) + \eta_i$. With $0 < k_i < 1$ for new links, the error system in SIR is a stable linear system driven by a bounded input η_i. Linear system theory (Lewis 1999) shows that the SIRs of the links reach their targets in a finite time.

5.2.4 Distributed Power Control: Application to Cellular Networks

The DPC is naturally applied to peer-to-peer networks in which communication exists between a transmitter and a receiver, whereas in the case of cellular networks, the base station would like to maintain a target SIR by acting as a receiver to all the mobile users. Normally, all the users in a cellular network have to achieve a target SIR value γ_i. The SIR value is a measure of quality of received signal and can be used to determine the control action that needs to be taken for the uplink process control. The SIR, γ_i can be expressed as (Rappaport 1999)

$$\gamma_i = ((E_b/N_0)/(W/R)) \qquad (5.18)$$

where E_b is the energy per bit of the received signal in watts, N_0 is the interference power in watts per Hertz, R is the bit rate in bits per second, and W is the radio channel bandwidth in Hertz. In our case, the metrics chosen to evaluate the performance of the power control scheme are outage probability and the total power consumed by the users. The *outage probability* is defined as the probability of failing to achieve adequate reception of the signal due to cochannel interference. It is defined as the ratio of the number of disconnected or handed over users to that of the total number of users in the system.

Example 5.2.1: Uplink Transmitter Power Control of Cellular Networks

The cellular network is considered to be divided into seven hexagonal cells covering an area of 10×10 km with 100 users randomly distributed in the cells. A base station, which is located at the center of the hexagonal cell, services a cell. It is assumed that there is no mobility in the network initially so that the performance of the DPC can be evaluated in the presence of path loss. The system parameters used for the simulation are taken from IS-95 system (Rappaport 1999). The receiver noise in the system κ_i, is taken as 10^{12}. The threshold SIR, γ is selected as 0.04. The ratio of energy per bit to interference power per hertz $E_b/_{N_0}$ is taken as 5.12 dB. Bit rate R_b, is 9600 bps. Radio channel bandwidth is B_c, and is considered to be 1.2288 MHz.

The system is simulated using the method of Bambos et al. (2000), optimal DPC, SSCD (Jagannathan et al. 2002), and constrained second-order power control (CSOPC) (Jantti and Kim 2000). Outage probability and total power are taken as metrics to evaluate the performance of the schemes. Plots of outage probability and total power consumed are generated for a given number of users in the network. A plot of outage probability with number of users is also generated to show the performance of the scheme when number of users in the cellular network varies.

Figure 5.2 shows the randomly placed location of users in the cellular network, which is divided into seven hexagonal cells. Figure 5.3 shows the plot of outage probability with time. From this simulation, it is clear that the outage probability in the system using SSCD or optimal DPC scheme is lower when compared to CSOPC and Bambos scheme, i.e., our approach can accommodate a greater number of users, rendering high system capacity. Figure 5.4 illustrates the plot of total power consumed by all the users in the network with respect to time. The result shows that SSCD and optimal DPC allow the users to attain their target SIRs with lower transmitter powers when compared to other schemes, while maintaining a lower outage probability. Simulation experiments are performed

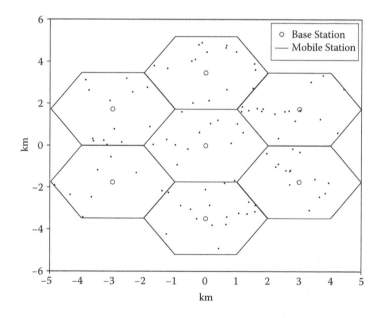

FIGURE 5.2
Cellular network with seven cells.

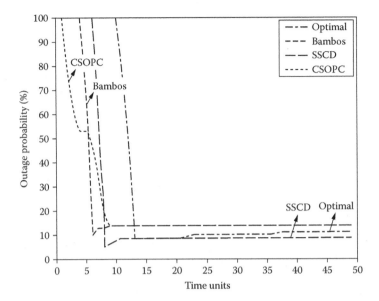

FIGURE 5.3
Outage probability as a function of time units.

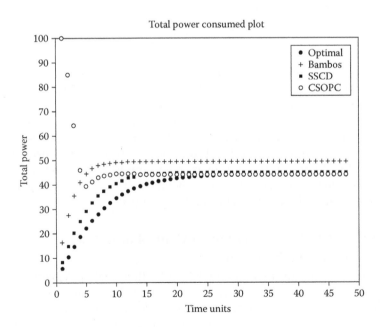

FIGURE 5.4
Total power as a function of time units.

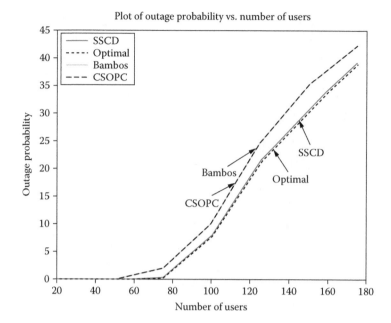

FIGURE 5.5
Outage probability with number of users.

with a varying number of nodes to evaluate the performance of DPC schemes. The plots of outage probability in Figure 5.5 and total power in Figure 5.6 demonstrate that the outage probability and total power consumed, using the SSCD and optimal DPC schemes, respectively, continue to be lower than when compared to Bambos and CSOPC schemes. When the number of users varies, the lower outage probability observed in Figure 5.3 is the result of low power consumption by our DPC schemes in comparison with others while the time to reach the target is longer than others, as seen in Figure 5.4. This is the trade-off.

Example 5.2.2: Transmitter Power Control of Cellular Networks with Mobile Users

To illustrate the performance of the DPC in cellular networks with mobile users, consider the scenario from Example 5.2.1. The requirements of the target SIR and design parameters were selected from Example 5.2.1. Mobility in the cellular environment has been simulated as follows: A user can move a maximum of 0.01 km per time unit in any one of the eight predefined directions chosen at random at the beginning of the

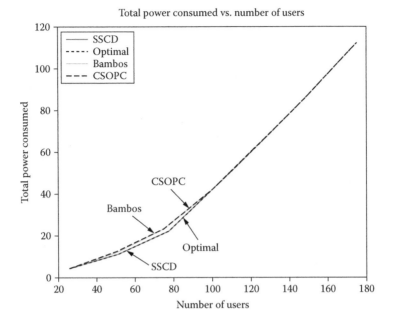

FIGURE 5.6
Total power consumed with number of users.

simulation. Given the small power update interval, a value of 0.01 km is a considerable amount of distance. The initial and final location of users in the cellular network is shown in Figure 5.7 and Figure 5.8. Soft handoff is utilized when a user moves from one cell to the other. From Figure 5.9 and Figure 5.10, we see that the outage probability and total power consumed by optimal DPC scheme is significantly lower compared to Bambos and CSOPC.

Example 5.2.3: DPC for Peer-to-Peer Networks

To demonstrate the performance of our DPC, a peer-to-peer network is selected. The wireless network is designed to span an area of 500 units as shown in Figure 5.2. The link's transmitter is placed randomly in a square region of dimensions 500 x 500 units. For all the receivers, the normalized noise floor, $\frac{\kappa_i}{g}$, is taken to be at 10^9. The link's receiver is placed at a distance of 50 units from the transmitter. Path loss only is considered, and the attenuation of power is assumed to follow the inverse fourth power law. The SIR at the transmitter of the link i is computed using Equation 5.2.

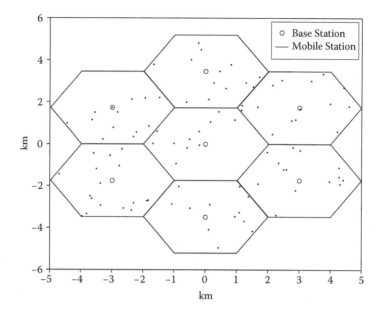

FIGURE 5.7
Initial placement of users in the cellular network.

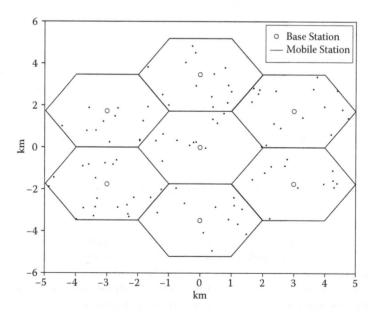

FIGURE 5.8
Final placement of users in the cellular network.

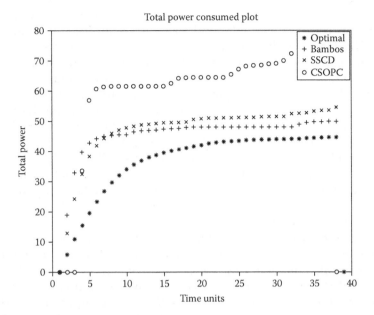

FIGURE 5.9
Total power consumed.

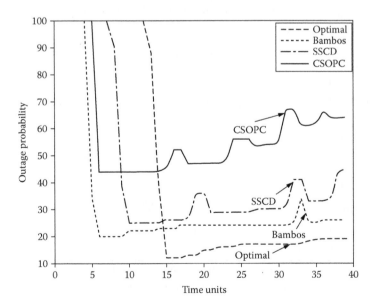

FIGURE 5.10
Outage probability.

All links are assumed to attain the same target value for SIR defined as $\gamma = 5$. Each new link, at a random time instant, tries to attain the target SIR starting with an initial power value of $p_i = 10^{-5}$ units. An upper limit on the power with which each link transmitter can transmit is set as $p_{max} = 5$ units. Each new link is given a predefined amount of admission time T_i to achieve the target SIR. If it is able to attain the target SIR during T_i, it is declared active, allowed to transmit to the receiver for a certain predefined amount of service time; and thereafter, it terminates transmission. Optimal power control and SSCD schemes allow the new links to gradually power up and attain the target SIR value. The schemes maintain the SIR value of the active links above the target γ while the new links are gaining access to the channel. A new link, according to the scheme, is admitted only if it maintains the system's stable state even after its addition.

Several scenarios are considered, and the proposed DPC/ALP schemes are evaluated. In SSCD and optimal control schemes, in addition to error in SIR Equation 5.12, the cumulative sum of errors (El-osery and Abdullah 2000) is also considered for the simulation. As a result, the feedback gains are selected as $k_i = [k_1 \ k_2]$. To compare the performance of our scheme with that of Bambos et al. (2000), the simulation environment is chosen as seen in Figure 5.11. Eleven nodes with

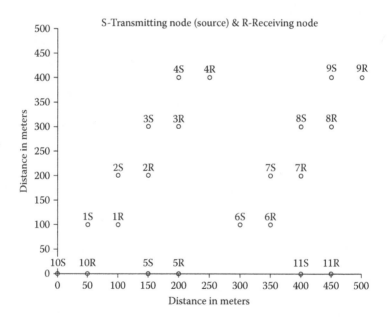

FIGURE 5.11
Link placements.

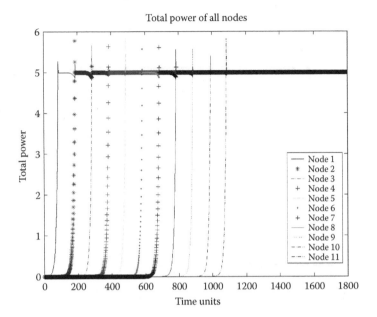

FIGURE 5.12
Response of Bambos DPC scheme.

transmitter/receivers were considered to be requesting admission into the network, one at a time, at predefined intervals of time. The network is then simulated, and the SIRs computed using various schemes discussed in the chapter and are plotted as shown in Figure 5.12, Figure 5.13, and Figure 5.14. From these figures, we can observe that the SSCD and optimal control schemes help the new links to attain target SIR in a fewer number of iterations when compared to Bambos. But the total power from each of these schemes, plotted at every time instant, shows that Bambos' scheme consumes slightly less power when compared with our schemes (see Figure 5.15). By varying the gains k_1 and k_2, we can allow the new links in the network to attain target SIR using less power with SSCD and optimal schemes compared to Bambos. But the number of iterations needed for convergence would be more. Therefore, there is a trade-off between the convergence speed and total power consumed in all the schemes discussed in this paper in peer-to-peer case. In other words, the selection of k_i and the weight matrices Q_i and T_i affect the power and convergence. This option is not available in the work of Bambos.

Further, within the proposed suite of DPC/ALP schemes, the simulations show that the total power of all the transmitters with the optimal DPC scheme (see Figure 5.16) is quite small compared to using SSCD

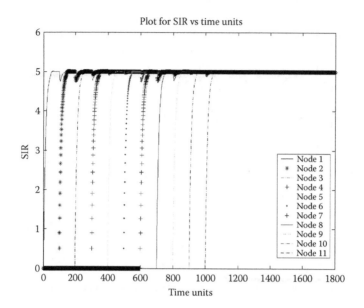

FIGURE 5.13
Response of SSCD scheme.

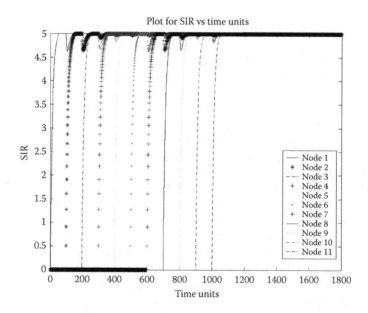

FIGURE 5.14
Response of optimal DPC.

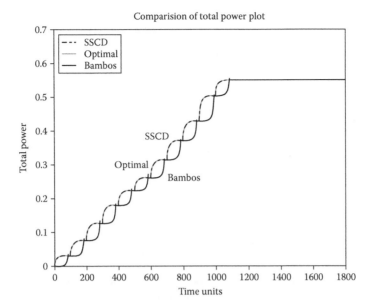

FIGURE 5.15
Total power consumed.

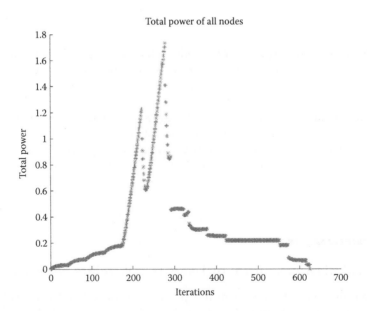

FIGURE 5.16
Total power with optimal DPC scheme.

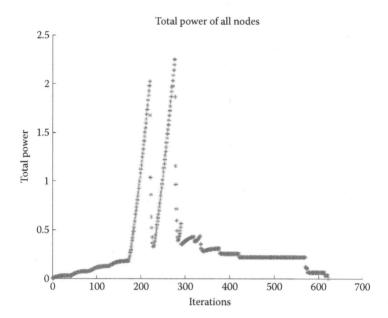

FIGURE 5.17
Total power with SCSD scheme.

scheme (see Figure 5.17). Here a node, which is dropped, gets added to the network after waiting for a period of "dormant time." The optimal DPC scheme for this scenario results in an optimal selection of the gain value for each link as $k_1 = 0.1654$ and $k_2 = 0.1330$, whereas for the SSCD scheme the gains, $k_1 = 0.3$ and $k_2 = 0.5$, are selected after careful analysis. In both the cases, the new links are updated with $k_1 = 0.1$ and $k_2 = 0.1$. As seen from Figure 5.16 and Figure 5.17, the total power consumed by all transmitters is small for the optimal DPC scheme compared to the SSCD scheme.

5.3   Admission Control of Users in a Wireless Network

In the previous section, transmitter power control is detailed for cellular and peer-to-peer networks. In this section, admission control of new users is treated with DPC. Then the DPC/ALP with admission controller algorithm is presented in a unified manner.

Defining $x_i(l) = e_i(l)$ for each link, the SIR update equation for the entire network can be represented in the vector form as $R(l+1)=R(l) + v(l)$, where $v(l) = -Kx(l)+\eta$, with $R(l)$, $v(l)$, K, $x(l)$, γ, η represented as

$$R(l) = \begin{bmatrix} R_1(l) \\ R_2(l) \\ . \\ . \\ R_i(l) \end{bmatrix}, \quad v(l) = \begin{bmatrix} v_1(l) \\ v_2(l) \\ . \\ . \\ v_i(l) \end{bmatrix}, \quad K = \begin{bmatrix} k_1 & 0 & 0 & 0 & . . & 0 \\ 0 & k_2 & 0 & 0 & . . & 0 \\ . & . & . & . & . . & . \\ . & . & . & . & . . & . \\ 0 & 0 & 0 & 0 & . . & k_i \end{bmatrix},$$

$$x(l) = \begin{bmatrix} x_1(l) \\ x_2(l) \\ . \\ . \\ x_i(l) \end{bmatrix}, \quad \gamma = \begin{bmatrix} \gamma_1 \\ \gamma_2 \\ . \\ . \\ \gamma_i \end{bmatrix}, \quad \eta = \begin{bmatrix} \eta_1 \\ \eta_2 \\ . \\ . \\ \eta_i \end{bmatrix} \tag{5.19}$$

Using the SSCD or optimal power control approach presented in this paper, the active links are always maintained active. Inactive links, after some time, get added into the network if they are admissible. They remain active until the end of their transmission.

We now consider a group of $N + M$ links, such that originally the ones in the set

$$A_0 = \{1, 2, 3, \ldots, N-1, N\} \tag{5.20}$$

are active, whereas the ones in the set

$$B_0 = \{N+1, N+2, N+3, \ldots, N+M-1, N+M\} \tag{5.21}$$

are inactive new links. We are mainly interested in whether the new links will eventually become active.

THEOREM 5.3.1 (TOTALLY INADMISSIBLE NEW LINKS)
Given that the network operates under the state-space-based or optimal power control scheme, if

$$A_l = A_0 \neq \phi \quad \text{and} \quad B_l = B_0 \neq \phi \tag{5.22}$$

for every $l \in \{1, 2, 3, \ldots\}$, (time slots) then the following limits exist:

$$\lim_{l \to \infty} R(l) = R^* < \infty \tag{5.23}$$

and

$$\lim_{l \to \infty} P(l) = D^* < \infty \tag{5.24}$$

for some positive constants $D_i^* R^*$ and for each $i \in A_0 \cup B_0$. Moreover, $R_i^* = \gamma_i$ for every initially active link $i \in A_0$, while

$$R_i^* \leq \gamma_i \quad \text{for every initially inactive link } i \in B_0. \tag{5.25}$$

Thus, if no link is ever admitted then: (1) the SIRs of the active links initially are squeezed down to their lowest acceptable values γ_i; (2) the SIRs of all new links saturate below their required thresholds, γ_i; and (3) the transmitter powers of all links geometrically increase to infinity.

PROOF We have $N + M$ links such that $A_0 = \{1, 2, 3, \ldots, N-1, N\}$ are active and $B_0 = \{N+1, N+2, N+3, \ldots, N+M-1, N+M\}$ are inactive links. Because links in B_0 remain forever inactive, we have from Theorem 5.3.1 that

$$P_n(l) = P_{\min} = D_i^* < \infty \quad \text{for every } i \in B_0 \tag{5.26}$$

and for each and every iteration of the power update scheme, we get

$$R_n(l) \leq R_i^* \leq \gamma_i \quad \text{for every } i \in B_0 \tag{5.27}$$

As the SIRs of the inactive links increase and because of the fact that they remain inactive forever, we need to study the behavior of active links. We define the set of powers of all the links to be a set of powers of active and inactive links

$$P = [\, P_a \; P_n \,]^T \tag{5.28}$$

where we define,

$$P_a(l) = (P_1(l), P_2(l), \ldots, P_i(l), P_{i+1}(l), \ldots, P_N(l))^T \tag{5.29}$$

as the vector of powers of active links and

$$P_n(l) = (P_{N+1}(l), P_{N+2}(l), ..., P_{N+i}(l), P_{N+i+1}(l), ..., P_{N+M}(l))^T \qquad (5.30)$$

the vector of powers of inactive links.

Also we have

$$P_a(l+1) = [R(l) + v(l)]I(l+1) \qquad (5.31)$$

Substituting $v(l) = -Kx(l) + \eta$ into Equation 5.31 results in

$$P_a(l+1) = [(I-K)R(l) + (K\gamma + \eta I)]I(l+1), \qquad (5.32)$$

where K, P, R, η, γ are defined in Equation 5.19. Moreover, $R(l) = \frac{P(l)}{I(l)}$ and $\varphi(l) = \frac{I(l+1)}{I(l)} < \beta(l)$ from Theorem 5.2.6.

Therefore Equation 5.31 yields

$$P_a(1) = (I-K)\varphi(0)P(0) + (K\gamma + \eta I)\varphi(0)I(0) \qquad (5.33)$$

$$P_a(2) = (I-K)\varphi(1)P(1) + (K\gamma + \eta I)\varphi(1)I(1). \qquad (5.34)$$

Substituting $P_a(1)$ in $P_a(2)$ yields

$$P_a(2) = (I-K)^2 \varphi(1)\varphi(0)P(0) + (K\gamma + \eta I)[(I-K)\varphi(1)\varphi(0)I(0) + \varphi(1)I(1)] \quad (5.35)$$

Consequently, we can write $P_a(l)$ as

$$P_a(l) = (I-K)^l \varphi(l-1)\varphi(l-2)\cdots\varphi(0)P(0) + (K\gamma + \eta I)[(I-K)^{l-1}\varphi(l-1)$$

$$\varphi(l-2)\cdots\varphi(0)I(0) + (I-K)^{l-1}\varphi(l-1)\varphi(l-2)\cdots\varphi(1)I(1)] + \cdots + \varphi(l-1)I(l-1)].$$
$$(5.36)$$

From Equation 5.36, we can clearly see that the $P_a(l)$ value converges to

$$\lim_{l \to \infty} P_a(l) \to D^*, \quad \text{which is a constant.} \qquad (5.37)$$

Recall, that the SIR of the ith link at the lth power update is given by

$$R(l+1) = R(l)[I-K] + K\gamma + \eta \qquad (5.38)$$

Now, at each power control scheme iteration, we have

$$R(1) = R(0)[I - K] + K\gamma + \eta \qquad (5.39)$$

$$R(2) = R(1)[I - K] + K\gamma + \eta \qquad (5.40)$$

Substituting the value of R(1) in R(2) to get,

$$R(2) = R(0)(I - K)^2 + (I - K)(K\gamma + \eta I) + (K\gamma + \eta I) \qquad (5.41)$$

Similarly, the value of SIR at the *l*th time instant is given by

$$R(l) = R(0)(I - K)^l + [(I - K)^{l-1} + (I - K)^{l-2} + \cdots + 1](K\gamma + \eta I) \qquad (5.42)$$

Taking limits, on both sides to get

$$\lim_{l \to \infty} R(l) = \gamma I + \frac{\eta}{K} = R^*, \qquad (5.43)$$

because $\|I - K\| < 1$; therefore, as $l \to \infty$ $\|(I - K)^l\| \to 0$ and the second term in $R(l)$ is in an infinite geometric progression. This concludes the proof of the theorem.

5.3.1 DPC with Active Link Protection and Admission Control

Inactive links try to gain access to the network. Some of these new links gain admission to the network whereas the rest never do so. New links, as their SIR saturates, can drop out from the network in two ways:

1. Timeout-based voluntary dropout
2. Adaptive timeout-based voluntary dropout

A dropped link from the channel can immediately try to gain access to the channel after remaining dormant for a backoff period. The link then starts to power up again as any other new link. The rest of this section describes these two ways of voluntary dropout.

CASE I: TIMEOUT-BASED VOLUNTARY DROPOUT The target time span for a new link to get admitted is fixed at T_i. New links are not admitted if they have reached the target time span T_i because SIR of the new links gets saturated. This dropout of the new link in fact decreases the interference experienced by other new links or previously dropped links trying to gain admission and increases their chances of getting added to the network.

CASE II: ADAPTIVE TIMEOUT-BASED VOLUNTARY DROPOUT   It would be unfair for a new link, which has almost reached (but couldn't get added to) the required target SIR in the predefined amount of time T_i target time span to drop out. Hence, optimal power control (OPC) scheme is modified to accommodate the issue of admission control. The following is the algorithm showing the behavior of the OPC scheme with admission control.

1. A Link goes to step 2 or step 3 depending upon whether it is active $R_i \geq \gamma_i \; i \in A_l$ or inactive $R_i < \gamma_i \; i \in B_l$.

2.

   a. If $l \in \{l_i^a, \; l_i^a+1, \; l_i^a+2, \; \dots, \; l_i^a+S_i\}$, active link i updates its power according to the OPC active link update, sets $l \leftarrow l+1$ and goes to step 1. Here, l_i^a is the first time instant when the link i became active, and S_i is its maximum allowable transmission time.

   b. If l has reached $l_i^a + S_i + 1$ iterations, link i drops off voluntarily. The power of this link is set to zero, removing it from the list of active links.

3.

   a. $l \in \{l_i^b, \; l_i^b+1, \; l_i^b+2, \; \dots, \; l_i^b+T_i+D_i-1\}$ inactive link i updates its power according to the OPC inactive link updates, sets $l \leftarrow l+1$ and goes to step 1. Here, l_i^b is the first time instant when the link i starts to gain admission into the network. The parameter T_i is the maximum allowable admission time. D_i is the extra admissible time before dropping out. The value of D_i is computed as follows: $D_i = \max[\, f_i(\gamma_i - R_i(T_i)), f_i(p_{\max} - P_i(T_i))]$ can be written as $\max([A_i \, e^{-\alpha_i(\gamma_i - R_i(T_i))}\,], [B_i \, e^{-\beta_i(P_{\max} - P_i(T_i))}\,])$ where $A_i, B_i, \; \alpha_i \geq 0, \beta_i \geq 0$ are design parameters. The function f_i is decreasing in this argument. The closer the link i gets to the target, the longer the link is allowed to get added to the network before dropping out.

   b. l reached $K_i^b + T_i + D_i$ iterations, link i voluntarily drops out, and sets its power to zero. Go to step 4.

4.

   a. $l \in \{l_i^c, \; l_i^c+1, l_i^c+2, \; \dots, \; l_i^c+B_i\}$, link i remains in dormant mode in the backoff period, and it sets $l \leftarrow l+1$ until the end of backoff period.

   b. l reached $l_i^c + B_i + 1$, the link starts with minimum power and tries to get added as any other new link, same as in 1.

A new link is not added to the system if one of these is satisfied: (1) when the transmitter power of an active link is in danger of exceeding p_{max}; (2) when the link takes a greater amount of time than the "admission time," T_i; and (3) when the link's rate of increase of power is more than a certain predefined value. This constraint rate of increase in power is necessary because (1) the drastic increase in power of the new link creates a large interference on the nearby active links, making it necessary for them to increase their power levels to maintain the predefined SIR level and (2) the increase of power levels by the active links drain their battery power. If a link wants to terminate, its transmitter power is made zero so that it does not cause any interference to other active links. A link, if it were not added because of these reasons, would try to get admission into the network after a certain predefined amount of backoff-time B_i. The DPC/ALP with admission control scheme is presented in the following subsection.

5.3.2 DPC/ALP and Admission Controller Algorithm

The proposed distributed algorithm with power control can be itemized as:

1. New links trying to gain access to the channel start with a minimum power p_{min}.
2. The interference experienced by all links existing in the network is calculated as $I_i(l) = \left(\sum_{j \neq i}^{n} P_j * \frac{G_{ij}}{G_{ii}} + \frac{\kappa_i}{G_{ii}} \right)$.
3. All links in the network calculate their SIR according to Equation 5.2 and check against threshold γ_i.
4. If SIR of a link is greater than the threshold value γ_i, the link is declared active. The link thus gains access to the network and starts communicating with the receiver.
5. A new link trying to gain access to the network is dropped if:
   a. Its transmitter power exceeds p_{max}.
   b. The rate of increase of power exceeds a certain threshold. (To drop new links that cause large amount of interference to already existing nodes.)
   c. The rate of increase of SIR is less than a given threshold. (To prevent SIR saturation.)
   d. The amount of time for a link to gain access to the network exceeds maximum admissible delay. (To prevent SIR saturation.)
6. A dropped link, if it has waited for dormant or backoff period B_i, is added to the set of existing new links.

7. An active link is dropped from the network if it has transmitted for its maximum allowable transmission time T_i.

8. If link's SIR is less than the threshold value γ_i (not active) and if it is not dropped, its power is updated by calculating the new transmitter powers using Theorem 5.2.1.

9. The above steps are repeated until all the links get added to the network and have transmitted for a maximum allowable transmission time.

Example 5.3.1: Admission Control of Links in a Peer-to-Peer Network

Efficiency of a power control scheme would depend on maintaining the stable state of the system after a new link is added. The scheme should maintain the SIR of all active nodes above a threshold, and power at which a node transmits should be maintained below the maximum power level. From this example, it will be shown that Bambos (2000) DPC scheme appears not to satisfy this condition during admission control. The power levels of the active links increase to maintain their actual SIRs because of the interference caused by the new node. To evaluate the DPC with active link protection, link 5 is introduced closer to the receiver of the link 1 as shown in Figure 5.18. The power update scheme of Bambos DPC allows

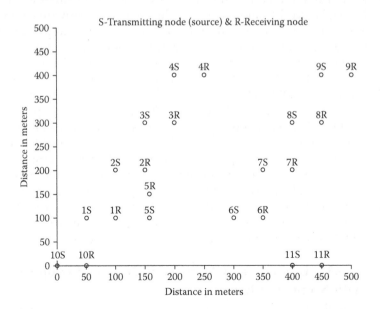

FIGURE 5.18
Node placement.

Wireless Ad Hoc and Sensor Networks

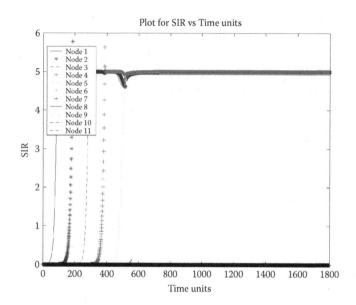

Plot for SIR vs Time units

FIGURE 5.19

Response of Bambo's update during admission of new links.

the link 5 to get admitted into the network causing severe interference to link 1. In fact, Figure 5.19 and Figure 5.20 depict that the individual powers of all active links increase significantly as link 5 is being added into the system, causing a surge in power consumption of all active links.

The SSCD and optimal schemes simulation displayed in Figure 5.21 and Figure 5.22 would not allow link 5 to be added into the system. In fact, SSCD and optimal DPC schemes use an extra requirement that has to be satisfied by the admission-seeking links. The admission links power up gradually, but whenever a link power update is greater than a predefined value (which implies it would cause high interference when added to the active links), the link is dropped from the network. The threshold level should be carefully selected to increase the efficiency of the network. The comparison of the total power consumed for the two schemes in Figure 5.23 illustrates clearly that the total power consumed using the optimal DPC update is way below that of Bambos scheme. Hence, the optimal DPC scheme provides efficient admission control.

Example 5.3.2: Evaluation of Admission Delay and Dropped Links with Number of Users

To study the effect of number of users gaining admission and delay, consider the scenario from Example 5.3.1 except the number of admission seeking links is increased. Figure 5.24 and Figure 5.25 show the dependence of

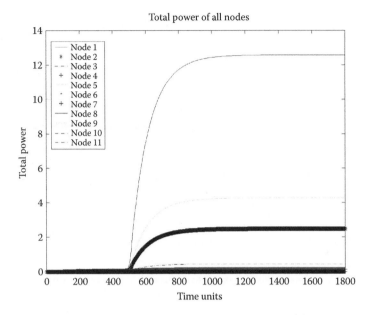

FIGURE 5.20
Individual link powers during admission.

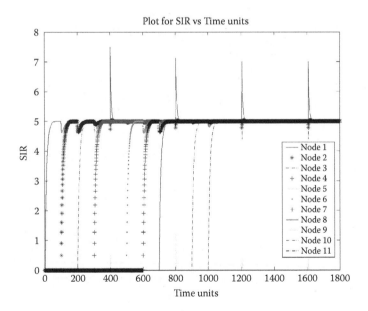

FIGURE 5.21
Response of optimal DPC during admission.

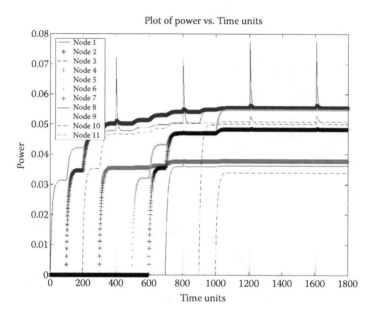

FIGURE 5.22
Individual powers during admission.

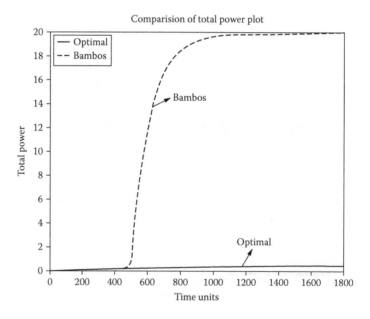

FIGURE 5.23
Total power consumption with time.

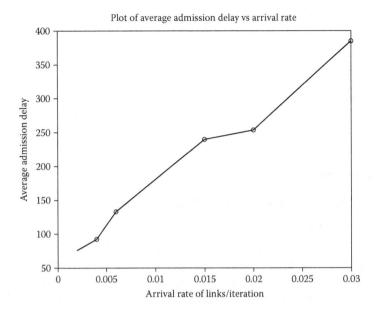

FIGURE 5.24
Average admission delay with arrivals.

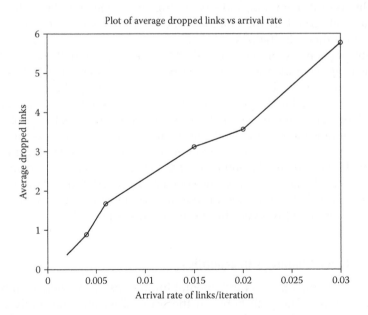

FIGURE 5.25
Average dropped links with arrivals.

average admission delay and dropped links on the arrival rate. As arrival rate of the users increases, interference in the network increases. The power control scheme would take more time to assign reasonable power levels for the large number of admission-seeking users. Some nodes due to this reason may not reach their target SIRs in the predefined admission time assigned; hence, they drop out. Therefore, as arrival rate of the users in the network increases, the admission delay and average number of dropped links increase.

5.4   Distributed Power Control in the Presence of Fading Channels

Earlier DPC works (Bambos 2000, Jantti and Kim 2000, Jagannathan et al. 2002, Dontula and Jagannathan 2004) presented in Section 5.2 neglect the changes observed in the radio channel. In fact, they all assume: (1) only the path loss component is present, (2) no other uncertainty exists in the channel, and (3) the interference is held constant. Consequently, the convergence speed of these algorithms and the associated power updates are of an issue in a highly dynamic wireless environment in which user mobility is quite common, and shadowing and Rayleigh fading effects are typically observed in the channel. The proposed work in this section overcomes these limitations.

In this section, we present a novel DPC scheme (Jagannathan et al. 2006) for the next generation wireless networks with channel uncertainties. This algorithm estimates the variations in the slowly varying channel, and it is subsequently used in the power update so that a desired SIR is maintained. This algorithm, being highly distributive in nature, does not require inter-link communication, centralized computation, and reciprocity assumption as required in a centrally controlled wireless environment. In addition, the modified DPC scheme (Jagannathan et al. 2006) converges faster compared to the other schemes in the presence of channel uncertainties. As the necessity of interlink communication is eliminated, network capacity increases, and easy controlled recovery from error events is possible.

5.4.1   Radio Channel Uncertainties

The radio channel places fundamental limitations on wireless communication systems. The path between the transmitter and the receiver can vary from simple line-of-sight to one that is severely obstructed by buildings, mountains, and foliage. Unlike wired channels that are stationary and predictable, radio channels involve many uncertain factors, so they are extremely random and do not offer easy analysis. In wireless networks, channel uncertainties such as path loss, the shadowing, and Rayleigh fading can attenuate the power of the signal at the receiver and thus cause variations in the received SIR and, therefore, degrading the performance of any DPC scheme. It is important to understand these uncertainties before the development of a DPC scheme.

5.4.1.1 Path Loss

As presented earlier, if only path loss is considered, the power attenuation is taken to follow the inverse fourth power law (Aein 1973):

$$g_{ij} = \frac{\overline{g}}{d_{ij}^n} \tag{5.44}$$

where $\overline{g}$ is a constant usually equal to 1 and d_{ij} is the distance between the transmitter of the jth link to the receiver of the ith link, and n is the path loss exponent. A number of values for n have been proposed for different propagation environments, depending on the characteristics of the communication medium. A value of $n = 4$ is taken in our simulations, which is commonly used to model path loss in an urban environment. Further, without user mobility, g_{ij} is a constant.

5.4.1.2 Shadowing

High buildings, mountains, and other objects block the wireless signals. A blind area is often formed behind a high building or in the middle of two buildings. This is often seen especially in large urban areas. The term $10^{0.1\varsigma}$ is often used to model the attenuation of the shadowing to the received power (Canchi and Akaiwa 1999, Hashem and Sousa 1998), where ς is assumed to be a Gaussian random variable.

5.4.1.3 Rayleigh Fading

In mobile radio channels, the Rayleigh distribution is commonly used to describe the statistical time-varying nature of the received envelope of a flat fading signal, or the envelope of an individual multipath component. The Rayleigh distribution has a probability density function (pdf) given by (Rappaport 1999)

$$p(x) = \begin{cases} \dfrac{x}{\sigma^2} \exp\left(-\dfrac{x^2}{2\sigma^2}\right) & (0 \le x \le \infty) \\ 0 & (x < 0) \end{cases} \tag{5.45}$$

where x is a random variable, and σ^2 is known as the fading envelope of the Rayleigh distribution.

Because the channel uncertainties can distort the transmitted signals, therefore, the effect of these uncertainties is represented via a channel loss (gain) factor, typically multiples of the transmitter power. Then, the channel gain or loss, g, can be expressed as (Canchi and Akaiwa 1999)

$$g = f(d, n, X, \zeta) = d^{-n} \cdot 10^{0.1\zeta} \cdot X^2 \tag{5.46}$$

where d^{-n} is the effect of path loss, $10^{0.1\zeta}$ corresponds to the effect of shadowing. For Rayleigh fading, it is typical to model the power attenuation as X^2, where X is a random variable with Rayleigh distribution. Typically, the channel gain, g, is a function of time.

5.4.2 Distributed Power Controller Scheme Development

The goal of transmitter power control now is to maintain a required SIR threshold for each network link while the transmitter power is adjusted so that the least possible power is consumed in the presence of channel uncertainties. Suppose there are $N \in Z_+$ links in the network. Let g_{ij} be the power loss (gain) from the transmitter of the jth link to the receiver of the ith link. It involves the free space loss, multipath fading, shadowing, and other radio wave propagation effects, as well as the spreading/processing gain of CDMA transmissions. The power attenuation is considered to follow the relationship given in Equation 5.46. In the presence of such uncertainties, our objective is to present a novel DPC and to compare its performance with others.

The channel uncertainties will appear in the power loss (gain) coefficient of all transmitter–receiver pairs. Calculation of SIR, $R_i(t)$, at the receiver of ith link at the time instant t (Jagannathan et al. 2006), is given by

$$R_i(t) = \frac{g_{ii}(t)P_i(t)}{I_i(t)} = \frac{g_{ii}(t)P_i(t)}{\displaystyle\sum_{j \neq i} g_{ij}(t)P_j(t) + \eta_i(t)} \tag{5.47}$$

where $i, j \in \{1, 2, 3, \ldots, n\}$, $I_i(t)$ is the interference, $P_i(t)$ is the link's transmitter power, $P_j(t)$ are the transmitter powers of all other nodes, and $\eta_i(t) > 0$ is the thermal noise at its receiver node. For each link i, there is a lower SIR threshold γ_i. Therefore, we require

$$\gamma_i \leq R_i(t) \leq \gamma_i^* \tag{5.48}$$

for every $i = 1, 2, 3, \ldots, n$. The lower threshold value for all links can be taken equal to γ for convenience, reflecting a certain QoS the link has to maintain to operate properly. An upper SIR limit is also set, to decrease the interference due to its transmitter power at other receiver nodes. In the literature, several DPC schemes have been proposed. The most recent DPC work includes Bambos et al. (2000), CSOPC by Jantti and Kim (2000), SSCD and optimal by Jagannathan et al. (2002), Dontula and Jagannathan (2004), respectively, and they are discussed in the previous section. This algorithm is given next.

5.4.2.1 Error Dynamics in SIR

In the previous DPC schemes presented in Section 5.2 and Section 5.3, only path loss uncertainty is considered. As a result, during fading conditions, high value of outage probability, as illustrated using simulations, is observed. The work, presented in this section, is aimed at demonstrating the performance in the presence of several channel uncertainties.

In the time domain, however, the channel is time varying when channel uncertainties are considered and therefore $g_{ij}(t)$ is not a constant. In Lee and Park (2002), a new DPC algorithm is presented where $g_{ii}(t)$ is treated as a time-varying function due to Rayleigh fading by assuming that the interference $I_i(t)$ is held constant. Because this is a strong assumption, in this paper, a novel DPC scheme is presented (Jagannathan et al. 2006) where $g_{ii}(t)$ and the interference $I_i(t)$ are time varying, and channel uncertainties are considered for all the mobile users. This relaxes the assumption of other works where both $g_{ij}(t)$ and $P_j(t)$ are held at a fixed value.

Considering SIR from Equation 5.47 where the power attenuation $g_{ij}(t)$ is taken to follow the time-varying nature of the channel and differentiating Equation 5.47 to get

$$R_i(t)' = \frac{(g_{ii}(t)P_i(t))'I(t) - (g_{ii}(t)P_i(t))I(t)'}{I_i^2(t)} \tag{5.49}$$

where $R_i(t)'$ is the derivative of $R_i(t)$, and $I_i(t)'$ is the derivative of $I_i(t)$.

To transform the differential equation into the discrete time domain, $x'(t)$ is expressed using Euler's formula as $\frac{x(l+1)-x(l)}{T}$, where T is the sampling interval. Equation 5.49 can be expressed in discrete time as

$$R_i(l)' = \frac{(g_{ii}(l)P_i(l))'I(l) - (g_{ii}(l)P_i(l))I(l)'}{I_i^2(l)}$$

$$= \frac{1}{I_i^2(l)}\left[(g_{ii}{}'(l)P_i(l))I(l) + (g_{ii}(l)P_i'(l))I_i(l) - (g_{ii}(l)P_i(l))\left(\sum_{j \neq i}g_{ij}(l)P_j(l) + \eta_i(t)\right)'\right] \tag{5.50}$$

In other words,

$$\frac{R_i(l+1)-R_i(l)}{T} = \frac{1}{I_i(l)}\frac{[g_{ii}(l+1)-g_{ii}(l)]}{T}P_i(l) + \frac{1}{I_i(l)}g_{ii}(l)\frac{[P_i(l+1)-P_i(l)]}{T}$$

$$- \frac{g_{ii}(l)P_i(l)}{I_i^2(l)}\sum_{j\neq i}\left(\frac{g_{ij}(l+1)-g_{ij}(l)}{T}P_j(l) + \frac{P_j(l+1)-P_j(l)}{T}g_{ij}(l)\right)$$

$$(5.51)$$

Canceling T on both sides and combining to get

$$R_i(l+1) = \left[\frac{g_{ii}(l+1)-g_{ii}(l)}{g_{ii}(l)} - \frac{\sum_{j\neq i}\{[g_{ij}(l+1)-g_{ij}(l)]P_j(l)+[P_j(l+1)-P_j(l)]g_{ij}(l)\}}{I_i(l)}\right]$$

$$\times R_i(l) + g_{ii}(l)\frac{P_i(l+1)}{I_i(l)}$$

$$(5.52)$$

Now, define

$$\alpha_i(l) = \frac{g_{ii}(l+1)-g_{ii}(l)}{g_{ii}(l)} - \frac{\sum_{j\neq i}\{[g_{ij}(l+1)-g_{ij}(l)]P_j(l)+[P_j(l+1)-P_j(l)]g_{ij}(l)\}}{I_i(l)}$$

$$= \frac{\Delta g_{ii}(l)}{g_{ii}(l)} - \frac{\sum_{j\neq i}\Delta g_{ij}(l)P_j(l)+\Delta P_j(l)g_{ij}(l)}{I_i(l)}$$

$$(5.53)$$

where

$$\beta_i(l) = g_{ii}(l) \tag{5.54}$$

and

$$v_i(l) = \frac{P_i(l+1)}{I_i(l)} \tag{5.55}$$

Equation 5.52 can be expressed as

$$R_i(l+1) = \alpha_i(l)R_i(l) + \beta_i(l)v_i(l) \tag{5.56}$$

with the inclusion of noise, Equation 5.52 is written as

$$R_i(l+1) = \alpha_i(l)R_i(l) + \beta_i(l)v_i(l) + r_i(l)\omega_i(l) \tag{5.57}$$

where $\omega(l)$ is the zero mean stationary stochastic channel noise with $r_i(l)$ as its coefficient.

The SIR of each link at time instant l is obtained using Equation 5.57. Carefully observing Equation 5.57, it is clear that the SIR at the time instant $l+1$ is a function of channel variation from time instant l to $l+1$. The channel variation is not known beforehand, and this makes the DPC scheme development difficult and challenging. Because α is not known, it has to be estimated for DPC development. As indicated earlier, all the available schemes so far ignore the channel variations during DPC development, and therefore, they render unsatisfactory performance.

Now define $y_i(l) = R_i(l)$, then Equation 5.57 can be expressed as

$$y_i(l+1) = \alpha_i(l)y_i(l) + \beta_i(l)v_i(l) + r_i(l)\omega_i(l) \tag{5.58}$$

The DPC development is given in two scenarios.

CASE 1 α_i, β_i, and r_i are known. In this scenario, one can select feedback as

$$v_i(l) = \beta_i^{-1}(l)[-\alpha_i(l)y_i(l) - r_i(l)\omega_i(l) + \gamma + k_v e_i(l)] \tag{5.59}$$

where the error in SIR is defined as $e_i(l) = R_i(l) - \gamma$. This implies that

$$e_i(l+1) = k_v e_i(l) \tag{5.60}$$

Appropriately selecting k_v by placing the eigenvalues within a unit circle, it is easy to show that the closed-loop SIR system is asymptotically stable in the mean or asymptotically stable, $\lim_{l\to\infty} E\{e_i(l)\} = 0$. This renders that $y_i(l) \to \gamma$.

CASE 2 α_i, β_i, and r_i are unknown. In this scenario, Equation 5.58 can be expressed as

$$y_i(l+1) = [\alpha_i(l) \quad r_i(l)]\begin{bmatrix} y_i(l) \\ \omega_i(l) \end{bmatrix} + \beta_i(l)v_i(l)$$
$$= \theta_i^T(l)\psi_i(l) + \beta_i(l)v_i(l) \tag{5.61}$$

where $\theta_i(l) = [\alpha_i(l) \quad r_i(l)]$ is a vector of unknown parameters, and $\psi_i(l)$ $= \left[\begin{smallmatrix} y_i(l) \\ \omega_i(l) \end{smallmatrix} \right]$ is the regression vector. Now selecting feedback for DPC as

$$v_i(l) = \beta_i^{-1}(l)[-\hat{\theta}_i(l)\psi_i(l) + \gamma + k_v e_i(l)] \tag{5.62}$$

where $\hat{\theta}_i(l)$ is the estimate of $\theta_i(l)$, then the SIR error system is expressed as

$$\begin{aligned} e_i(l+1) &= k_v e_i(l) + \theta_i^T(l)\psi_i(l) - \hat{\theta}_i^T(l)\psi_i(l) \\ &= k_v e_i(l) + \tilde{\theta}_i^T(l)\psi_i(l) \end{aligned} \tag{5.63}$$

*TPC: Transmitter power control *DPC: Distributed power control

where $\tilde{\theta}_i(l) = \theta_i(l) - \hat{\theta}_i(l)$ is the error in estimation. From Equation 5.63, it is clear that the closed-loop SIR error system is driven by the channel estimation error. If the channel uncertainties are properly estimated, then estimation error tends to zero. In this case, Equation 5.63 becomes Equation 5.60. In the presence of error in estimation, only the boundedness of SIR error can be shown. We can show that the actual SIR approaches the target provided the channel uncertainties are properly estimated. Figure 5.26 illustrates the block diagram representation of the proposed DPC where channel estimation and power selection are part of the receiver. To proceed further, the Assumption 5.4.1 is required and therefore stated.

5.4.2.2 Adaptive Scheme Development

ASSUMPTION 5.4.1
The channel changes slowly compared to the parameters' updates.

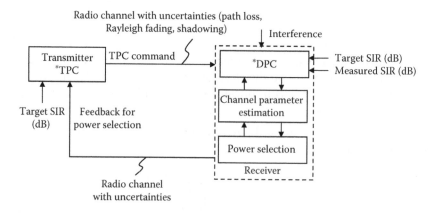

FIGURE 5.26
Block diagram representation of distributed power control with channel uncertainties.

THEOREM 5.4.1
Given the DPC in the preceding text (with channel uncertainty), if the feedback from the DPC scheme is selected as Equation 5.62, then the mean channel estimation error along with the mean SIR error converges to zero asymptotically, if the parameter updates are taken as

$$\hat{\theta}_i(l+1) = \hat{\theta}_i(l) + \sigma \psi_i(l) e_i^T(l+1) \tag{5.64}$$

provided

$$\sigma \|\psi_i(l)\|^2 < 1 \tag{5.65}$$

$$k_{v\max} < \frac{1}{\sqrt{\delta}} \tag{5.66}$$

where $\delta = \frac{1}{1-\sigma\|\psi_i(l)\|^2}$, and σ is the adaptation gain.

PROOF Define the Lyapunov function candidate

$$J_i = e_i^T(l)e_i(l) + \frac{1}{\sigma}\kappa\left[\tilde{\theta}_i^T(l)\tilde{\theta}_i(l)\right] \tag{5.67}$$

whose first difference is

$$\Delta J = \Delta J_1 + \Delta J_2 = e_i^T(l+1)e_i(l+1) - e_i^T(l)e_i(l) + \frac{1}{\sigma}\kappa\left[\tilde{\theta}_i^T(l+1)\tilde{\theta}_i(l+1) - \tilde{\theta}_i^T(l)\tilde{\theta}_i(l)\right] \tag{5.68}$$

Consider ΔJ_1 from Equation 5.68 and substituting Equation 5.63 to get

$$\begin{aligned}
\Delta J_1 &= e_i^T(l+1)e_i(l+1) - e_i^T(l)e_i(l) \\
&= \left(k_v e_i(l) + \tilde{\theta}_i^T(l)\psi_i(l)\right)^T\left(k_v e_i(l) + \tilde{\theta}_i^T(l)\psi_i(l)\right) - e_i^T(l)e_i(l)
\end{aligned} \tag{5.69}$$

Taking the second term of the first difference from Equation 5.68 and substituting Equation 5.64 yields

$$\begin{aligned}
\Delta J_2 &= \frac{1}{\sigma}\kappa\left[\tilde{\theta}_i^T(l+1)\tilde{\theta}_i(l+1) - \tilde{\theta}_i^T(l)\tilde{\theta}_i(l)\right] \\
&= -2[k_v e_i(l)]^T \tilde{\theta}_i^T(l)\psi_i(l) - 2\left[\tilde{\theta}_i^T(l)\psi_i(l)\right]^T\left[\tilde{\theta}_i^T(l)\psi_i(l)\right] \\
&\quad + \sigma\psi_i^T(l)\psi_i(l)\left[k_v e_i(l) + \tilde{\theta}_i^T(l)\psi_i(l)\right]^T\left[k_v e_i(l) + \tilde{\theta}_i^T(l)\psi_i(l)\right]
\end{aligned} \tag{5.70}$$

TABLE 5.3

Distributed Power Controller during Fading Channels with Estimation Error Being Zero

SIR system state equation	$y_i(l+1) = [\alpha_i(l) \quad r_i(l)]\begin{bmatrix} y_i(l) \\ \omega_i(l) \end{bmatrix} + \beta_i(l)v_i(l)$
	$= \theta_i^T(l)\psi_i(l) + \beta_i(l)v_i(l)$
where	$y_i(l) = R_i(l)$
SIR error	$e_i(l) = R_i(l) - \gamma_i$
Feedback controller	$v_i(l) = \beta_i^{-1}(l)[-\hat{\theta}_i(l)\psi_i(l) + \gamma_i + k_v e_i(l)]$
Channel parameter update	$\hat{\theta}_i(l+1) = \hat{\theta}_i(l) + \sigma\psi_i(l)e_i^T(l+1)$
Power update	$P_i(l+1) = (v_i(l)I_i(l) + P_i(l))$

where k_v, γ_i, σ and η_i are design parameters

Combining Equation 5.69 and Equation 5.70 to get

$$\Delta J = -e_i^T(l)\left[I - \left(1 + \sigma\psi_i^T(l)\psi_i(l)k_v^T k_v\right)\right]e_i(l)$$

$$+ 2\sigma\psi_i^T(l)\psi_i(l)[k_v e_i(l)]^T\left[\tilde{\theta}_i^T(l)\psi_i(l)\right] - \left(1 - \sigma\psi_i^T(l)\psi_i(l)\right)\left[\tilde{\theta}_i^T(l)\psi_i(l)\right]^T\left[\tilde{\theta}_i^T(l)\psi_i(l)\right]$$

$$\leq -\left(1 - \delta k_{v\max}^2\right)\|e_i(l)\|^2 - \left(1 - \sigma\|\psi_i(l)\|^2\right)\left\|\tilde{\theta}_i^T(l)\psi_i(l) - \frac{\sigma\|\psi_i(l)\|^2}{1 - \sigma\|\psi_i(l)\|^2}k_v e_i(l)\right\|^2$$

$$(5.71)$$

where δ is given after Equation 5.66. Taking expectations on both sides yields

$$E(\Delta J) \leq -E$$

$$\times \left(\left(1 - \delta k_{v\max}^2\right)\|e_i(l)\|^2 - (1 - \sigma\|\psi_i(l)\|^2)\left\|\tilde{\theta}_i^T(l)\psi_i(l) + \frac{\sigma\|\psi_i(l)\|^2}{1 - \sigma\|\psi_i(l)\|^2}k_v e_i(l)\right\|^2\right)$$

$$(5.72)$$

Because $E(J) > 0$ and $E(\Delta J) \leq 0$, this shows the stability in the mean via sense of Lyapunov provided the conditions (Equation 5.65) and (Equation 5.66) hold, so $E[e_i(l)]$ and $E[\tilde{\theta}_i(l)]$ (and, hence, $E[\hat{\theta}_i(l)]$) are bounded in the mean if $E[e_i(l_0)]$ and $E[\tilde{\theta}_i(l_0)]$ are bounded in a mean. Summing both sides of Equation 5.72 and taking limits $\lim_{l\to\infty} E(\Delta J)$, the SIR error can be shown to converge $E[\|e_i(l)\|] \to 0$.

Consider now the closed-loop SIR error system with channel estimation error, $\varepsilon(l)$, as

$$e_i(l+1) = k_v e_i(l) + \tilde{\theta}_i^T(l)\psi_i(l) + \varepsilon(l) \tag{5.73}$$

using the proposed DPC. The DPC scheme, presented in Table 5.4, will be employed when the channel estimation error is nonzero. In fact, in the following theorem, it is demonstrated that the SIR and channel parameter estimation errors are bounded when the channel estimation error is nonzero.

THEOREM 5.4.2
Assume the hypothesis as given in Theorem 5.4.1, with the channel uncertainty (path loss, shadowing, and Rayleigh fading) now estimated by

$$\hat{\theta}_i(l+1) = \hat{\theta}_i(l) + \sigma\psi_i(l)e_i^T(l+1) - \left\| I - \psi_i^T(l)\psi_i(l) \right\| \hat{\theta}_i(l) \tag{5.74}$$

where $\varepsilon(l)$ is the error in estimation that is considered bounded above $\|\varepsilon(l)\| \leq \varepsilon_N$, with ε_N a known constant. Then the mean error in SIR and the estimated parameters are bounded provided Equation 5.65 and Equation 5.66 hold.

PROOF See Jagannathan et al. (2006).

TABLE 5.4

Distributed Power Controller during Fading Channels: Nonideal Case (Estimation Error Nonzero)

SIR system state equation	$y_i(l+1) = [\alpha_i(l) \quad r_i(l)]\begin{bmatrix} y_i(l) \\ \omega_i(l) \end{bmatrix} + \beta_i(l)v_i(l)$
	$= \theta_i^T(l)\psi_i(l) + \beta_i(l)v_i(l)$
where	$y_i(l) = R_i(l)$
SIR error	$e_i(l) = R_i(l) - \gamma_i$
Feedback controller	$v_i(l) = \beta_i^{-1}(l)[-\hat{\theta}_i(l)\psi_i(l) + \gamma_i + k_v e_i(l)]$
Channel parameter update	$\hat{\theta}_i(l+1) = \hat{\theta}_i(l) + \sigma\psi_i(l)e_i^T(l+1) - \|I - \psi_i^T(l)\psi_i(l)\|\hat{\theta}_i(l)$
Power update	$p_i(l+1) = (v_i(l)I_i(l) + p_i(l))$

where k_v, γ_i, σ, and η_i are design parameters

5.4.2.3   Simulation Examples

Example 5.4.1: DPC Evaluation in the Presence of Fading Channel

In the simulations, all the mobiles in the network have to achieve a desired target SIR value γ_i in the presence of channel uncertainties. The SIR value is a measure of quality of the received signal, and can be used to determine the control action that needs to be taken. The SIR, γ_i can be expressed as

$$\gamma_i = ((E_b / N_0) / (W / R)) \tag{5.75}$$

where E_b is the energy per bit of the received signal in watts, N_0 is the interference power in watts per Hertz, R is the bit rate in bits per second, and W is the radio channel bandwidth in Hertz.

The cellular network is considered to be divided into seven hexagonal cells covering an area of 10 x 10 km (Figure 5.2). Each cell is serviced via a base station, which is located at the center of the hexagonal cell. Mobile users in each cell are placed at random. It is assumed that the power of for each mobile user is updated asynchronously. Consequently, the powers of all other mobile users do not change when the power of the ith link is updated. The receiver noise in the system η_i is taken as 10^{-12}. The threshold SIR, γ, which each cell tries to achieve is 0.04 (13.9794 dB). The ratio of energy per bit to interference power per hertz $\frac{E_b}{N_0}$ is 5.12 dB. Bit rate R_b, is chosen at 9600 bps. Radio channel bandwidth B_c is considered to be 1.2288 MHz. The maximum power for each mobile P_{max} is selected as 1 mw. Two cases are considered: channel changing sharply at a certain time instant and channel changing smoothly. The system is simulated with different DPC schemes with 100 users. In the first few simulations, the users in each cell are placed at random and are considered stationary. Later, the users are mobile.

5.4.2.3.1   Stationary Users

CASE I: CONSTANT BUT ABRUPTLY CHANGING CHANNEL   In this scenario, we select the parameters as $k_v = 0.01$ and $\sigma = 0.01$. Figure 5.27 shows the change of g_{ii} with time as a result of channel fluctuation, which obeys the Rayleigh fading and shadowing. Though channel changes sharply at every ten time-units only, the channel gain g_{ii} is changed once in every ten time-units, and it is held constant otherwise. Figure 5.28 illustrates the plot of SIR of a randomly selected mobile user. From this figure, it is clear that the proposed DPC scheme is the only scheme that maintains the target SIR in the presence of channel variations. Figure 5.29 presents the plot of total power consumed by all the mobiles in the network. The result shows that all the schemes consume similar power values (about 90 mW) for the entire network. Figure 5.30 shows the plot of outage

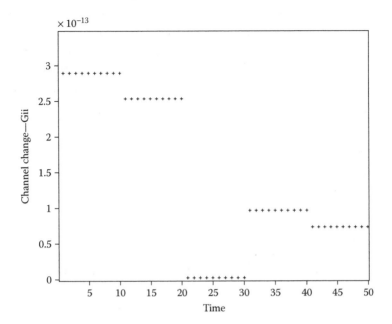

FIGURE 5.27
Channel variations over time.

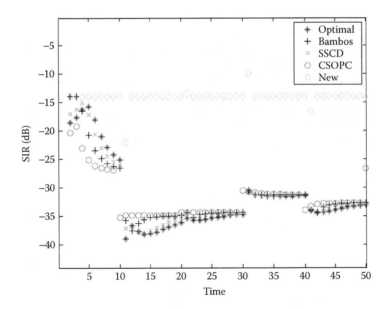

FIGURE 5.28
SIR of a randomly selected user.

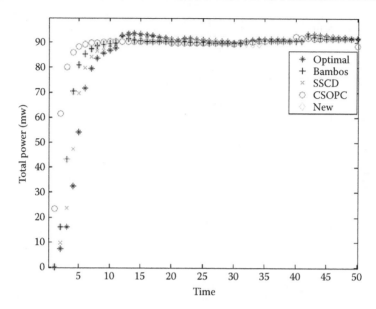

FIGURE 5.29
Total power consumed.

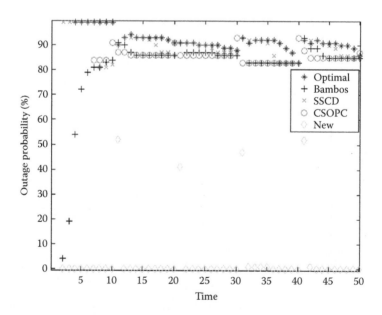

FIGURE 5.30
Outage probability.

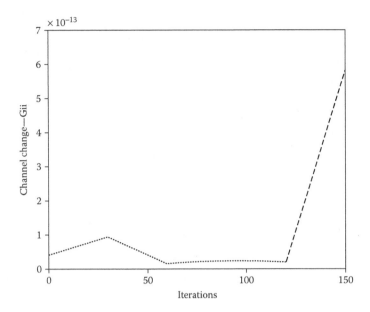

FIGURE 5.31
Smoothed channel variation.

probability with time wherein the outage probability using the proposed DPC scheme is significantly less (near zero) when compared to all the other schemes (about 85%), i.e., our approach can accommodate more number of mobile users rendering high channel utilization or capacity in the presence of channel variations. This, in turn, implies that the power consumed per active user is less compared to other schemes.

CASE II: SLOWLY VARYING CHANNEL In this scenario, $k_v = 0.01$ and $\sigma = 3$. In this case, though the channel changes every ten time-units, the channel variation illustrated in Figure 5.31 follows the Rayleigh fading and shadowing behavior. The channel variation is smoothed out using a linear function between the changes. In this case also, the proposed DPC scheme renders a low outage probability (about 30% as observed in Figure 5.32) while consuming less power per active mobile because the proposed DPC scheme maintains the SIR of each link closer to its target compared to others. Other schemes result in about 85% outage probability. The low outage probability for the proposed DPC scheme is the result of faster convergence, and low SIR error while it consumes satisfactory power per active user.

Example 5.4.2: Performance Evaluation with Number of Users
When the total number of mobile users in the cellular network varies, we compare how the total power and outage probability vary. In this

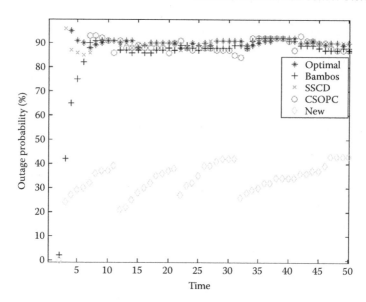

FIGURE 5.32
Outage probability.

simulation scenario, the number of mobile users trying to gain admission increases by 25 from the previous value, and the corresponding outage probability is calculated. Figure 5.33 and Figure 5.34 present the performance of the DPC in terms of total power consumed and outage probability, respectively, when the channel changes slowly. As expected, the proposed scheme renders significantly low outage probability while ensuring low power per active mobile user compared to other schemes.

Example 5.4.3: Mobile User Scenario

Because the users are mobile and in the presence of channel uncertainties, Figure 5.35 illustrates the initial location of mobile users, whereas Figure 5.36 presents the final location. Users in the cellular network try to move in any one of the eight predefined directions chosen at random at the beginning of the simulation. A user can move a maximum of 0.01 km per time unit. Because the time unit is small, the 0.01 km is a considerable amount of distance for any mobile user. Figure 5.37 and Figure 5.38 display the total power consumed and the corresponding outage probability. As evident from the result, the proposed DPC renders a lower outage probability (average of about 30%) compared to others (about 90%) while consuming low power per active mobile.

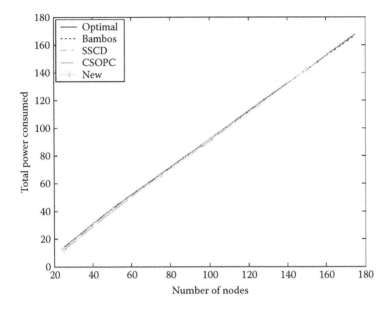

FIGURE 5.33
Total power consumed.

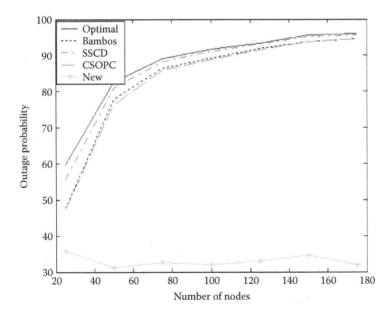

FIGURE 5.34
Outage probability.

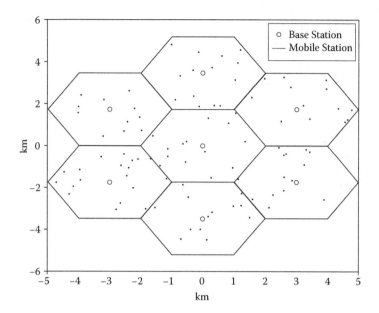

FIGURE 5.35
Initial state of mobile placement.

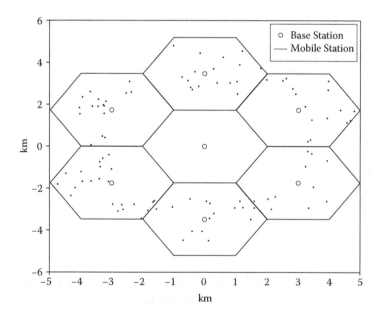

FIGURE 5.36
Final state of mobile placement.

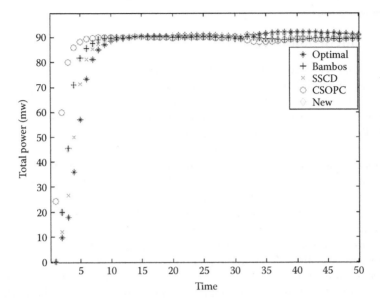

FIGURE 5.37
Total power consumed.

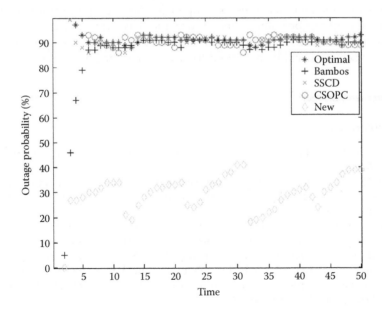

FIGURE 5.38
Outage probability.

5.5 Conclusions

In this chapter, a suite of DPC/ALP schemes is presented capturing the essential dynamics of power control. It was observed that the proposed DPC/ALP schemes allow fully distributed power and admission control, supporting ALP. The key idea introduced and leveraged here is to use a protection margin (matched with gradual power-up) in the dynamics of the SIR. Using a novel power update, it is shown that the overall system maintains a desired target SIR value for each link analytically. The analytical results are verified via simulation in several scenarios where the link arrivals are simulated in a uniform distribution. Simulation results show that our DPC/ALP schemes offer a superior performance in terms of convergence and maximizing the network capacity compared to the available ones in the literature both for peer-to-peer and cellular network scenario.

Subsequently, the DPC scheme is modified to render satisfactory performance in the presence of radio channel uncertainties. The simulation results show that the modified DPC scheme converges faster than others, maintains a desired target SIR value for each link and can adapt to the channel variations in the radio channel better. In the presence of channel uncertainties, the DPC scheme can render lower outage, using significantly less transmitter power per active user compared to other DPC schemes. As a result, the DPC schemes of Chapter 5 offer superior performance in terms of convergence, and it maximizes the network capacity compared to the available ones in the literature. In the next chapter, the applicability of these schemes will be exploited for wireless ad hoc networks.

References

Aein, J.M., Power Balancing in Systems Employing Frequency Reuse, *COMSAT Technical Review*, 1973, pp. 277–299.

Alavi, H. and Nettleton, R.W., Downstream power control for a spread spectrum cellular mobile radio system, in *Proceedings of IEEE GLOBECOMM*, 1982, pp. 84–88.

Bambos, N., Chen, S.C., and Pottie, G.J., Channel access algorithms with active link protection for wireless communication networks with power control, *IEEE/ACM Transactions on Networking*, Vol. 8, 583–597, October 2000.

Bambos, N., Toward power-sensitive network architectures in wireless communications: concepts, issues and design aspects, *IEEE Personal Communications*, Vol. 5, June 1998, pp. 50–59.

Canchi, R. and Akaiwa, Y., Performance of adaptive transmit power control in $\pi/4$ DQPSK mobile radio systems in flat Rayleigh fading channels, *Proceedings of the IEEE Vehicular Technology Conference*, Vol. 2, 1999, pp. 1261–1265.

Dontula, S. and Jagannathan, S., Active link protection for wireless peer-to-peer and cellular networks with power control, *Proceedings of the World Wireless Congress*, May 2004, pp. 612–617.

El-Osery, A. and Abdullah, C., Distributed Power Control in CDMA Cellular Systems, *IEEE Antennas and Propagation Magazine*, Vol. 42, No. 4, August 2000, pp. 152–159.

Foschini, G.J. and Miljanic, Z., A simple distributed autonomous power control algorithm and its convergence, *IEEE Transactions on Vehicular Technology*, Vol. 42, November 1993, pp. 641–646.

Grandhi, S., Vijayan, R., Goodman, D.J., A distributed algorithm for power control in cellular radio systems, *IEEE Transactions on Communications*, Vol. 42, No. 2, 226–228, April 1994.

Hanly, S., Capacity and power control in a spread spectrum macro diversity radio networks, *IEEE Transactions on Communications*, Vol. 44, 247–256, February 1996.

Hashem, B. and Sousa, E., Performance of cellular DS/CDMA systems employing power control under slow Rician/Rayleigh fading channels, *Proceedings of the International Symposium on Spread Spectrum Techniques and Applications*, Vol. 2, September 1998, pp. 425–429.

Jagannathan, S., Chronopoulos, A.T., and Ponipireddy, S., Distributed power control in wireless communication systems, *IEEE International Conference on Computer Communications and Networks*, November 2002, pp. 493–496.

Jagannathan, S., Zawdniok, M., and Shang, Q., Distributed power control of transmitters in wireless cellular networks in the presence of fading channels, *IEEE Transactions on Wireless Communications*, Vol. 5, No. 3, 540–549, March 2006.

Jantti, R. and Kim, S.L., Second-order power control with asymptotically fast convergence, *IEEE Journal on selected areas in communications*, Vol. 18, No. 3, March 2000.

Lee, G. and Park, S.-C., Distributed power control in fading channel, *Electronics Letters*, Vol. 38, No. 13, 653–654, June 2002.

Lewis, F.L., *Optimal Control*, John Wiley and Sons, 1999.

Mitra, D., An asynchronous distributed algorithm for power control in cellular radio systems, *Proceedings of the 4th WINLAB workshop*, Rutgers University, New Brunswick, NJ, 1993.

Rappaport, T.S., *Wireless Communications: Principles and Practice*, Prentice Hall, 1999.

Wu, Q., Optimum transmitter power control in cellular systems with heterogeneous SIR thresholds, *IEEE Transactions on Vehicular Technology*, Vol. 49, No. 4, July 2000, pp. 1424–1429.

Zander, J., Distributed cochannel interference control in cellular radio systems, *IEEE Transactions on Vehicular Technology*, Vol. 41, August 1992, pp. 305–311.

Problems

Section 5.2

Problem 5.2.1: Evaluate the performance of the SSCD scheme using a ten-cell network by placing the users in a random manner.

Plot the outage probability and power consumed by each user. Use the parameters from the Example 5.2.1.

Problem 5.2.2: Evaluate the performance of the optimal scheme using a ten-cell network by placing the users in a random manner. Plot the outage probability and power consumed by each user. Compare the outcome with that of SSCD. Use the parameters from the Example 5.2.1.

Section 5.3

Problem 5.3.1: Evaluate the performance of the SSCD scheme now with admission control included and using a ten-cell network by placing the users in a random manner. Plot the outage probability and power consumed by each user. Use the parameters from the Example 5.2.1.

Problem 5.3.2: Evaluate the performance of the optimal scheme now with admission control included and using a peer-to-peer network. Plot the outage probability, average admission delay power consumed by each user, and average dropped links. Compare the outcome with that of SSCD. Use the parameters from the Example 5.3.1.

Section 5.4

Problem 5.4.1: Evaluate the performance of the DPC scheme with embedded channel estimator for a ten-cell network by placing the users in a random manner. Consider shadowing with slow and abrupt fading. Plot the outage probability and power consumed by each user. Use the parameters from the Example 5.4.1.

Problem 5.4.2: Evaluate the performance of the DPC using a ten-cell network by placing the users in a random manner with mobility. Plot the outage probability and power consumed by each user. Use the parameters from the Example 5.4.3, except the node mobility is increased to 0.02 km per unit time.

6

Distributed Power Control and Rate Adaptation for Wireless Ad Hoc Networks

In the last chapter, the distributed power control (DPC) scheme was demonstrated for cellular networks both analytically and via simulation. Ad hoc wireless networks have gained great importance in recent years. Wireless communication technology has come into wide use with the advent of the IEEE 802.11 standard. An ad hoc network is a group of wireless mobile nodes dynamically forming a temporary network without any fixed infrastructure or centralized administration. The applications for ad hoc networks have grown tremendously with the increase in the use of wireless sensor networks. For wireless networks, it was demonstrated that energy efficiency is an important QoS metric besides throughput, loss rate, and end-to-end delay. Therefore DPC is necessary even for ad hoc wireless and sensor networks. The DPC derived for the cellular networks is extended to ad hoc wireless networks.

In this chapter, DPC scheme from the previous chapter is reviewed and a medium access control (MAC) protocol based on the DPC is introduced for wireless ad hoc networks (Zawodniok and Jagannathan 2004) in the presence of radio channel uncertainties such as path loss, shadowing, and Rayleigh fading. The DPC quickly estimates the time-varying nature of the channel and uses the information to select a suitable transmitter power value, even for ad hoc wireless and sensor networks, to maintain a target signal-to-interference ratio (SIR) at the receiver. To accommodate the sudden changes in channel state, a safety factor is incorporated in the power selection. The standard assumption of a constant interference during a link's power update used in other research works is relaxed.

The performance of the DPC for ad hoc wireless networks can be demonstrated analytically because it is exactly the same as that of the cellular networks. Moreover, the power used for all RTS-CTS-DATA-ACK frames is selected using the proposed DPC; hence, energy savings and modest improvement in spatial reuse can be achieved. The hidden-terminal problem, which is commonly encountered in ad hoc wireless networks, can be overcome by periodically increasing the power. The NS-2 simulator is used to compare the proposed scheme with that of 802.11. The proposed MAC protocol renders significant increase in throughput in the presence of channel variations compared with 802.11 although consuming low energy per bit.

6.1 Introduction to DPC

The objectives of transmitter power control include minimizing power consumption while increasing the network capacity and prolonging the battery life of mobile units by managing mutual interference so that each mobile unit can meet its SIR and other quality of service (QoS) requirements. Rigorous work on DPC was performed for cellular networks (Bambos 1998, Dontula and Jagannathan 2004, Jagannathan et al. 2002, Jagannathan et al. 2004, Hashem and Sousa 1998). A few DPC schemes (Park and Sivakumar 2002, Jung and Vaidya 2002, Gomez et al. 2001, Karn 1990, Pursley et al. 2000) were developed for wireless ad hoc networks where the topology constantly changes because of node mobility and communication link failures. Similarly, see Maniezzo et al. (2002), Woo and Culler (2001), Ye et al. (2002), Singh and Raghavendra (1998).

Unlike wired networks, radio channel uncertainties in a wireless network, for instance path loss, shadowing, and Rayleigh fading, can attenuate the power of the transmitted signal causing variations in the received SIR and thus degrading the performance of any DPC. Low SIR means high bit error rate (BER), which is unsatisfactory. Reported DPC schemes (Park and Sivakumar 2002, Jung and Vaidya 2002, Gomez et al. 2001, Karn 1990, Pursley et al. 2000) for ad hoc networks assume that: (1) only path loss is present, (2) no other channel uncertainty exists, and (3) the mutual interference among the users is held constant during power update of each user. Moreover, improvement of spatial reuse factor is not adequately addressed. Our previous effort (Jagannathan et al. 2004) was to develop new protocols to accommodate channel uncertainties for cellular networks. No work is currently reported for ad hoc wireless networks that accommodate channel uncertainties.

Furthermore, in an ad hoc network as mentioned in Chapter 1, request to send (RTS) and clear to send (CTS) messages are used to establish a connection for data transmission between a transmitter and a receiver. In (Jung and Vaidya 2002, Gomez et al. 2001), authors propose using maximum transmitter power only for RTS-CTS, whereas DATA and ACK transmission is accomplished at a much lower power. This lower power is calculated according to RTS and CTS reception conditions. However, the channel state can change between the RTS and the DATA transmissions resulting in an inaccurate power selection. On the other hand,

Jung and Vaidya (2002) suggested that calculating the transmitter power using a DPC scheme for the DATA and ACK frames in 802.11 results in the degradation of QoS. More collisions will occur, thus causing a large increase in the number of retransmissions. Consequently, this results in higher power consumption, lower throughput per node, and lower network utilization. Additionally, previous DPC schemes for wireless ad hoc networks (Park and Sivakumar 2002, Jung and Vaidya 2002, Gomez et al. 2001, Karn 1990, Pursley et al. 2000) ignore performance guarantees using analytical methods. The work from Zawodniok and Jagannathan (2004) overcomes these limitations.

In this chapter, the DPC scheme with an embedded channel prediction scheme from the previous chapter is extended to ad hoc wireless networks and a medium access control (MAC) protocol is designed using the DPC for wireless ad hoc networks in the presence of radio channel uncertainties such as path loss, shadowing, and Rayleigh fading. This embedded scheme predicts the time-varying fading channel state for the next transmission, in contrast to (Jung and Vaidya 2002, Gomez et al. 2001) where the delayed channel parameters from the previous transmission are used to select the power for the subsequent transmission. The MAC protocol from Zawodniok and Jagannathan (2004) uses the DPC scheme to update the power so that a target SIR is maintained at the receiver. To minimize the effect of sudden changes in the channel state in the power selection, a safety factor is incorporated. This DPC scheme is shown to converge analytically to any target SIR value in the presence of channel uncertainties. Moreover, the proposed MAC protocol assigns power adaptively for all MAC frames (RTS, CTS, DATA, and ACK) while overcoming the hidden-terminal problem. As a result, a modest improvement in spatial reuse factor is observed. Finally, a comparison with the standard 802.11 protocol is also included.

6.2 Channel Uncertainties

Unlike wired channels that are stationary and predictable, radio channels involve many uncertain factors, and as a result they are difficult to analyze. We focus our effort on these main channel uncertainties, such as path loss, shadowing, and Rayleigh fading.

6.2.1 Signal-to-Interference Ratio (SIR)

Each receiving node on a given link measures the interference present in the channel and communicates this information to the transmitter. In addition, each link autonomously decides how to adjust its transmitter power.

Therefore, the decision-making is fully distributed at the link level. Consequently, the overhead due to the feedback control is minimal in DPC compared to its counterpart for centralized operations. Because each receiver provides feedback to the transmitter, the DPC presented in the previous chapter can be extended to ad hoc wireless and sensor networks.

The goal of transmitter power control is to maintain a target SIR threshold for each network link although the transmitter power is adjusted so that the least possible power is consumed in the presence of channel uncertainties. Suppose there are $N \in Z_+$ links in the network. Let g_{ij} be the power loss (gain) from the transmitter of the *j*th link to the receiver of the *i*th link. The power attenuation is considered to follow the relationship given in the following paragraph.

Calculation of SIR, $R_i(t)$, at the receiver of *i*th link at the time instant t, is given by

$$R_i(t) = \frac{g_{ii}(t)P_i(t)}{I_i(t)} = g_{ii}(t)P_i(t) \Big/ \sum_{j \neq i} g_{ij}(t)P_j(t) + \eta_i(t) \qquad (6.1)$$

where i, j {1, 2, 3,..., n}, $I_i(t)$ is the interference, $P_i(t)$ is the link's transmitter power, $P_j(t)$ are the transmitter powers of all other nodes, and $_i(t) > 0$ is the variance of the noise at its receiver node. For each link *i*, there is a lower SIR threshold $_i$ and upper threshold $_i^*$. Therefore, we require

$$\gamma_i \leq R_i(t) \leq \gamma_i^* \qquad (6.2)$$

for every $i = 1, 2, 3,...$, n the lower threshold value for all links can be taken equal to for convenience, reflecting a certain QoS that the link has to maintain to operate properly. An upper SIR limit is also set, to manage the interference.

6.2.2   Radio Channel Model with Uncertainties

The radio channel places fundamental limitations on wireless communication systems. The path between the transmitter and the receiver can vary from clear line-of-sight to one that is severely obstructed by buildings, mountains, and foliage. In ad hoc wireless networks, channel uncertainties such as path loss, shadowing, and Rayleigh fading can attenuate the signal power during transmission and thus cause variations in the received SIR which degrades the performance of any DPC scheme. The effect of these uncertainties (see Chapter 5) is represented via a channel loss (gain) factor that typically multiplies the transmitter power. Therefore, the channel loss or gain, g, can be expressed as (Rappaport 1999, Canchi and Akaiwa 1999)

$$g = f(d, n, X, \zeta) = d^{-n} \cdot 10^{0.1\zeta} \cdot X^2 \qquad (6.3)$$

where d^{-n} is the effect of path loss and $10^{0.1.\zeta}$ corresponds to the effect of shadowing. For Rayleigh fading, it is typical to model the power attenuation as X^2, where X is a random variable with Rayleigh distribution. Typically the channel gain, g, is a function of time.

6.3 Distributed Adaptive Power Control

In the previous DPC schemes (Bambos 1998, Janti and Kim 2000, Dontula and Jagannathan 2004, Jagannathan et al. 2002), only path loss uncertainty is considered. Moreover, the DPC algorithm proposed in (Bambos 1998) appears to be slow in convergence compared to (Dontula and Jagannathan 2004) for cellular networks and the outage probability is slightly higher. Nevertheless, in the presence of other channel uncertainties, the performances of these DPC schemes fail to render satisfactory performance as shown in (Jagannathan et al. 2004) for cellular networks. The work, presented in this chapter, is aimed at demonstrating the performance of DPC in the presence of several channel uncertainties for wireless ad hoc networks, because there are significant differences between cellular and ad hoc networks.

However, the channel is time-varying when uncertainties are considered and therefore $g_{ij}(t)$ is not a constant. In Lee and Park (2002), a new DPC algorithm is presented where $g_{ii}(t)$ is treated as a time-varying function due to Rayleigh fading by assuming that the interference $I_i(t)$ is held constant. Because this is an invalid assumption, in this paper, a novel DPC scheme is given where both $g_{ii}(t)$ and the interference $I_i(t)$ are time-varying, and channel uncertainties are considered for all the mobile users. In other words, in all existing works (Bambos et al. 2000, Janti and Kim 2000, Dontula and Jagannathan 2004), both $g_{ij}(t)$ and $I_j(t)$ are considered to be held constant, whereas in our work, this assumption is relaxed. Moreover, the persistence of excitation condition requirement on the input signals in (Jagannathan et al. 2004) is relaxed in this work.

Following the work and similar notation from Chapter 5, the DPC calculation for an ad hoc wireless network can be given for two scenarios.

CASE 1 α_i, β_i, and r_i are known. In this scenario, one can select the feedback control as

$$v_i(l) = \beta_i^{-1}(l)[\gamma - \alpha_i(l)y_i(l) - r_i(l)\omega_i(l) + k_v e_i(l)] \tag{6.4}$$

where the error in SIR is defined as $e_i(l) = R_i(l)$. This results in

$$e_i(l+1) = k_v e_i(l) \tag{6.5}$$

By appropriately selecting k_v via placing the eigenvalues within a unit circle, it is easy to show that the closed-loop SIR system is asymptotically stable in the mean or asymptotically stable, $\lim_{l \to \infty} E\{e_i(l)\} = 0$. This renders that $y_i(l) \to \gamma$.

CASE 2 α_i, β_i, and r_i are unknown. In this scenario, SIR error equation (see Chapter 5) can be expressed as

$$y_i(l+1) = [\alpha_i(l) \quad r_i(l)]\begin{bmatrix} y_i(l) \\ \omega_i(l) \end{bmatrix} + \beta_i(l)v_i(l) = \theta_i^T(l)\psi_i(l) + \beta_i(l)v_i(l) \quad (6.6)$$

where $\theta_i^T(l) = [\alpha_i(l) \quad r_i(l)]$ is a vector of unknown parameters, and $\psi_i(l) = [\begin{smallmatrix} y_i(l) \\ \omega_i(l) \end{smallmatrix}]$ is the regression vector. Now, selecting feedback control for DPC as

$$v_i(l) = \beta_i^{-1}(l)[-\hat{\theta}_i(l)\psi_i(l) + \gamma + k_v e_i(l)] \quad (6.7)$$

where $\hat{\theta}_i(l)$ is the estimate of $\theta_i(l)$, then the SIR error system is expressed as

$$e_i(l+1) = k_v e_i(l) + \theta_i^T(l)\psi_i(l) - \hat{\theta}_i^T(l)\psi_i(l) = k_v e_i(l) + \tilde{\theta}_i^T(l)\psi_i(l) \quad (6.8)$$

where $\tilde{\theta}_i(l) = \theta_i(l) - \hat{\theta}_i(l)$ is the error in estimation. From Equation 6.8, it is clear that the closed-loop SIR error system between a transmitter–receiver pair in an ad hoc wireless network is driven by channel estimation error. If the channel uncertainties are properly estimated, then estimation error tends to be zero. In this case, Equation 6.8 becomes Equation 6.5. In the presence of error in estimation, only boundedness of error in SIR can be shown. We can show that the actual SIR approaches the target provided the channel uncertainties are properly estimated. To proceed further, Assumption 6.3.1 is used in Jagannathan et al. (2004).

ASSUMPTION 6.3.1
The channel should change slowly in comparison to parameter updates.

REMARK 1
The channel estimation scheme works well when the channel changes slowly compared to estimation. However, to accommodate the sudden changes in channel conditions, an additional safety factor has to be introduced in this work.

Consider now the closed-loop SIR error system with channel estimation error, (l), as

$$e_i(l+1) = k_v e_i(l) + \tilde{\theta}_i^T(l) \cdot \psi_i(l) + \varepsilon(l) \quad (6.9)$$

where $\varepsilon(l)$ is the error in estimation which is considered bounded above $\|\varepsilon(l)\| \le \varepsilon_N$, with ε_N a known constant.

THEOREM 6.3.1 (CONVERGENCE ANALYSIS)

Given the DPC scheme above with channel uncertainties for a wireless ad hoc network and under the assumption of persistence of excitation of input signals, if the feedback from the DPC scheme is selected as Equation 6.4, then the mean channel estimation error along with the mean SIR error converges to zero asymptotically, if the parameter updates are taken as Equation 5.64.

PROOF See Zawodniok and Jagannathan (2004).

6.4 DPC Implementation

A preliminary version of the proposed DPC algorithm was implemented for cellular networks (Jagannathan et al. 2004), which was relatively straightforward. Full duplex links and time-synchronized communication simplifies the estimation of channel gain. Additionally, the packets are sent at predefined intervals. This allows calculating the change of channel conditions between packets accurately, with a small estimation error. Moreover, the power for all connections is updated synchronously assuming that the interference from other users is experienced by a particular connection at the same time. Hence, the interference will not change during the packet transmission.

On the other hand, the proposed MAC protocol is derived from 802.11 standards, where the CSMA/CA random access method is used. This introduces additional uncertainties with different challenges. Therefore the proposed DPC algorithm has to be modified and has to be implemented carefully in the MAC protocol.

6.4.1 DPC Feedback

The basic communication of a user packet employs a four-way handshake exchange that consists of four frames: request to send (RTS) — from source to destination; clear to send (CTS) — from destination to source; DATA frame — from source to destination; acknowledgment (ACK) — from destination to source. All these frames are transmitted over a single radio channel for wireless ad hoc networks. Hence, the communication between any two nodes is carried out over a shared half-duplex medium. Additionally, a particular handshake can be initiated at any time instance. In consequence, the interference experienced by a particular connection can change at any time even during the transmission of a particular frame. Hence, the estimated power feedback has to overcome such an interference uncertainty.

Moreover, the time interval between any consecutive transmissions occurring in the same direction will vary from frame to frame because of

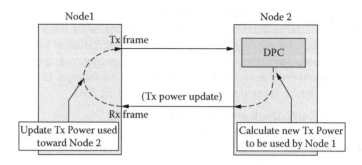

FIGURE 6.1
DPC feedback loop.

the following reasons. First, the frames within the four-way handshake are of different sizes — ranging from few bytes for ACK to over 2500 bytes for DATA. Second, the delay between two consecutive handshakes can vary significantly because of packet transmission intervals and channel contention. As a result, the channel estimation error will vary depending on these time gaps, if feedback from the receiver is used to determine the transmitter power for any DPC scheme. Therefore, the target SIR has to be chosen suitably to overcome the worst-case scenario due to these uncertainties. In our implementation, the target SIR value is obtained by using a certain safety factor.

6.4.2   DPC Algorithm for 802.11 Type Ad Hoc Networks

In the proposed implementation shown in Figure 6.1, a feedback loop is used between the transmitter and the receiver to successfully implement the DPC. The detailed description of the proposed DPC algorithm implementation is presented in Subsection 6.4.4.

6.4.3   Retransmissions and Power Reset

In real scenarios, it is expected that the channel conditions at certain time instants can change too quickly to be correctly estimated by the DPC algorithm, by violating the Assumption 6.3.1. Then frame losses will occur. Two mechanisms are introduced to overcome and mitigate such problems: increase of transmitter power in case of retransmission and reset of power for a long idle connection.

A retransmission implies that the transmitted signal strength is not sufficient for decoding. A simple approach would be to retransmit the frame with the same power as in first transmission, and hope that channel state will improve. However, the channel attenuation and/or interferences could also become worse. On the other hand, in an active approach for each retransmission, the power is increased by a certain factor. Unfortunately, this will increase interference as well as power consumption. However, experiments done with both methods have indicated that the active adaptation of power yielded higher throughput and resulted in fewer retransmissions when compared to the passive method. As a result, the active retransmission method is better in terms of throughput and energy efficiency, and it was chosen for the DPC scheme (Zawodniok and Jagannathan 2004).

Additionally, as the estimated power value for a given destination becomes less and less accurate, either the feedback or time delay increases between any two receptions or consecutive frame transmissions respectively. After a certain idle period, the channel estimate will not reflect the future channel state accurately. To overcome this, after a certain idle time interval, the proposed algorithm will reset the transmission power to the maximum value defined for the network. Then the DPC process described in previous subsection is restarted from the beginning.

6.4.4 DPC Algorithm

In the proposed implementation shown in Figure 6.1 a feedback loop is used between the transmitter and the receiver to successfully implement the DPC. The source node 1 and destination node 2 are used in this description.

1. The source node 1 sends the RTS frame to the destination node 2 using the maximum transmission power defined for the network.

2. When the destination node 2 receives the RTS:
   - The interference is measured during the RTS reception, and a suitable transmission power increment required to successfully transmit the next packet by the source node is calculated.

3. The first CTS frame from the destination node 2 is transmitted using the maximum power value as follows:
   - The CTS frame will embed the power value calculated in Step 2 for the source node 1.

4. When the source node 1 receives the CTS:
   - The interference is measured at the node 1, and the suitable transmission power increment required by the node 2 is calculated for the next packet.
   - The transmission power at the source node 1 for the next packet is obtained by using the power increment value embedded within CTS.

5. When the source sends the DATA frame:

   • The transmission power just calculated by using the information from the CTS is used.

   • The power increment for the destination is included.

6. When the destination receives the DATA frame:

   • The interference is measured, and the suitable transmission power increment for the next packet at the source node 1 is calculated.

   • The transmission power of the destination B is updated according to the power value carried in the DATA frame.

7. When the destination sends the ACK frame:

   • The transmission power required to transmit the ACK frame is obtained by using the power increment information carried by the DATA frame.

   • Power increment value for the source node 1 is included.

8. When the source receives the ACK frame:

   • The interference is measured, and the suitable transmission power increment for the destination node 2 is calculated.

   • The transmission power of the source node 1 is updated by using the power information carried in the ACK.

9. Then the nodes wait for next user packet to be sent from the source to the destination.

10. After a new user packet is ready to be sent from the source, it sends RTS frame to the destination node:

    • The power increment value carried by the last ACK frame is used to obtain the transmitter power.

    • The power increment value for the destination node is embedded.

11. The above steps are repeated for all the frames: transmitter power is calculated by DPC for each frame and passed as feedback in the following frame both at the source and the destination.

6.5 Power Control MAC Protocol

For this DPC implementation, the original MAC protocol for 802.11 has been modified. These modifications will occur at several different layers. Furthermore, a novel idea of train of pulses, which is adapted from

(Jung and Vaidya 2002), is used to overcome the hidden-terminal problem and it is combined with the proposed power control scheme.

The proposed MAC protocol uses the novel DPC algorithm to calculate the transmitter power for every MAC frame, whereas the other protocols (Jung and Vaidya 2002, Gomez et al. 2001) modify the power only for the DATA and ACK frames. Additionally, the algorithm predicts the channel state for the next transmission, whereas the other protocols use the old measurements to select a power value. Finally, a safety factor is utilized to adjust the transmission power in the case of delayed feedback. Thus, the proposed power control MAC protocol selects a more suitable transmission power for fading channels whereas others do not. Moreover, the train of pulses is used for every MAC frame including the RTS and CTS, whereas the protocol in Jung and Vaidya (2002) applies it to DATA frames only. Consequently, the proposed approach, as explained in the following subsection, results in higher channel utilization and energy efficiency.

6.5.1 Hidden-Terminal Problem

The hidden-terminal problem occurs in a wireless network when a third node causes a collision with an ongoing communication between any two nodes. The problem is illustrated in Figure 6.2, where node A transmits data to node B. Node F is outside the sensing range of transmitter A, and it will not detect transmission of the DATA frame that is sent from node A.

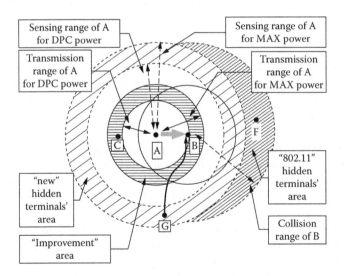

FIGURE 6.2
Hidden-terminal problem in ad hoc networks.

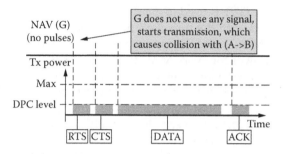

FIGURE 6.3
Transmission using DPC.

Therefore, F can initiate transmission, which will collide with DATA frame at the receiving node B. In Figure 6.2, nodes located inside a marked region, called "hidden terminals' area," are potential sources for this problem. The DPC schemes, generally, use much lower power for the transmission of frames, and therefore result in increased collisions, as shown in Figure 6.3. A solution similar to that found in other schemes (Jung and Vaidya 2002) is proposed.

Generally, when a lower transmission power is used, the transmission and sensing range of a node will decrease as shown in Figure 6.2. Node G will fail to detect transmission from node A and it will initiate transmission by assuming that the channel is idle. If node G uses maximum power, then a collision will occur at receiver node B. Hence, the probability of hidden-terminal problem occurring in the network will increase when a low transmitter power is used.

To overcome this problem, in our proposed methodology, a train of short pulses with increased transmitted power, periodically, is used during transmission. The RTS, CTS, DATA, and ACK frames are transmitted using the power dictated by the proposed DPC, along with the train of pulses. The pulses use maximum transmission power defined for the network. This ensures that all the nodes in the sensing range of the transmitter will detect the pulses and update their NAV vectors accordingly. Thus, the nodes in the sensing range of the transmitter node will not cause collision. Figure 6.3 and Figure 6.4 display the difference in handling of NAV vector in the available DPC schemes with and without the train of pulses.

The generation of the train of pulses should be implemented in the hardware. In details, the circuitry of RF amplifier should periodically increase the transmission power for every transmission. Additionally, this solution is simpler to implement than the one proposed in Jung and Vaidya (2002), because an amplifier module in this scheme does not need to know the type of a transmitted frame.

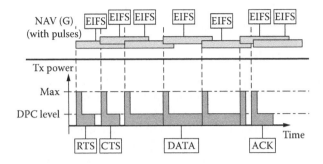

FIGURE 6.4
Periodic increase in power during frame transmission.

6.5.2 Protocol Design

In the proposed protocol, only the initial RTS-CTS frames for a new link have to be transmitted using the maximum power defined for the network. Subsequently, all frames, including RTS-CTS-DATA-ACK frames, will use the power calculated according to the proposed DPC scheme. The MAC header needs to be changed to allow power information to be sent between the communicating nodes.

In other words, the MAC frames have to embed the power information that is used for the current packet, as well as for any subsequent response. This process repeats for all transmissions occurring between any two nodes A and B resulting in an increase in overhead, which could cause a decrease in throughput. However, an increase in throughput observed because of better channel utilization overcomes the penalty introduced by the additional overhead. Moreover, this overhead can be further reduced by using discrete power levels. Consequently, the number of bits necessary for storing the power value in the header will decrease, thus reducing the overhead. Additionally, these fields can be defined as an option in the frame header by using a one-bit flag.

Whenever the power level changes, this flag will be set indicating the receiving node to calculate its transmitter power for the subsequent transmission. Otherwise, the flag will be cleared and the node will continue to use its previous value.

Because a lower power is selected for transmission of the MAC frames, during severe fading, certain frames will be unsuccessfully decoded because of poor reception and they will be dropped. It is important to note that predicting the channel during severe fading conditions is quite difficult and impossible. To mitigate this problem, the proposed protocol increases the power by a predefined safety factor before each retransmission, to reduce the packet drops.

6.5.3   Channel Utilization

Though hidden-terminal problem occurrence increases with lower transmitter powers, an improvement in channel utilization and throughput can be seen by using the proposed MAC protocol. In fact, Figure 6.2 depicts the enhancement in utilization that will occur when a lower transmitter power is used for subsequent RTS-CTS transmissions whenever the RTS-CTS handshake between any two given nodes A and B has failed. In this scenario, node B will not respond to request from node A. This will occur, for example, if node F is transmitting at the same time as node A is trying to send the RTS frame. Therefore, node B is unable to receive the RTS frames because of a collision. After the predefined number of retransmissions, node A will cease to send the packet. In such a case, node C will be able to start transmission earlier than in the case of using maximum power for RTS-CTS. As a result, the contention time for frames from certain nodes, such as C, decreases.

Consider the scenario when maximum power is utilized for all RTS-CTS frames. Node C will decode the RTS frame because it was sent with the maximum transmitter power defined for the network. Consequently, node C will update its NAV vector using the RTS frame. No transmission occurs; hence, the channel is idle. On the other hand, if the RTS frame is sent at a power level calculated by the DPC, the node C will only detect the RTS frame and will set its NAV vector to the EIFS time. Hence, shortly after EIFS, node C is free to initiate communication. Because of the availability of a channel to C, the throughput increases.

This improvement applies to all nodes within the improvement area depicted in Figure 6.2. Given the high density of nodes in the case of wireless ad hoc networks, the probability of a node accessing the channel is quite high. Therefore, an increase in aggregated throughput is observed with the proposed protocol.

Because of higher channel utilization, the *spatial reuse factor*, which is defined as the number of successful transmissions within a given time interval for a given area, will increase for the proposed DPC scheme. For 802.11, the NAV vector will be set for an entire expected duration of flow transmission; hence, there will be time intervals when no transmissions take place. As the result, there will be fewer transmissions for a given time interval in comparison with the theoretical capacity of the radio channel. In our scheme, these idle periods are detected, nodes are allowed to transmit sooner and, thus, the total number of successful transmissions within a given time interval increases. Consequently, the spatial reuse factor increases for the proposed DPC when compared to 802.11.

6.5.4 Contention Time

The change in contention time for the proposed DPC scheme is due to two major factors: (1) more retransmissions during fading channel conditions and (2) improved channel utilization. During fading channel conditions, retransmissions will increase with the proposed DPC because of the possibility of insufficient power for the reception of a packet. As a result, the average contention time increases. Additionally, higher utilization due to the proposed DPC will cause an increase in the throughput causing congestion. Under these conditions, the proposed protocol will cause certain frames to be delayed longer compared to the 802.11 standard. Therefore, the contention time will increase with the DPC from Zawodniok and Jagannathan (2004).

6.5.5 Overhead Analysis

The proposed MAC protocol requires additional data to be incorporated into the 802.11 frames for transmission. This additional information will include the current and the new transmitter power value to be used for the response. All RTS, CTS, DATA, and ACK frames will embed this information. The following analysis is used to evaluate the efficiency of the proposed protocol and to compare it with 802.11. In particular, we have analyzed the case where the RTS/CTS messages are followed by a single DATA/ACK exchange.

6.5.5.1 RTS/CTS Followed by a Single DATA/ACK Frame Scenario

In this scenario, there will be a total of four frames transmitted: RTS, CTS, DATA, and ACK. This is a typical sequence used for an Ethernet/IP based packets (length up to 2500 octets). Each frame includes two power values; thus overhead per data packet will include a total of eight power values. Let the size of power value in octets be expressed by S_{power}; the overhead (OH) size in octets per data packet is equal to:

$$OH = 4\,frames \times (S_{power} * 2) = 8 * S_{power} \qquad (6.10)$$

6.5.5.2 Minimizing Overhead Impact

In the simulations, the power values are stored as real numbers and they are sent in the MAC frame. However, in actual implementation, the overhead can be minimized by allowing discrete values for power levels and lowering the OH in terms of number of bits used for power. Second, the power values can be embedded in the frame only when the transmitter power changes between the power levels. This can be accomplished by using a one-bit flag to indicate whether the power values are added to the header or not. The one-bit flag field will be included in all the frames. If the power value does not change from its previous value, the flag is cleared, and no additional data is sent. Otherwise, the bit is set, and the new power value is included in the header.

Let us assume that the power value will change between frames with a probability p. Then the OH per data packet — in case of RTS/CTS followed by a single DATA/ACK — will be expressed as:

$$OH_{save} = 4\,frames \times (2 \times 1bit\_flag + p \times 2 \times S_{power}) = 8 * (1bit\_flag + p * S_{power})$$

(6.11)

where p is the probability with which the change in power level will occur for a frame, a one-bit flag is used to indicate whether the power value is included in the header or not.

6.5.5.3 Protocol Efficiency for RTS/CTS/DATA/ACK Sequence

The efficiency of the protocol in terms of OH size can be evaluated as the ratio of user data portion to total data transmitted (data + frame headers + backoff) (Wei et al. 2002). The efficiency can be expressed as

$$\eta = \frac{S_{packet}}{S_{packet} + S_{RTS} + S_{CTS} + S_{DATA} + S_{ACK} + S_{BACKOFF}}$$

(6.12)

where S_{packet} is the size of data packet in octets; S_{RTS}, S_{CTS}, and S_{ACK} represent the size of RTS, CTS, and ACK frames, respectively, S_{DATA} denotes the size of DATA frame header (without data packet), and $S_{BACKOFF}$ represents the backoff time given in octets.

Because of the implementation of the DPC in the MAC protocol, the frame size of RTS/CTS/DATA/ACK will increase by an amount equal to OH from Equation 6.11 and Equation 6.12, respectively. To understand the OH, the efficiency of the proposed implementation has been compared with that of the standard 802.11 protocol. Different size fields used for the power levels have been compared: 4-bit, allows 32 different power levels, and 8-bit (one octet) allows 255 power levels and so on. Also, probability levels are used to assess the power change between frames: p = 0.5 implies that the change occurs at every second frame and p = 0.1 represents the change occurring at every tenth frame.

In the worst-case scenario all the frames will contain the power fields. Because of the additional OH resulting from the incorporation of power levels for the proposed DPC, a 2.5% decrease in efficiency calculated using Equation 6.12 is observed when compared to 802.11. Thus the impact of OH introduced by the proposed MAC protocol appears to be negligible.

6.5.6 NS-2 Implementation

The NS-2 simulator was used for evaluating the proposed DPC scheme. The modifications made to incorporate the proposed DPC algorithm and protocol development, mainly focus on two layers of 802.11: Physical layer — modified to collect necessary data, for example interference; and Medium Access Control (MAC) layer — modified to implement the DPC algorithm and protocol. Furthermore, floating-point variables are used for communicating the power values.

The standard 802.11 and the proposed MAC protocols are evaluated under similar channel conditions, identical node placements, node movement and data flows (type, rate, start time, source, destination, etc.), and SIR thresholds, with a deterministic propagation model. The path loss effect is calculated as in *Propagation/Shadowing* object from NS-2 simulator. Additionally, shadowing and Rayleigh fading effects have been implemented. The calculations use the model from Equation 6.3 and store effects in sample files. This ensures that the channel uncertainties for the simulations are the same for all protocols. Figure 6.5 illustrates an example of such attenuation. The simulations were repeated for a number of different, randomly generated scenarios, and the results were averaged.

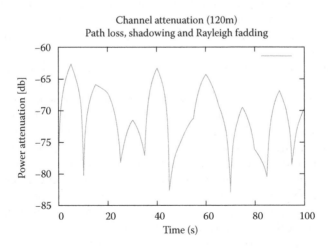

FIGURE 6.5
Power attenuation.

6.6   Simulation Parameters

The AODV routing protocol with 2-Mbps radio channel rate was used. The random topology is used to evaluate the schemes. The maximum power used for the 802.11 and proposed DPC is equal to 0.2818 W. The proposed DPC should maintain a target SIR of 10, which is 2.5 times higher than the minimum SIR for error-free reception that is equal to 4 (~6 db), as in Singh and Raghavendra (1998). The target SIR is increased to overcome a possible change in channel condition between the time when the power was calculated and the time when this power is used. For the proposed DPC, the design parameters are selected as $Kv = 0.01$, and = 0.01. The power safety factor in the case of retransmissions is set to 1.5 for the proposed DPC scheme.

The random topology consists of a 1000×1000 m square area with 100 nodes placed at random. Node mobility is also randomly generated with a maximum speed of 3 m/sec and with 2-sec pauses between the moves. The simulations are executed for 50 sec. CBR traffic is used for 50 flows, starting randomly during the first 2 sec. Each data flow generates steady traffic using 512-B-long packets. The results were averaged over simulation trials using different fading effects, node placement, and movement. The simulations were executed by varying the per flow rates. The radio channel with 2-Mbps bandwidth was used.

In the random topology scenario, each flow can use different number of hops (miniflows) between the source and the destination pairs. This can lead to different end-to-end throughput depending on number of hops used in a particular scenario (determined by node placement). Hence, the miniflow transmissions were used instead of end-to-end transmissions.

The total data transmitted for all miniflows is presented in Figure 6.6. Compared to the 802.11, the proposed DPC transmits more data regardless of traffic flow rate, because the higher utilization is achieved for the proposed protocol, as explained in Section 6.3. Similarly, the proposed protocol outperforms the 802.11 protocols in terms of energy efficiency, as shown in Figure 6.7, indicating lower energy consumption than 802.11. Regardless of the traffic load, the proposed protocol allows transmission of more data per joule when compared to 802.11. Higher energy efficiency of the proposed protocol is a result of transmission power control scheme, which selects a more suitable power value needed for correct decoding of the frames. Additionally, the proposed protocol consumes energy more efficiently in case of congestion, because it performs better in terms of throughput when compared to the 802.11.

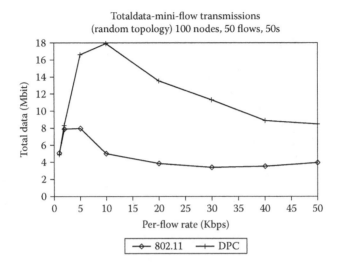

FIGURE 6.6
Total data transmitted (miniflows).

In the case of the 802.11, the maximum throughput is achieved at 5-kbps per flow rate, but the energy efficiency is highest for 2.5-kbps per flow rate. This indicates that with higher throughput more packets are dropped, with simultaneously increasing the energy consumption. By contrast, in

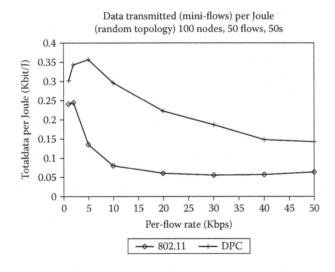

FIGURE 6.7
Total data per joule transmitted.

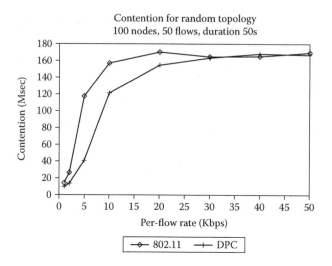

FIGURE 6.8
Contention time.

the proposed protocol, the maximum throughput and energy-efficiency is achieved at the higher per flow rate (10 and 5 kbps, respectively), because the proposed protocol achieves higher channel utilization when compared to the 802.11 protocol.

The average contention time is presented in Figure 6.8. The proposed protocol yields lower contention time for the per flow rate (up to 30 kbps), because of higher channel utilization and shorter interval between packet transmissions, than in the case of the 802.11. As per flow rate increases further, the contention time increases for the proposed protocol. This neutralizes the advantage of higher channel utilization in the proposed protocol. Consequently, both protocols achieve similar contention time for per flow rates above 30 kbps.

The simulations in random topology were repeated for constant a per flow rate of 10 kbps with varying packet size. The energy efficiency is presented in Figure 6.9. As expected, the energy efficiency increases with packet size. The MAC protocol introduces constant bit OH for every packet transmitted; thus, for the same rate, the OH decreases as the packet size increases. In consequence, a higher channel capacity is utilized for transmission of bigger user data packets. Additionally, the DPC protocol outperforms the 802.11 scheme for all packet sizes, which conforms the previous results that are displayed in Figure 6.10.

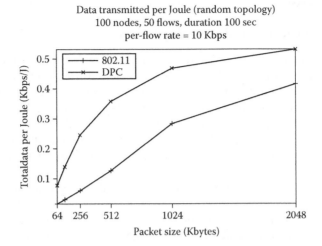

FIGURE 6.9
Data transmitted per joule with packet size.

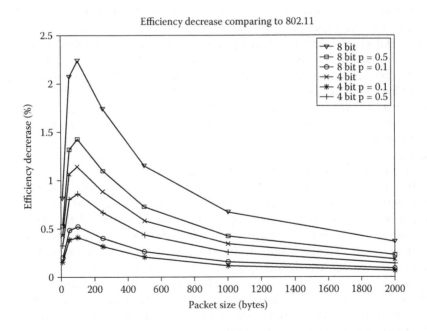

FIGURE 6.10
Protocol efficiency.

6.7 Background on Rate Adaptation

Resource constraints as pointed out in Chapter 1 require that ad hoc wireless and sensor networks are energy efficient during transmission and rate adaptation. In this chapter, we present two novel energy-efficient rate adaptation schemes from Zawodniok and Jagannathan (2005) to select modulation schemes online to maximize throughput based on channel state while saving energy. These protocols use the DPC algorithm from previous sections to predict the channel state and determine the necessary transmission power that optimizes the energy consumption. The first proposed rate adaptation scheme heuristically alters the transmission rate using energy efficiency as a constraint to meet the required throughput, which is estimated with the queue fill ratio. Moreover, the backoff scheme is incorporated to mitigate congestion and reduce packet losses due to buffer overflows, thus minimizing corresponding energy consumption. The backoff scheme implemented recursively becomes a back-pressure signal. Consequently, the nodes will conserve energy when the traffic is low, offer higher throughput when needed, and save energy during congestion by limiting transmission rates.

The second rate adaptation scheme uses the burst mode described in the 802.11 standard to provide a flow control mechanism. The dynamic programming (DP) principle is employed to provide an analytical method to select the modulation rate and a burst size to be transmitted over the radio link. The proposed quadratic cost function minimizes the energy consumption. Additionally, buffer occupancy is included in the cost function for the purpose of congestion control. The proposed DP solution renders a Riccati equation ultimately providing an optimal rate selection. The simulation results, shown later in this chapter, indicate that an increase in throughput by 96% and energy-efficiency by 131% is observed when compared to the receiver-based auto rate (RBAR) protocol (Holland et al. 2001).

Because of the need for higher throughputs in the next generation wireless networks, modulation schemes that render higher data rates have been introduced; for example 54 Mbps capability of the 802.11g standard. However, the communication range decreases as the transmission rate increases. Hence, connectivity is reduced for modulation schemes that provide higher throughput. A simple remedy is to increase the transmission power. However, a node's energy is drained quickly and the energy-efficiency of transmission, which is measured as the number of bits transmitted per joule, decreases with rate reducing the overall lifetime of the nodes and the network.

To address the rate adaptation in wireless networks based on the 802.11 standard, several schemes were proposed in the literature (Holland et al. 2001, Kamerman and Monteban 1997). However, these protocols focus mainly on maximizing the throughput, regardless of transmission power, channel state, and network congestion. For instance, the auto rate fallback (ARF) (Kamerman and Monteban 1997) protocol incrementally alters the transmission rate after experiencing a number of consecutive correct or erroneous packet receptions. As a result, ARF slowly converges to the more appropriate rate dropping significant number of packets because of low signal-to-noise ratio (SNR) or SIR. In some cases, transmission rate is much lower than acceptable reducing the throughput. By contrast, the RBAR protocol proposed in (Holland et al. 2001) uses the predefined lower and upper SNR thresholds to select an appropriate modulation scheme and hence the rate. By using the measured SNR value from the previous MAC frame, a more suitable modulation scheme is selected using SNR thresholds. However, the channel measurements used to select a given rate are from the previous transmission, and they do not accurately describe the channel state for the subsequent transmission. A common problem that is found in both ARF and RBAR, and with many available protocols, is the transmission of data at the maximum power reducing energy efficiency. Additionally, the protocols (Holland et al. 2001, Kamerman and Monteban 1997) ignore the effect of congestion on throughput and energy efficiency.

A more appropriate energy-efficient rate adaptation would be to use several modulation schemes and dynamically selecting a suitable one online, based on the channel state and the network traffic. The basic concept was analyzed in Schurgers et al. (2001) and proved to be effective. Relevant parameters for the selection of the modulation scheme include bit-error-rate (BER) and SNR. The former indicates the probability of error occurrence during transmission for a given SNR. The latter defines the received signal quality. Hence, for a given modulation scheme, a threshold SNR can be calculated according to a desired BER level. In general, for small target BER (low level of errors) a high SNR is typically required.

The heuristic rate adaptation protocol from (Zawodniok and Jagannathan 2005) uses the DPC scheme (Zawodniok and Jagannathan 2004) to predict the channel state and to meet the target SNR. Thus, this scheme can select a more appropriate rate when compared to available protocols (Holland et al. 2001, Kamerman and Monteban 1997). In addition, DPC in rate adaptation reduces energy consumption by selecting a minimal power needed for successful transmission. Additionally, this scheme selects a wide range of suitable rates by taking into account the demanded throughput and the energy efficiency to accommodate the network congestion and to conserve energy through the selection of a backoff interval.

The protocol from Zawodniok and Jagannathan (2005) uses the backoff mechanism to mitigate the impact of congestion. As a result, packet forwarding is prevented either if the next-hop node cannot buffer them or the channel state is not conducive. Under such circumstances, the maximum possible rate will be selected in the case of high congestion to free the channel quicker. In short, the proposed protocol maximizes throughput and saves energy by both modifying the transmission power and selecting an appropriate rate while minimizing the packet losses in the case of congestion. Thus, the inclusion of DPC in rate adaptation is extremely important to assess the channel state to select a suitable rate and transmission power.

The second scheme from Zawodniok and Jagannathan (2006) is based on dynamic programming (DP) principle (Bertsekas 1987, Angel and Bellman 1972, White 1969, Bambos and Kandukuri 2000), which utilizes a burst mode transmission to control incoming flow rate by varying admissible burst size. This method of flow control is more precise than a backoff scheme used in the heuristic protocol, because it can specify an exact amount of data that will be transmitted to the receiver. In consequence, queue utilization can be maintained close to the target value. Moreover, the usage of burst mode transmission increases overall network efficiency, thus providing higher data rates for the end users. The IEEE 802.11 burst mode transmits a number of data packets within a single RTS/CTS/DATA/ACK exchange thereby reducing the number of transmitted RTS/CTS frames and minimizing an associated bit overhead. In the IEEE 802.11, a burst size is limited by the modulation rate because of the way the time of a single RTS/CTS/DATA/ACK exchange is selected.

The second rate adaptation scheme is an improvement of the power controlled multiple access (PCMA) protocol (Bambos and Kandukuri 2000). The proposed algorithm minimizes energy consumption while providing required the QoS. This scheme, in contrast to the heuristic one, can provide the desired level of service by selecting the appropriate target queue utilization. Maximizing throughput normally reduces the lifetime of the node, overall network, and throughput while increasing congestion levels. Therefore, a suitable outgoing rate has to be identified so that an optimal trade-off results between the increasing modulation rate to maximize throughput and the decreasing modulation rate to maximize energy-efficiency. A quadratic cost function is introduced to obtain this trade-off between throughput and energy-efficiency. The analytical solution, which is derived using a dynamic programming (Bertsekas 1987) approach, provides a performance guarantee, unlike the simplified heuristic scheme, where the performance of the rate adaptation scheme is not guaranteed. The DP-based scheme eliminates packet losses due to buffer overflows, which will result in energy efficiency due to reduced retransmissions.

6.7.1 Rate Adaptation

A given application requires that a BER is maintained below a certain level. For wireless networks, the BER translates into a minimum SNR for which the packet is considered to be successfully decoded. The relation between

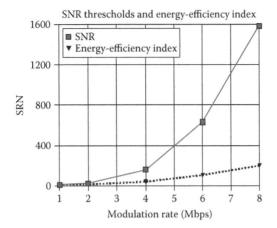

FIGURE 6.11
SNR thresholds and energy-efficiency index.

the BER and the SNR threshold is well defined for a given modulation scheme (Holland et al. 2001). In general, a SNR threshold increases with the rate as illustrated in Figure 6.11 for the set of modulation rates used in simulations (1, 2, 4, 6, and 8 Mbps). Consequently, the minimal required transmission power has to be increased with the rate. On the other hand, the power value is limited by the hardware capabilities. Thus, there is a maximum SNR value and corresponding modulation scheme that could be selected for a given power constraint. Given this maximum power constraint, the rate will result in highest throughput and higher energy efficiency as explained next. If the maximum power will yield SNR = 800, then the maximum possible rate that can be achieved is equal to 6 Mbps, as illustrated in Figure 6.11. It is important to note that channel state impacts the rate of transmission because the power will vary with interferences and signal attenuation. Hence, it is important for the rate adaptation scheme to accurately assess the channel state to identify a suitable rate that results in the highest throughput under the channel conditions.

Figure 6.11 depicts an energy-efficiency index that results in the lowest modulation rate. It shows the SNR levels for all rates assuming a constant energy-efficiency value. Comparing the energy-efficiency index and the actual SNR levels, it can be noticed that the energy-efficiency decreases with an increase in modulation rate.

6.7.2 Protocol Comparison

In the previous works (Holland et al. 2001, Kamerman and Monteban 1997), the problem of rate adaptation was attempted either by considering the past transmission history, as in ARF protocol, or assuming that the channel state does not change significantly between consecutive transmissions, as in the case of RBAR. Hence, the selected rate was often not optimal. Additionally, these protocols neither take energy-efficiency into consideration nor modify the transmission power to save energy during rate adaptation.

The operation of the ARF protocol (Kamerman and Monteban 1997) is presented in Figure 6.12. The four rates (R1, R2, R3, and R4) are considered with corresponding SNR thresholds. In this example, the first packet is sent with the maximum rate allowed for the channel state. The following packets are sent using the same rate, though the SNR could have increased thus lowering the throughput for the current channel state. After three consecutive, successfully received packets, the rate is then increased even though the channel state could have changed significantly during this time. The new rate (R2) used for the fourth packet could be still lower than the maximum possible rate. On the other hand, when the SNR decreases, the selected rate will always be higher than the acceptable throughput possible, resulting in problems of decoding of packets at the receiver.

In short, the problems observed in the ARF protocol are the result of the lack of information about the radio channel state, because, no measurements of signal reception are considered. In consequence, the throughput achieved by ARF is lower than possible for a given channel state. Furthermore, the energy is consumed inefficiently.

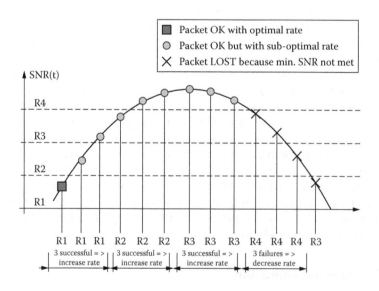

FIGURE 6.12
ARF rate selection.

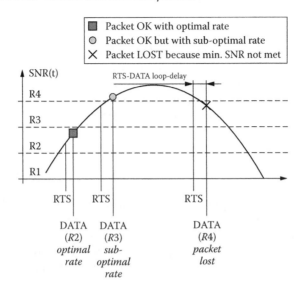

FIGURE 6.13
RBAR rate selection.

By contrast, the RBAR protocol (Holland et al. 2001) improves the throughput via suitable rate adaptation by measuring the SNR ratio, at the receiver, for the RTS MAC frame. This information is piggybacked in the CTS frame, which is sent to the transmitting node. Then a suitable rate is selected for the DATA frame transmission as shown in Figure 6.13. Because the RTS and DATA frames are sent one after the other, the rate selection is more accurate than in ARF. Additionally, a more appropriate rate is selected in RBAR, while in ARF, the rate changes one step at a time. Consequently, fewer packets are received incorrectly, and the achieved throughput is always higher than the ARF protocol. However, the rate adaptation is determined using outdated information that was measured one loop-delay prior to the DATA transmission. Hence, some packets will be transmitted with an inaccurate rate that is either too low or too high. Moreover, the packets are transmitted at the maximum power. As a result, both the throughput and the energy consumption are not optimal.

6.8 Heuristic Rate Adaptation

The proposed heuristic protocol uses the previously mentioned DPC algorithm (Zawodniok and Jagannathan 2004) to predict the channel state one step ahead, which is subsequently used to select the rate and the transmission power. This rate is adjusted according to the achieved energy-efficiency levels and local queue utilization. In consequence, a more accurate rate is selected while transmitting the DATA frame when compared to ARF or RBAR. Because of the DPC, minimal power required to transmit data packets under the current channel state is used, thus saving energy. Finally, to minimize buffer overflows during network congestion, a backoff interval is altered based on the buffer utilization of the receiver.

6.8.1 Overview

Figure 6.14 presents the dataflow for the proposed rate adaptation scheme, which is applied at a transmitting node. First, the SNR thresholds are pre-defined for all modulation schemes in accordance with the desired BER value, thus reducing the computation overhead. During communication, the DPC algorithm is used to continuously assess the channel state and to calculate the transmission power for the target SNR_0 of the lowest supported rate. In Subsection 6.8.2, the details of the channel state estimation is given.

Next, a set of possible rates is calculated. The upper limit of the set corresponds to the maximum rate that can be achieved due to the transmission

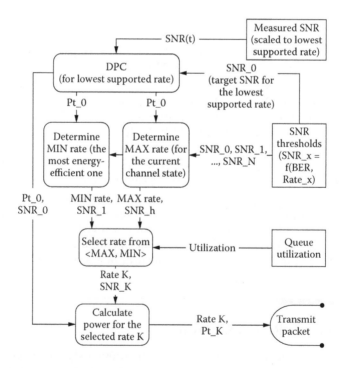

FIGURE 6.14
Energy-efficient rate adaptation data flow.

power limitations of the hardware. In Subsection 6.8.2, we present the details of how the scheme uses the channel-state estimator provided by the DPC algorithm. Next, the lower limit of the set is identified as the most energy-efficient rate. The underlying DPC implementation is taken into account, because the train of pulses used during transmission of the MAC frames introduces the energy overhead.

Subsequently, the selection of the most suitable modulation rate from the set is presented in Subsection 6.8.5. The buffer utilization at the transmitter dictates the selection of a particular modulation rate from the set of rates calculated in Subsections 6.8.6. Next, the transmission power is adjusted for the selected rate such that a corresponding target SNR is achieved. Here, the buffer utilization at the receiving node dictates a suitable value for the backoff interval. Finally, changes needed to the MAC layer as the result of the proposed protocol implementation are discussed in Subsection 6.8.8.

REMARK 2
In this scheme, we assume that all nodes are capable of handling the same set of N modulations. In the case of a heterogeneous network, the nodes will have to exchange the rate information by using the control packets transmitted at the lowest rate. To simplify the description of the problem, the modulation schemes are sorted by rate.

6.8.2 Channel-State Estimation

The proposed protocol uses a threshold method similar to Holland et al. (2001) to determine the maximum rate possible for the given SNR. The maximum SNR that corresponds to the maximum transmission power is also to be determined. In the proposed scheme, the DPC is used to predict this maximum SNR for transmitting the next frame.

THEOREM 6.8.1 (MAXIMUM SNR PREDICTION)
Given the maximum transmission power defined for the network, and information (Zawodniok and Jagannathan 2005) for the lowest supported rate through DPC, the maximum possible SNR for subsequent transmission can be predicted using

$$SNR_{MAX}(t) = SNR_0 * Pt_{MAX}/Pt_0(t) \qquad (6.21)$$

where SNR_0 is the target SNR for the lowest supported rate, $Pt_0(t)$ is the estimated power value for the lowest supported rate, and Pt_{MAX} is the maximum transmission power defined for the network.

PROOF First, let us consider the SNR ratio of a received signal expressed as

$$SNR(t) = Pr(t)/I(t) = Pt(t) \cdot G_i(t)/I(t) \qquad (6.22)$$

where Pr(t) is a received signal strength, Pt(t) is the transmission power, $G_i(t)$ is a gain (loss) experienced by the signal, and $I(t)$ is an interference plus a noise level.

It is important to note that the SNR depends upon transmitted power. Furthermore, we can calculate the ratio of two SNR values using the corresponding transmission power values. This ratio is equal to

$$\frac{SNR_l(t)}{SNR_k(t)} = \frac{Pt_l(t)*Gi(t)/I(t)}{Pt_k(t)*Gi(t)/I(t)} = \frac{Pt_l(t)}{Pt_k(t)} \tag{6.23}$$

where $SNR_l(t)$ and $SNR_k(t)$ are the measured SNRs, and $Pt_1(t)$ and $Pt_k(t)$ are the corresponding transmission powers used in each case, respectively.

Let the value of $Pt_0(t)$ needed to meet the target SNR_0 along with the maximum power Pt_{MAX} be known. The maximum SNR, when the maximum power is used, can be calculated from Equation 6.23 as

$$SNR_0 / SNR_{MAX}(t) = Pt_0(t) / Pt_{MAX} \tag{6.24}$$

Equation 6.24 can be rewritten as Equation 6.21.

6.8.3   Maximum Usable Rate

Next, the maximum achievable rate is selected using a thresholding method similar to the case of RBAR protocol (Holland et al. 2001). However, in the proposed scheme, the current SNR value is an estimated value of the SNR for the next transmission, while in the case of RBAR, the SNR is an outdated value measured from the reception of the previous frame (i.e., a half loop-delay before the DATA frame). Hence, the proposed protocol will be able to accurately select the rate taking into account the channel state. The highest rate, m, is selected from

$$SNR_m/SNR_{MAX}(t) < SNR_{m+1} \tag{6.25}$$

where $SNR_{MAX}(t)$ is the estimated SNR ratio if the maximum power is used for transmission, and SNR_m is the lower threshold for the mth rate. All modulation schemes that render rates up to the mth rate are considered usable because, the transmission power for these rates, is below the maximum network threshold.

6.8.4   Minimum Usable Rate

The most energy-efficient rate will become the optimal rate that can be used. Under the ideal circumstances, energy efficiency decreases with an increase in rate. Hence, the lowest rate should be the most energy-efficient. However, in the proposed protocol, the train of pulses in DPC (Zawodniok and Jagannathan 2004) used to overcome the hidden-terminal problem introduces an additional energy OH and becomes an important factor to be considered for rate adaptation. Hence, the lowest rate is not always energy efficient. Consequently, the most energy-efficient rate is found by comparing the energy consumed at each rate for a given packet under the current channel state. The rate with lowest energy consumption that takes into account the channel state and energy overhead due to the hidden-terminal problem will be selected as the minimum usable rate as discussed next.

Normally, the energy consumed for transmission is equal to

$$E = D \cdot Pt \tag{6.26}$$

where E is the energy consumed, D is the duration of transmission, and Pt is the current transmission power. For the case of the DPC algorithm with the train of pulses illustrated in Figure 6.14, the energy consumed during transmission is equal to

$$E_{TOTAL} = E_{DPC} + E_{PULSES} \tag{6.27}$$

where E_{TOTAL} is the total energy consumed on transmission of the packet with pulses, E_{DPC} represents the energy consumed for the transmission of the packet using the power selected by the DPC algorithm, and

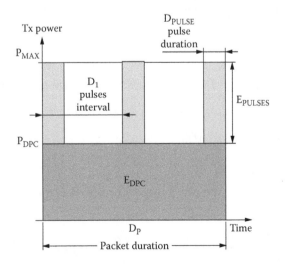

FIGURE 6.15
Energy consumption during packet transmission.

E_{PULSES} denotes the additional energy consumed for the generation of pulses. Therefore, E_{DPC} and E_{PULSES} can be expressed using Equation 6.26 as

$$E_{DPC} = Pt_{DPC} * Pkt_{size}/R \tag{6.28}$$

and

$$E_{PULSES} = (Pkt_{size}/R) * (D_{PULSE}/D_I) * (Pt_{MAX} - Pt_{DPC}) \tag{6.29}$$

After substituting Equation 6.28 and Equation 6.29 into Equation 6.27 we get,

$$E_{TOTAL} = (Pkt_{size}/R) * Pt_{DPC} + (Pkt_{size}/R) * (D_{PULSE}/D_I) * (Pt_{MAX} - Pt_{DPC}) \tag{6.30}$$

It is important to note that for the given packet size and rate, the total energy consumed varies linearly with transmission power. Also, the DPC scheme increases the transmission power for a given rate proportional to the target SNR as per Equation 6.23. Thus for a rate k, Equation 6.30 is expressed as

$$E_{TOTAL,k} = (Pkt_{size}/R_k)[(D_{PULSE}/D_I) * Pt_{MAX} + (Pkt_{size}/R_k)$$
$$\times (SNR_k/SNR_0) * (1 - (D_{PULSE}/D_I))Pt_0(t)] \tag{6.31}$$

The upper and lower thresholds of the DPC power (calculated for the lowest supported rate) need to be found so that the most energy-efficient rate can be selected. By equating Equation 6.31 for the two successive rates, the threshold values for the rate adaptation are determined as

$$E_{TOTAL,k} = E_{TOTAL,k+1}$$

$$(Pkt_{size}/R_k)[(D_{PULSE}/D_I) * Pt_{MAX} + (SNR_k/SNR_0) * (1 - (D_{PULSE}/D_I))Pt_0(t)]$$

$$= (Pkt_{size}/R_{k+1})[(D_{PULSE}/D_I) * Pt_{MAX} + (SNR_{k+1}/SNR_0)$$

$$* (1 - (D_{PULSE}/D_I))Pt_0(t)]$$

$$\tag{6.32}$$

Using Equation 6.31, the rate's upper threshold using power as a constraint is given by

$$Pt_{O,k} = \frac{\varphi}{1-\varphi} * \frac{1-\alpha_k}{\gamma_k - \alpha_k * \gamma_{k+1}} * Pt_{MAX} \tag{6.33}$$

where $\alpha_k = R_k/R_{k+1}$, $\gamma_k = SNR_k/SNR_0$ and $\varphi = D_{PULSES}/D_I$.

TABLE 6.1

Power Thresholds for Selecting the Most
Energy-Efficient Rate

SNR [dB]	SNR	Rate [Mbps]	Upper Thresholds of Tx Power [mW]
10	10.00	1	> 20.09
14	25.12	2	20.09
22	158.49	4	10.64
28	630.96	6	1.49
32	1584.89	8	0.30

REMARK 3

The power threshold given by Equation 6.33 depends upon DPC calculations for the lowest supported rate. The power thresholds can be calculated in advance, reducing online calculation overhead of the protocol. Table 6.1 presents the thresholds calculated for a set of five rates used in the simulations. For example, if the transmission power calculated by the DPC algorithm is equal to 15 mW, 2 Mbps is the most energy efficient rate, because 15 mW belongs to a set defined between 20.09 mW (2 Mb) and 10.64 mW (4 Mbps).

6.8.5   Modulation Rate to Overcome Congestion

In the previous subsections, the maximum throughput and energy-efficient rates were calculated. The throughput increases with rate; however, the lower the rate used, larger the number of bits that can be transmitted per joule. Therefore, the rate adaptation has to generate a trade-off between the throughput and energy-efficiency. Hence, the selection of a modulation scheme that results in the best throughput while satisfying the energy constraint is discussed next. Moreover, in the proposed protocol, the buffer utilization at the transmitter is used as an indication of the required throughput, and will impact the rate selection method.

Because a high transmission rate implies high buffer occupancy at the receiving node, provided the channel is conducive, the buffer occupancy at each node should be taken into consideration during rate adaptation. The basic idea is presented in Figure 6.16. Here the lowest supported rate is selected when the queue at the transmitting node has low utilization. Increased queue occupancy at this node indicates a higher traffic demand. Hence, the modulation rate that results in higher throughput is selected.

This rate adaptation based on buffer occupancy can be explained as follows. The lowest supported rate will be chosen if the congestion is low. As the congestion in the network increases, the queue occupancy will increase. Consequently, a higher rate will be selected to clear the congestion, rendering a

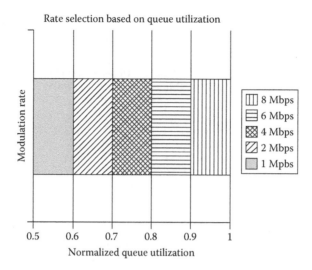

FIGURE 6.16
Rate selection as a function of queue utilization.

higher throughput in contrary to the belief that increased congestion levels should be controlled by decreasing transmission. As the congestion decreases, the lower rate will be selected resulting in higher energy efficiency.

6.8.6 Power Selection for the Rate

The DPC algorithm calculates the transmission power only for the lowest supported rate. In the case of proposed rate adaptation scheme, the necessary power has to be selected to reflect the used rate. Hence, an appropriate transmission power for a given rate is calculated as follows.

THEOREM 6.8.2 (POWER CONTROL WITH RATE ADAPTATION)
Given the lower and upper threshold values of SNR for a given rate, and the information for the lowest supported rate through DPC, the transmission power for the proposed energy-efficient rate adaptation scheme is given by

$$Pt_k(t) = Pt_0(t) * SNR_k / SNR_0 \tag{6.34}$$

where SNR_0 is the target SNR for the lowest supported rate, SNR_k is the target SNR for the kth modulation scheme, $Pt_0(t)$ is the estimated power value for the lowest supported rate, and $Pt_k(t)$ is the power value for the kth modulation scheme. Moreover, the error in estimation of power is bounded.

PROOF Recall the ratio of SNR values depending on the transmission power selected is expressed from Equation 6.22 as

$$SNR_l(t)/SNR_k(t) = Pt_l(t)/Pt_k(t) \tag{6.35}$$

where $SNR_l(t)$ and $SNR_k(t)$ are the measured SNR values, and $Pt_l(t)$ and $Pt_k(t)$ are the transmission powers used in each case, respectively. Assume that value of $Pt_0(t)$ is known, which is needed to meet the target SNR_0. Let the target SNR_k for a different rate be provided beforehand. The transmission power $Pt_k(t)$ needed to meet the target SNR_k can be obtained from Equation 6.35 as

$$SNR_0/SNR_k = Pt_0(t)/Pt_k(t) \tag{6.36}$$

Equation 6.36, after rewriting becomes Equation 6.34. Error in estimation for the kth rate is bounded provided the power calculated by DPC is bounded. Because the power for the lowest supported rate is estimated using the DPC algorithm, the power for the kth rate is also an estimate. Assume that ε_0 is the maximum error in estimation of power by the DPC algorithm. Applying this to Equation 6.34, the maximum error in estimation for the kth rate ε_k is given as

$$SNR_0/SNR_k = Pt_0(t)/Pt_k(t) \tag{6.37}$$

It is important to note that the error in estimation for the kth rate is proportional to the error in estimation of the basic rate and the ratio of the two SNR values. Consequently, the proposed DPC protocol can be used to calculate transmission power although noting that the error in estimation is bounded.

6.8.7 Backoff Mechanism

The desired backoff mechanism should dictate the backoff interval. Higher queue utilization in the next-hop node indicates that a longer backoff interval is needed for all the nodes in the region so that the next-hop node will access the channel more often. A quadratic relationship for backing off is considered in this algorithm, with the intention that for high buffer occupancy the backoff interval will be longer preventing buffer overflowing, and that for low buffer occupancy the delay caused by backoff will be low, thus not undermining throughput. The backoff interval is calculated using

$$BI = \rho \cdot SF \cdot \left(1 + \alpha \cdot Q_{NEXT\text{-}HOP}(t)^2\right) \tag{6.38}$$

where BI is the backoff interval, ρ is a random number, SF and α are the scaling factors, $Q_{NEXT\text{-}HOP}{}^{(t)}$ and is the utilization of queues at the next-hop node.

6.8.8 MAC Protocol Design

The DPC protocol presented in (Zawodniok and Jagannathan 2004) between a transmitting and receiving node is modified to accommodate the rate adaptation. The DPC information is added to the MAC frames similar to DPC for ad hoc wireless networks. To implement the rate adaptation, it must be incorporated with the MAC protocol along with the necessary transformation required for the power and SNR values shown in Equation 6.34 and Equation 6.39. Additionally, the MAC frames have to be modified to include the queue utilization for the proposed backoff mechanism.

The RTS, CTS, and ACK frames are transmitted using the lowest supported rate whereas the DATA frames are transmitted at the rate selected by the algorithm. The 802.11 frame header contains a standard field, which indicates the rate used for payload transmission. We use this field to indicate the rate of transmission to the receiver so that it can correctly decode the payload.

The DPC requires that the input value of the SNR corresponds to the transmission power calculated by the DPC algorithm. However, in the proposed scheme, the power value is modified according to Equation 6.34. Hence, the received SNR will be invalid for the DPC algorithm. In such a case, the measured $SNR_k(t)$ needs to be scaled to correspond to the lowest supported rate. Multiplying both sides, of Equation 6.36 by $G_i(t)/I(t)$ we get,

$$SIR_0(t) = SIR_k(t) * SNR_0/SNR_k \qquad (6.39)$$

where SNR_0 and SNR_k are target values for the lowest supported rate and kth rate, respectively, $SNR_k(t)$ is the measured SNR ratio, and $SNR_0(t)$ is the SNR value that would be measured if the lowest supported rate (and adequate power) was used. Now, the $SNR_0(t)$ value can be used in the DPC algorithm to calculate the transmission power for the lowest supported rate.

6.9 Dynamic-Programming-Based Rate Adaptation

The second proposed rate adaptation scheme based on dynamic programming minimizes energy consumption, while providing required level of service. Recall that this scheme, in contrast to the heuristic one, is able to provide the desired level of service by selecting an appropriate target value of queue utilization. The proposed DP solution will utilize a burst mode transmission to control an incoming flow rate by varying an admissible burst size. Though the IEEE 802.11 burst mode transmits a number of data packets within a single RTS/CTS/DATA/ACK exchange, it limits the time a single RTS/CTS/DATA/ACK exchange can last. Thus, the maximum size of a burst is limited by the selected modulation rate.

The proposed scheme will be recursively applied at every link that connects transmitting and receiving nodes. The rate selection is performed at the receiving node and then transmitted to the receiver, where rate adaptation is performed and a suitable burst size can be selected and applied. The proposed DP-based scheme uses the DPC algorithm (Zawodniok and Jagannathan 2004) to calculate the transmission power required for the lowest supported rate. Similar to DPC, the receiving node performs the necessary calculation as to what modulation rate and what size of a burst should to be used by the transmitting node.

The rate adaptation algorithm uses the state equation for the buffer dynamics at the receiving node. To calculate optimal policy, a cost function is proposed that includes a cost of queuing packets and a cost of transmitting a given burst of data. The latter is equal to an energy required for a transmission of the burst using a selected modulation rate. The solution dictated by the proposed DP policy is subject to subsequent adjustments, for instance, the selected modulation rate may be reduced because of the maximum transmission power supported by the node.

The section is organized as follows. First, the state equation is presented. Next, the cost function is introduced and discussed. Subsequently, a dynamic programming solution that uses the Riccati equation is presented. Then, supplementary alterations of rate and burst size are discussed. Finally, the implementation essentials are given.

6.9.1 Buffer Occupancy State Equation

Consider the queue utilization equation given as

$$q_i(k+1) = q_i(k) + u_i(k) + w_i(k) \tag{6.40}$$

where $k = 0, 1, 2, \ldots, N-1$, N is a time instance, with N being the last step of the DP algorithm, $q_i(k)$ is a queue utilization at node i and at time instance k; $w_i(k)$, is the outgoing traffic at time k, and $u_i(k)$ is the incoming traffic at time k. The term $w_i(k)$ is dictated by the next-hop node $(i + 1)$ and is considered to be a random variable with a known distribution and expected value. The proposed scheme controls incoming traffic to minimize a cost function presented next.

6.9.2 Cost Function

The proposed cost function includes both the cost of queue occupancy and the cost of transmitting a given burst of data. The latter is then approximated with a quadratic form to obtain an optimal control law, which can be derived using the standard Riccati equation. Moreover, this solution is proven to be stable and will converge.

6.9.2.1 Queue Occupancy Cost

Consider a known, ideal queue utilization, q_{ideal}, which renders desired system performance in terms of throughput and delay. The buffer cost in such a case can be selected as a quadratic function of error between ideal and actual queue utilization:

$$B(q_i(k)) = \gamma \cdot (q_i(k) - q_{ideal})^2 \qquad (6.41)$$

where γ is a scaling factor and $q_i(k)$ is the current queue utilization. In such a case, the cost of queue utilization is the lowest for $q_i(k) = q_{ideal}$, and increases when the utilization is lower or higher than desired.

To get a quadratic cost function as in the Riccatti equation, the state variable is substituted by transiting queue utilization by a constant q_{ideal}

$$x_i(k) = q_i(k) - q_{ideal} \qquad (6.42)$$

Now, the state equation can be rewritten as

$$x_i(k+1) = x_i(k) + u_i(k) + w_i(k) \qquad (6.43)$$

and the cost function is expressed as

$$B(x_i(k)) = \gamma \cdot (x_i(k))^2 \qquad (6.44)$$

The selection of the parameter gamma will influence the convergence of the queue utilization dynamics to a target value. The bigger the gamma parameter, the more effort (cost) will be spent on converging to the target queue utilization and corresponding performance level. However, this will render higher cost for transmissions, thus consuming more energy than in the case of a lower gamma.

6.9.2.2 Transmission Cost

The wireless nodes consume energy to transmit packets. This energy depends on transmission power needed for successful transmission and duration of transmission. The selection of the modulation rate adds complexity to this term, because duration of transmission will vary with modulation rate, and the required power changes nonlinearly with change of modulation rate. The exact cost is expressed as:

$$C_{TX}(u_i(k), r_i(k)) = P_0(k) \frac{SNR(r_i(k))}{SNR(0)} \cdot \frac{u_i(k)}{R(r_i(k))} \qquad (6.45)$$

where $u_i(k)$ is a burst size, $r_i(k)$ is a modulation, $R(r)$ is a transmission rate for modulation r, $SNR(r)$ and $SNR(0)$ are SNRs for modulation r, and 0 (the lowest, available modulation), respectively, $P_0(k)$ is the power calculated for the modulation 0 (zero), and k is a time instant.

REMARK 4

The radio access standards have a limitation on how long a channel can be used by a given node. Hence, the burst size is limited by the modulation rate used as $u_i(k)/R(r_i(k)) \leq \max\_tx\_duration$.

REMARK 5

The DP is minimizing cost functions over the u_i and r_i. Thus the optimal value of r_i for a given u_i can be inferred *a priori*, as a lowest rate that supports a given burst size. Hence, the transmission cost can be expressed as function of only u_i.

$$C_{TX}^*(u_i(k)) = u_i(k) \cdot P_0(k) \cdot CM(u_i(k)) \qquad (6.46)$$

where $CM(u_i(k)) = SNR(r_i^*(k))/[SNR(0) \cdot R(r_i^*(k))]$ is the cost of transmitting a burst $u_i(k)$ using optimal modulation $r_i^*(k)$ for the burst size.

6.9.2.3 Approximating Cost Function

Overall, the cost function for the dynamic programming optimization problem is expressed as

$$J_k(x_i(k)) = B(x_i(k)) + C_{TX}^*(u_i(k)) + J_{k+1}(x_i(k+1)) \qquad (6.47)$$

where $J_k(x_i(k))$ is a cost function from time k to N (last step of algorithm) with initial state $x_i(k)$, $B(x_i(k))$ is the cost function of queuing, $C_{TX}^*(x_i(k))$ is a function of transmitting burst of size $u_i(k)$, and $J_{k+1}(x_i(k+1))$ is a to-go cost from time $k+1$.

REMARK 6

The cost function as in Equation 6.38, when used in DP will yield an optimal control law. However, the calculation of such a law is computationally intensive and highly sensitive to a number of possible values of $u_i(k)$.

Instead, an approximated quadratic function is proposed in form of

$$CQ(u_i(k)) = \alpha \cdot P_0(k) \cdot [u_i(k)]^2 \qquad (6.48)$$

where parameter a is selected such that the least-square error for approximation is minimized.

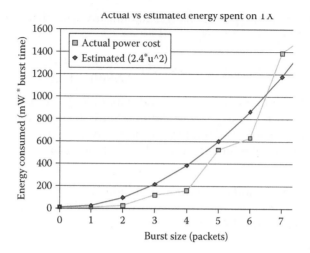

FIGURE 6.17
Rate selection as a function of queue utilization.

REMARK 7

Because the cost of transmission $CQ(u_i(k))$ is not exact, the calculated control law will be suboptimal.

Figure 6.17 presents an example for the set of modulations presented in Table 6.1. In this case, α is calculated to be equal to 2.4.

Now, the final cost function is expressed as

$$J_k(x_i(k)) = Q_k(x_i(k))^2 + R_k(u_i(k))^2 + J_{k+1}(x_i(k+1)) \qquad (6.49)$$

where $Q_k = \gamma$ and $R_k = \alpha P_0(k)$ are parameters from Equation 6.41 and Equation 6.45. The cost function in Equation 6.49 is in a quadratic form; hence, we can calculate an optimal control law using the standard Riccatti equation (Bertsekas 1987) because of the linear nature of the buffer dynamics. Next, the proposed solution is presented.

6.9.3 The Riccatti Equation

First, we notice that the rate adaptation problem should match closely the state of the system, x_i, to the outgoing flow, w_i, rather than the queue utilization, because keeping an adequate flow of the data is more important than keeping the queue at a certain level. Hence, we consider a new state variable that is equal to the sum of state x_i and outgoing flow w_i (negative value).

$$z_i(k) = x_i(k) + w_i(k) \qquad (6.50)$$

Now substituting a new state of Equation 6.41 in Equation 6.49, and including it in the last Nth iteration, a corresponding outgoing traffic component $^w n$, we get

$$J_k(z_i(k)) = Q_k(z_i(k))^2 + R_k(u_i(k))^2 + J_{k+1}(z_i(k+1)) \tag{6.51}$$

Applying the DP approach to Equation 6.42 we have

$$J_N(z(N)) = Q_N(z_i(N))^2$$
$$J_k(z(k)) = \min_{u_k} E\{Q_k(z_i(k))^2 + R_k(u_i(k))^2 + J_{k+1}(z_i(k) + u_i(k))\} \tag{6.52}$$

First, we expand the one before the last iteration

$$J_{N-1}(z(N-1)) = \min_{u_{N-1}} E\{Q_{N-1}(z_i(N-1))^2 + R_{N-1}(u_i(N-1))^2$$
$$+ Q_N(z_i(N-1) + u_i(N-1))^2\}$$
$$= Q_{N-1}(z_i(N-1))^2 + \min_{u_{N-1}} E\{R_{N-1}(u_i(N-1))^2 + Q_N(z_i(N-1))^2$$
$$+ Q_N z_i(N-1) u_i(N-1) + Q_N(u_i(N-1))^2\} \tag{6.53}$$

The minimization of Equation 6.44 with respect to $u_i(N-1)$ is performed by differentiating Equation 6.44 and equating it to zero which yields

$$u_i^*(N-1) = -z_i(N-1) \cdot Q_N / (Q_N + R_{N-1}) \tag{6.54}$$

By substituting $u_i(N-1)$ in Equation (6.44) with Equation 6.45 we get,

$$J_{N-1}(z(N-1)) = Q_{N-1}(z_i(N-1))^2 + R_{N-1}(z_i(N-1))^2(-Q_N/[Q_N + R_{N-1}])^2$$
$$+ Q_N(z_i(N-1))^2(1 - Q_N/[Q_N + R_{N-1}])^2 = G_{N-1}(z_i(N-1))^2 \tag{6.5}$$

where

$$G_{N-1} = Q_{N-1} + Q_N \left(R_{N-1}^2 - R_{N-1}Q_N\right) / (R_{N-1} + Q_N)^2$$
$$= Q_{N-1} + Q_N[1 - Q_N/(Q_N - R_{N-1})] \tag{6.56}$$

Following the preceding calculations we can calculate optimal input for $k = N - 2, N - 3, ..., 0$. In such a case, an optimal law for every k is equal to

$$u_i^*(k) = -z_i(k) \cdot G_{N+1} / (G_{k+1} + R_k) \tag{6.57}$$

where

$$G_n = Q_N$$

$$G_k = G_{k+1}\left(1 - \frac{G_{k+1}}{G_{k+1} + G_k}\right) + Q_k \tag{6.58}$$

However, because the transmission duration is unknown, it is desirable to calculate a steady-state solution by assuming an infinite flow. The steady-state solution is more useful in implementation, because most of the calculations can be performed offline before network deployment and only limited calculations have to be done online. In such a case, Equation 6.58 becomes

$$u_i^*(k) = -z_i(k) \cdot G / (G + R_k) \tag{6.59}$$

where G is stable state solution ($k \to \infty$) of Equation 6.49, $R_k = \alpha \cdot P_0(k)$ is a parameter of a cost function with $P_0(k)$ being the transmission power value calculated by the DPC for the next transmission.

By substituting Equation 6.33 and Equation 6.41 in Equation 6.59 we can calculate the stationary law that depends directly on the queue utilization as

$$u^*(k) = -G(q_i(k) - q_{ideal} + E\{w_i(k)\})/(G + R_k) \tag{6.60}$$

This control law is applied before each transmission to calculate a desired burst size u^*. Next, the optimal modulation rate is selected as in Subsection 6.9.2.2, as the lowest rate that can support a given burst size.

6.9.4 Additional Conditions for Modulation Selection

Transmission power is physically limited by the node's hardware. Hence, the rate adaptation has to exclude modulation rates that require transmission power higher than the maximum available value. This threshold is calculated according to Equation 6.24. The maximum power will dictate the maximum modulation rate that can be applied for a given channel state. If the modulation rate is reduced, then the burst size will be reduced as well.

In the case of high congestion, the radio channel experiences demand for access from numerous sources. Thus, it is necessary to increase the modulation rate for completing the transmission quicker. This will clear the channel sooner allowing other nodes to transmit data reducing congestion. Notably, this subsequent increase of modulation rate should not cause increase in burst size because it will defy the purpose of releasing the channel quicker. Moreover, the receiving node has limited amount of data it can accept because of buffer constraints and, thus, increasing burst size will cause uncontrolled increase in queue size at the receiver.

6.9.5 Implementation Consideration

In the DP-based scheme, the nodes use the standard backoff interval, and thus they will gain access to the channel in an orderly manner. The rate selection is performed at each RTS-CTS exchange. Next, the selected modulation rate and burst size are communicated from the receiving node using a CTS frame with additional fields for data. Then, the transmitting node can additionally modify the modulation rate and, if necessary, burst size according to conditions described in a subsection.

6.10 Simulation Results

The NS-2 simulator was used for evaluating the proposed rate adaptation protocols. The one-hop, two-hop, and random topologies are used to evaluate the proposed schemes. However, because of the analytical results obtained for the proposed technique, any network topology can be utilized, and similar results can be seen. The one-hop topology is used to evaluate performance of the protocols in the presence channel fading. The two-hop topology sets up two flows between two pairs of source–destination nodes, where a common relay node is forwarding data for both flows. The relay node becomes a bottleneck for communication and presents an excellent benchmarking case for rate adaptation. The last topology uses 50 nodes randomly located in the area of 1000×1000 m with 25 flows set up through the network. The proposed heuristic protocol is compared with the RBAR (Holland et al. 2001).

In case of the DP-based scheme, usage of the burst-mode transmissions results in increased data throughput because an overhead of backing off and transmitting RTS/CTS is reduced. Consequently, a comparison between protocols with and without burst mode support will be difficult.

Wireless Ad Hoc and Sensor Networks

For example, if the RBAR without the burst mode is compared with the proposed DP-based protocol which uses burst mode, then it will be difficult to identify whether the changes in throughput or energy-efficiency are a result of the algorithms themselves or due to the introduction of the burst mode. For that reason, the authors made a straightforward modification of the RBAR protocol to enable the burst mode. Therefore, the modified RBAR protocol is used when comparing the performance of the proposed DP-based protocol.

The standard AODV routing protocol was employed for testing the proposed work whereas any routing protocol can be employed with the rate adaptation scheme because the proposed work is independent of a routing protocol. The following values were used for all the simulations, unless otherwise specified: 5 modulation schemes were used with data rates equal to 1, 2, 4, 6, and 8 Mbps, and target SNR values of 10, 14, 22, 28, 34 dB, respectively. The maximum power used for transmission was selected as 0.2818 W. For the proposed DPC, the design parameters were selected as $K_v = 0.01$ and $\sigma = 0.01$. The safety margin in the case of retransmissions was set to 1.5 for the proposed DPC scheme. The fading channel with path loss, shadowing, and Rayleigh fading from Zawodniok and Jagannathan (2004) was used in the simulations.

6.10.1   One-Hop Topology

Table 6.2 presents the average throughput achieved by the protocols with flow rates of 0.5, 2, and 4 Mbps. Table 6.3 presents the energy efficiency of both protocols for different rates. Both protocols can transmit at similar throughput. However, the efficiency of the DP-based protocol outperforms the RBAR, such that the proposed protocol can transmit 3.5 times more data for the same energy consumed, because of the addition of DPC in rate adaptation.

6.10.2   Two-Hop Topology

Figure 6.18 illustrates the total throughput achieved for this network with varying per flow rates. The proposed protocol outperforms the

TABLE 6.2

Throughput (kbps)

Protocol Type	0.5 Mbps	1.2 Mbps	4 Mbps
RBAR	499	1083	1424
DP	499	1082	1434

TABLE 6.3

Energy Efficiency (kb/J)

Protocol	0.5 Mbps	1.2 Mbps	4 Mbps
RBAR	196.73	217.77	233.65
DP	670.58	780.05	803.58

RBAR for all the traffic rates because it intelligently selects the rate based on channel state and congestion. Here the higher throughput observed for the proposed protocol is the result of rate adaptation and not due to DPC which clearly shows that the channel state has to be considered during rate selection. Moreover, when observing energy-efficiency in Figure 6.19, the proposed protocol can transmit up to three times more data for the same amount of energy consumed, because of the inclusion of DPC.

Figure 6.20 displays the drop rate at the intermediate node. The number of dropped packets increases for the RBAR as traffic increases, whereas for the proposed protocol, this rate is constantly low because, in the proposed rate adaptation scheme, the buffer occupancy of the receiving node is fed back to the transmitter. This information is used at the transmitting node to delay transmissions, thus preventing packet drops and retransmissions, whereas the RBAR transmits constantly dropping

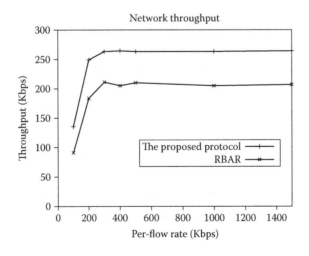

FIGURE 6.18
Throughput for varying per flow rate.

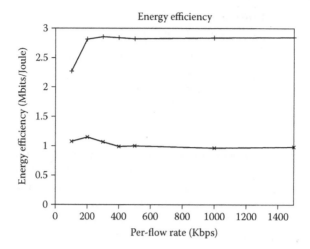

FIGURE 6.19
Energy efficiency for varying per flow rate.

packets at the relay node. This improvement, which is an effect of using the backoff mechanism, provides energy-efficient transmission as observed in the one-hop topology (though we cannot compare them directly because of differences in topology) giving higher throughput when compared to the RBAR.

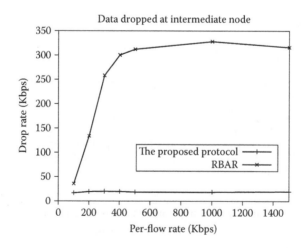

FIGURE 6.20
Drop rate for varying per flow rate.

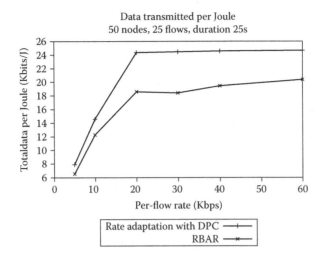

FIGURE 6.21
Data transmitted per joule for varying per flow rate.

6.10.3   Random Topology with 50 Nodes

Figure 6.21 shows the total data transmitted by 25 constant bit-rate (CBR) sources and received at the destinations, in the presence of fading channels, whereas the energy efficiency is presented in Figure 6.19. As expected, the proposed protocol transmits more data and consumes less energy per bit when compared to the RBAR protocol for all the traffic rates because of the proposed rate adaptation. These results reaffirm the conclusions from the previous simulations.

6.10.4   Two-Hop Results

The two-hop topology has been used while simulating the proposed DP-based scheme (PDP). For comparison, a modified RBAR protocol (BURST), which supports a burst mode, is used. The same parameters used in the previous simulations have been used. In addition, the information for the other parameters needed to simulate the DP-based algorithm include: cost function parameters $Q = 0.5$ and $\alpha = 2.4$ (from Figure 6.17), and an expected power value was taken as 100 mW. The stable value of G was calculated as 0.667. The target queue utilization is set at 30 packets per flow. The burst size, outgoing, and incoming traffic are expressed in number of packets (or number of packets per second). Furthermore, the burst duration was set to one packet duration when using the lowest rate. Consequently, the burst at the lowest rate of 1 Mbps can

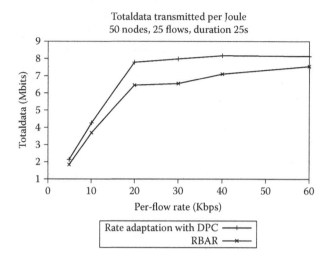

FIGURE 6.22
Total data transmitted for varying per flow rate.

accommodate one packet, the second rate of 2 Mbps can accommodate burst of maximum 2 packets, etc.

In case of low congestion with per flow traffic of up to 300 kbps, both protocols can transmit all generated packets without any difficulty as illustrated in Figure 6.23. However, as the congestion increases the PDP

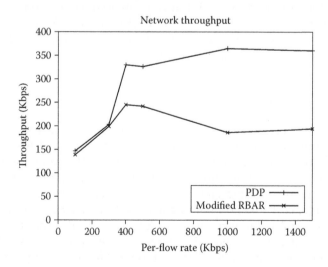

FIGURE 6.23
Throughput transmitted for varying per flow rate.

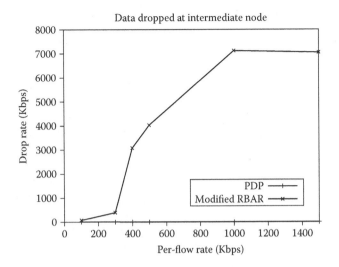

FIGURE 6.24
Number of packets dropped at the intermediate node for varying per flow rate.

scheme is able to increase the effective throughput more than RBAR, because the PDP prevents packet drops at the intermediate node as depicted in Figure 6.24, by limiting incoming flow using the feedback signal. The traffic emitted from sources is reduced as the queue utilization at the router increases, thus preventing buffer overflows. On the other hand, the modified RBAR with burst enabled cannot control the incoming traffic and, as a result, the packets are dropped at the intermediate node, and the end-to-end throughput suffers. An improvement of throughput up to 96% is observed for the proposed scheme, when compared with the modified RBAR protocol.

The buffer overflow in the case of modified RBAR protocol is shown in Figure 6.24 in terms of dropped packets at the intermediate nodes. The proposed PDP scheme eliminates any queue-related packet drops at the intermediate node regardless of the generated traffic, because it can precisely control the incoming flow by specifying a maximum number of admissible packets.

Figure 6.25 illustrates the energy efficiency (amount of data transmitted per joule of consumed energy) of the protocols. The proposed scheme outperforms the modified RBAR for all values of per flow rate, because the PDP reduces transmission power when possible, whereas the RBAR always transmits with maximum power. Additionally, the PDP avoids buffer overflow thus preventing retransmissions, which, in turn, increases energy

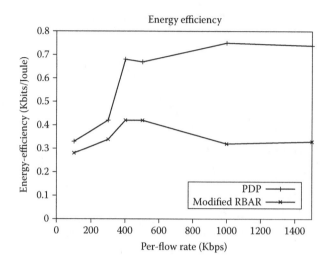

FIGURE 6.25
Data transmitted per joule for varying per flow rate.

consumption and decreases the end-to-end throughput. The proposed scheme achieves up to 131% higher energy-efficiency than the modified RBAR protocol.

6.11   Hardware Implementation of DPC

Implementation of novel wireless network protocols is traditionally evaluated through the use of network simulators such as NS2, OPNET, or MATLAB (Zawodniok and Jagannathan 2004, Holland et al. 2001, Kamerman and Monteban 1997). Although simulations provide a basis for comparison of protocol performances on a large scale, they lack the ability to evaluate the protocol against hardware constraints. Additionally, such approaches provide necessary feedback of hardware redesign to meet the needs of the protocol. In fact, very few protocols are implemented in hardware because it is quite difficult to access and modify the physical layer. Therefore, in this work, a hardware implementation is shown for the distributed adaptive power control (DAPC) protocol or, simply, DPC protocol. Though this is applied to an ad hoc type network, the idea can be demonstrated for cellular networks, wireless sensor networks, and RFID networks.

In wireless systems and networks, the radio channel can be highly unpredictable and can fundamentally limit the performance of any wireless communication system. Channel uncertainties such as path loss, shadowing, and Rayleigh fading can attenuate the power of the signal at the receiver, and thus cause variations in the received SNR or SIR. In addition, other wireless devices working in the same frequency band interfere, thereby increasing the noise level at the receiver. The objective of the DPC is to overcome these channel uncertainties and maintain a desired SNR at the receiver.

In cellular networks, the uncertain dynamic wireless channel results in intermittent connectivity and high energy consumption. DAPC schemes can be applied between the transmit tower and mobile user to reduce the effect of channel fluctuation and decrease the connection outage probability (Jagannathan et al. 2006). On the other hand, for wireless ad hoc and sensor networks, DAPC schemes applied between paired connections can increase the energy efficiency of the nodes, and therefore extend the lifetime of the network and improve the QoS (Zawodniok and Jagannathan 2004).

Similarly, in radio frequency identification (RFID) systems, the detection range and read rates will suffer from interference among high-power reading devices. The problem grows severe and degrades system performances in dense RFID networks. DAPC scheme can be deployed at the RFID reader to mitigate the interference issue among readers, and ensure the overall coverage area of the system while maintaining a desired read rate (Cha et al. 2006).

Work presented in this section focuses on the implementation of DAPC protocol on a generic wireless test platform developed at the University of Missouri — Rolla (UMR) (Cha et al. 2006). Hardware experiments are set up to evaluate the DAPC operation under various channel conditions in hardware because the authors are not aware of any known hardware implementations in this area. The results from the experiments illustrate that the protocol performs satisfactorily, as expected.

For the sake of simplicity, only the mathematical equation used for implementation is described here. In the discrete-time domain, the feedback control for the DAPC is selected as (Zawodniok and Jagannathan 2004)

$$p_i(l+1) = \frac{I_i(l)}{g_{ii}(l)}[-\hat{\theta}_i(l)R_i(l) + R_{required} + k_v(R_i(l) - R_{required})] \qquad (6.61)$$

where $\hat{\theta}_i(l)$ is the estimation of the unknown parameters defined as $\theta_i(l)$ and k_v is a control parameter.

The mean channel estimation error along with the mean SNR error converges to zero asymptotically, if parameter updates are taken as

$$\hat{\theta}_i(l+1) = \hat{\theta}_i(l) + \sigma R_i(l)(R_i(l) - R_{required}) \qquad (6.62)$$

where σ is the adaptation gain. The selection of k_v and σ should obey the following:

$$\sigma \|R_i(l)\|^2 < 1 \qquad (6.63)$$

$$k_{vmax} < \sqrt{1 - \sigma \|R_i(l)\|^2} \qquad (6.65)$$

For different network types, implementations of the DAPC can be different. However, the hardware implementation presented in this section is not specific to any existing type of wireless networks. It is only intended to demonstrate the working principles of the DAPC on a generic wireless test platform. The objective is to show that the desired SNR can be obtained in the presence of channel uncertainties. Moreover, this section from Cha et al. (2006) discusses in detail about the design specifications and requirements.

The proposed DAPC should be implemented at the MAC layer because it is specific to the connection and requires physical access to certain baseband parameters, such as RSSI reading and output power. A detailed description of the DAPC MAC is discussed in Subsection 6.5.2. We will now discuss the implementation in terms of hardware and software issues.

6.11.1   Hardware Architecture

In this section, an overview of the hardware implementation of the DAPC protocol is given from Cha et al. (2006). First, a customized wireless communication test platform for evaluating wireless networking protocols is presented. A detailed description of capabilities and limitations of the test platform is presented.

6.11.1.1   Wireless Networking Test Platform

To evaluate various networking protocols, a UHF wireless test platform is designed based on the UMR/SLU Generation-4 Smart Sensor Node (G4-SSN) (Fonda et al. 2006). Silicon Laboratories® 8051 variant microprocessors was selected for its ability to provide fast 8-bit processing, low power consumption, and ease of interfacing to peripheral components. ADF7020 ISM band transceiver was employed as the underlying physical radio for its ability to provide precise control in frequency, modulation, power, and data rate. A Zigbee compliant Maxstream XBee™ RF module was also employed as a secondary radio unit providing alternative wireless solutions. The former is suitable for low-level protocol development

FIGURE 6.26
Hardware block diagram.

at the MAC or Baseband level whereas the latter is great for implementing high-level routing and scheduling protocols. Using either the ADF7020 or the Zigbee radio interface, wireless networks can be formed, and various networking protocols can be implemented for evaluation. A block diagram of the hardware setup is shown in Figure 6.26.

6.11.1.2 Generation-4 Smart Sensor Node

The G4-SSN was originally developed at UMR and subsequently updated at the St. Louis University. The G4-SSN has various abilities in sensing and processing. The former include strain gauges, accelerometers, thermocouples, and general A/D sensing. The latter include analog filtering, CF memory interfacing, and 8-bit data processing at a maximum of 100 MIPS. These features provide a solid application level variability and have been utilized in previous works (Fonda et al. 2006). Moreover, the stackable connection easily allows for new hardware development. As seen in Figure 6.27, the Zigbee radio and ADF7020 radio stack can be used together, therefore allowing multiple radio interfaces.

As shown in Table 6.4, the G4-SSN provides powerful 8-bit processing, a suitable amount of RAM, and a low-power small form-factor.

FIGURE 6.27
Gen-4 SSN with Zigbee layer (left), ADF7020 Layer (right).

TABLE 6.4

G4-SSN Capabilities

	I_c @ 3.3 V [mA]	Flash Memory [bytes]	RAM [bytes]	ADC Sampling Rate [kHz]	Form-Factor	MIPS
G4-SSN	35	128k	8448	100 @ 10/12-bit	100-pin LQFP	100

6.11.1.3 ADF7020 ISM Band Transceiver

ADF7020 ISM band transceiver is used as the physical layer for the implementation of the DAPC protocol. The major advantage given by the ADF7020 is the freedom in controlling various physical layer properties, including operating frequency, output power, data rate, and modulation scheme. These features are essential in evaluating new wireless protocols, which require physical level access (Table 6.5). In the implementation of DAPC, direct access RSSI reading and output power are used. In addition, the low power consumption of this transceiver is suitable for embedded sensor network applications.

6.11.1.4 Limitations

Hardware implementation of any algorithm is constrained by the limitations of the hardware. With a single-chip software layered architecture, the microprocessor must simultaneously handle data communication with radio transceivers, internal processing, and applications. Therefore, the 8-bit processing power limits the data rate at which the radio transceivers can operate. Currently, a maximum data rate of 48 kbps has been successfully tested. Quantization is another issue faced in hardware and cannot be avoided. Quantization means that the hardware does not provide enough precision as desired by the algorithm, such as in calculation, A/D or digital-to-analog converter. In the implementation of the DAPC, the signal strength reading is only accurate up to 0.5 dB and power control is limited to 0.3-mW steps. These limitations must be treated to reduce the effects on the algorithm.

TABLE 6.5

AD7020 Capabilities

Feature	Capabilities
Frequency bands	431 MHz ~ 478 MHz and 862 MHz ~ 956 MHz
Data rates	0.15 kbps to 200 kbps
Output power	16–+13 dBm in 0.3 dBm steps
RSSI	6-bit digital readback
Modulation	FSK, ASK, GFSK
Power consumption	19 mA in receive, 28 mA in transmit (10 dBm)

6.11.1.5 RF Setup

The wireless channel for the DAPC implementation is chosen to be similar to the one in RFID systems. The nodes will operate at the central frequency of 915 MHz with 20-kHz channel bandwidth. To test the performance of only the DAPC, no other MAC is used. The data rate is set up at 12 kbps using FSK modulation with no encoding method. The output power at the transmitter can vary from 16 to +13dBm by 0.3-dB increments.

6.11.2 Software Architecture

A layered networking architecture is considered for the Gen-4 SSN wireless test platform. This would allow easier future implementations and protocol evaluations. A block diagram of the layered software architecture is shown in Figure 6.28. In this section, a detailed description of the baseband controller and DAPC MAC design is given.

6.11.2.1 Frame Format

Frame format used for DAPC implementation is shown in Figure 6.29. The physical layer header is composed of a series of SYNC bytes and a preamble sequence. The SYNC bytes, which are used to synchronize the transmitter and receiver clock, should be a DC-free pattern, such as 10101010 … pattern. The preamble sequence is a unique pattern indicating the beginning of a packet and must be universal to all nodes in the

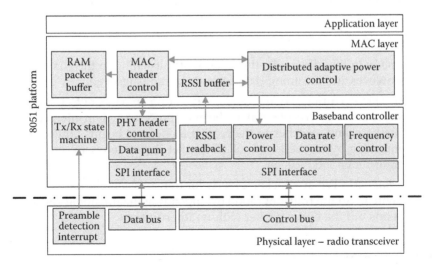

FIGURE 6.28
Software architecture.

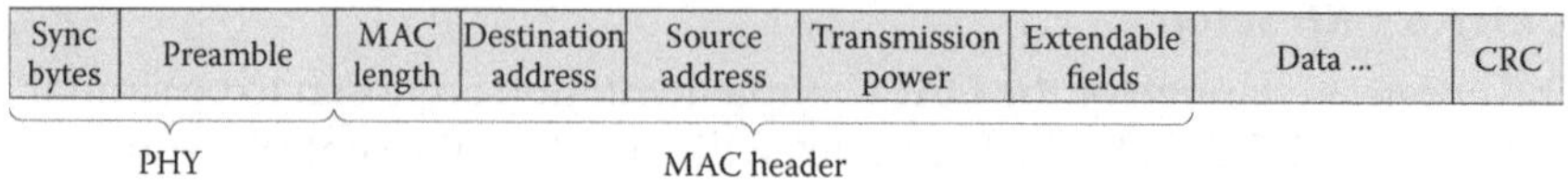

Sync bytes	Preamble	MAC length	Destination address	Source address	Transmission power	Extendable fields	Data ...	CRC

PHY   MAC header

FIGURE 6.29
Protocol frame format.

network. The ADF7020 provides the hardware preamble detection and interrupt source to the microprocessor.

The preamble is followed by the MAC header. The length of the MAC header can be programmable using its first byte, therefore allowing multiple extensions for the future. For the DAPC, only transmission power field is required. After the MAC header, data and CRC are transmitted.

6.11.2.2 Baseband Controller

A baseband controller is implemented to interface with the physical layer. It also provides an API for higher layers to access all functionalities offered by the radio transceiver. In the implementation of the DAPC MAC, only RSSI read-back and power control are used. Other options are available and can be utilized easily for future implementations of different protocols.

6.11.2.3 Operation Modes

The baseband controls the radio in three operational modes, *transmit, receive, and idle*, which are handled by the *Tx/Rx* state machine. The radio should always operate in the idle mode unless a packet is ready for transmission or a preamble is detected indicating the beginning of a packet reception.

Idle mode: In the idle mode, the radio is still listening to the channel however any incoming data from the radio is ignored.

Receive mode: During the idle mode, when a preamble is detected by the radio, an interrupt is sent to the microprocessor. Upon interrupt, the baseband switches to the receive mode, and begins buffering incoming bytes; the length of the packet is prefixed between the transmitter and receiver.

Transmit mode: When a packet is ready for transmission, the baseband switches to the transmit mode by appending the preamble and it sends out the entire packet with no interruptions.

6.11.2.4 RSSI Reading

The implementation of DAPC requires RSSI readings to calculate the SNR for every packet. To provide accurate SNR values, RSSI readings are taken

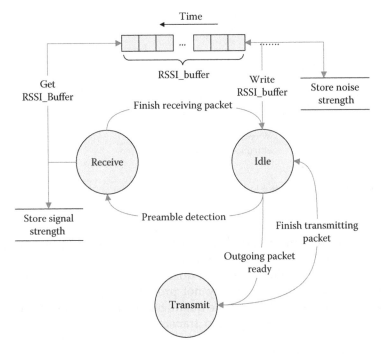

FIGURE 6.30
Baseband flowchart.

at the reception of every byte. When the radio is in the idle mode, any incoming data is discarded; however, the RSSI value is still recorded every 8 bits. To separate preamble from noise, a small "RSSI_buffer" stores the past N values of RSSI, where N equals to the length of preamble bytes. Any reading beyond N is averaged as the "noise_power." After the radio enters the receive mode, RSSI is recorded and averaged along with the values in the RSSI_buffer to provide "signal_power." A flowchart diagram of the mode switching and RSSI reading is shown in Figure 6.30.

6.11.2.5 DAPC MAC controller

Figure 6.31 illustrates the block diagram representation of the proposed DAPC control loop inside a transmitter and receiver.

At the receiver side, signal strength P_i, noise level I_i and, therefore, the SNR R_i are measured at the reception. Ouput power at the transmitter P_t is known from the previous calculation. Given P_t and P_i, the channel attenuation g_{ii} for the previous transmission can be calculated. Now, update θ_i and P_t using Table 5.3. P_t is then embedded into the MAC header of the next outgoing packet to the corresponding transmitter. At the reception of the next packet, the cycle begins again.

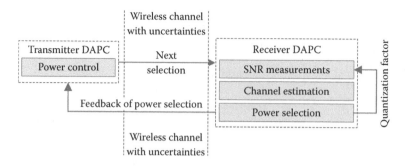

FIGURE 6.31 DAPC in control loop.

At the transmitter side, DAPC must extract the power information from the MAC header and inform the baseband to transmit P_t for the next outgoing packet to the corresponding receiver. In hardware implementation, especially in digital systems, a quantization factor should be introduced because the hardware may not provide the precision for output power that the DAPC desires. The quantization factor is simply the ratio between the actual $P_{t'}$ and desired transmission power P_t. This ratio is divided by the next power calculation to improve estimation accuracy and maintain system stability.

6.11.3   Experimental Results

In this section, hardware implementation results for the DAPC are presented. Various experiments are executed to create channel interferences to thoroughly evaluate the performance of the DAPC. Note that the experiments are conducted under normal office environments where other ISM band devices exist along with other channel uncertainties. Because of range and power limitations, the SNR for the test platform can reach a maximum of 80 dB. Therefore, the system control parameter kv and are very small and selected as 1e15 and 0.01, respectively.

In general, a paired connection between a transmitter and receiver is established. The transmitter sends a 100-byte packet to the receiver every 500 msec. The receiver sends the reply with a 100-byte packet immediately after reception. This also indicates that the power update rate is only 2 times per second. Essentially, the nodes act as transmitters and receivers, and the DAPC is implemented on both of them. The working range for the experiments is usually within 5 m.

Example 6.11.1: Path Loss Effect

In this setup, a paired connection is established. The receiver was slowly moved towards the transmitter and then taken away. The desired SNR

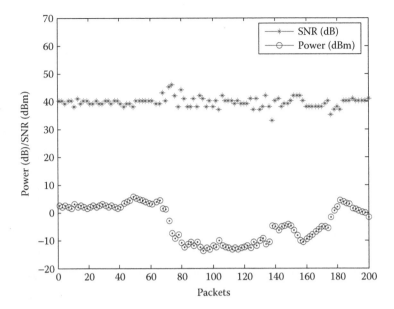

FIGURE 6.32
Receiver SNR and transmitter power level corresponding to channel uncertainties.

for the receiver is set at 40 dB. Figure 6.31 demonstrates the performance of the DAPC. In red, the SNR at the receiver is plotted. In blue, the output power of the transmitter is plotted (Figure 6.32). The receiver SNR was kept very close to the target SNR. We can clearly see that at packet number 65, the receiver starts moving close to the transmitter, resulting in a reduction in the power level. At the 180th packet, the radio had been moved back to its original location, and the output power for the transmitter has increased to provide required SNR. This experiment shows that the DAPC accurately estimates the channel loss g_{ii} in a noninterfered environment.

6.11.4 Slowly Varying Interference

In this experiment, a paired connection between a transmitter and a receiver is set up. At the same time, a constant interfering source is introduced to alter the channel with small variations per step. The time-varying transmission power for the interfering source is displayed in Figure 6.33. The transmission power for the interferer varies from 16dBm to 13 dB at a slow rate. Note that the rate for power update at the receiver is three times faster then rate of change of output power on the interferer.

At the desired SNR value equal to 45 dB from Figure 6.34, we can observe that the SNR seen at the receiver is obtained very close to the desired value.

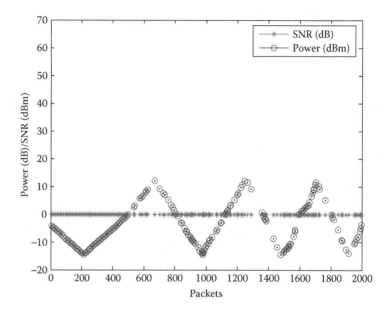

FIGURE 6.33
Power variation of a slow-changing interferer.

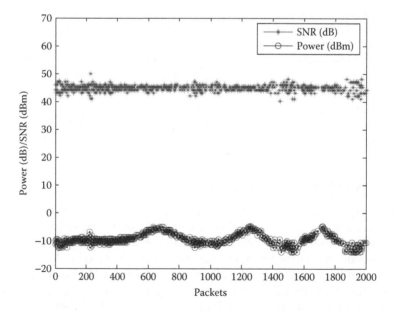

FIGURE 6.34
Receiver SNR and transmitter power level from a slow-changing interferer.

In blue, the output power at the transmitter is plotted. It shows that the change in transmitter power follows the power pattern of the interferer.

6.11.5 Abruptly Changing Channel with Slow Update

The setup here is the same as the previous experiment, except that the interferer varies the transmission power randomly. The rate for the power update is three times faster than the rate of the interferer. This is considered as a very brutal interferer. The interference level is shown here in Figure 6.35.

In Figure 6.36, we can observe that the SNR at the receiver is not very well leveled compared to a slowly varying channel because of the vast brutal interferer. However it is still kept at an acceptable margin around 45 dB.

6.11.6 Abruptly Changing Channel with Fast Update

In this setup, the same interferer is used as the previous experiment. However the rate for the power update is now 10 times faster than the interferer. With the desired SNR equal to 45 dB, we can observe from Figure 6.37 that the SNR at the receiver performs very well with a faster update rate.

Example 6.11.2: DAPC for Four Nodes

Next the DAPC is implemented on four nodes. One can treat this as an RFID system (see Chapter 10) where passive tags harvest energy from the

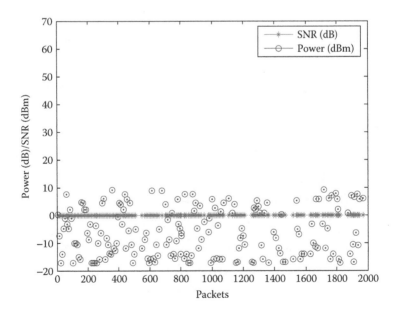

FIGURE 6.35
Power variation of an interferer with random output power.

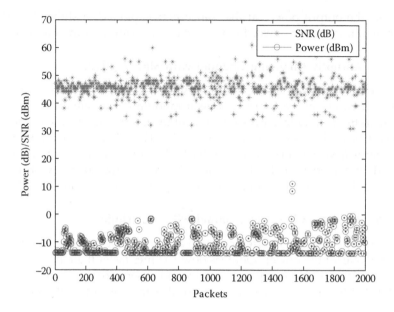

FIGURE 6.36
Receiver SNR and transmitter power level from a brutal interferer — slow power update.

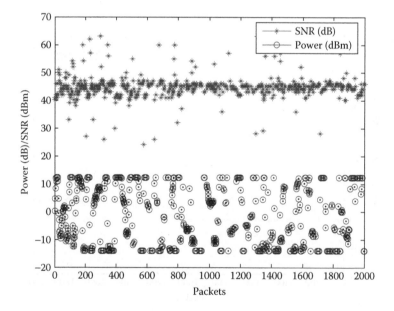

FIGURE 6.37
Receiver SNR and transmitter power level from a brutal interferer — fast power update.

RFID readers to power internal circuits and obtain communication. Readers operating in the same frequency may interfere with the others resulting in a reduced detection range and read rate. In addition, because the tags are at low cost, any intelligent power control must be designed on the reader side only. Because the reader and tag range are relatively stationary and short in distance, interference by others is considered as the main source for channel uncertainties in RFID systems. Therefore, by assuming g_{ii} in Equation 6.1 to be a constant, the DAPC feedback loop can be internal to the reader, and only interference measurements are necessary. Received SNR can be directly converted into detection range and measure system performances. For wireless ad hoc and sensor networks, meeting the SNR target is important to decode a packet successfully. Otherwise, retransmissions will result.

An ad hoc wireless network (or a RFID reader network with 4 readers) is implemented using the Gen-4 SSN setup. The desired SNR for the readers is at 10 dB and a channel attenuation between the tag and reader is assumed to be 40 dBm (g_{ii}). First, a system with no power control scheme is tested, and the output power of all four nodes is set to be 2 dBm. In Figure 6.38, the performance of all four nodes is illustrated, and it is clear

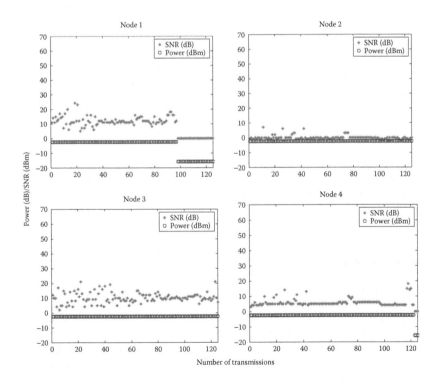

FIGURE 6.38
Network performances of 4 nodes with no power update scheme.

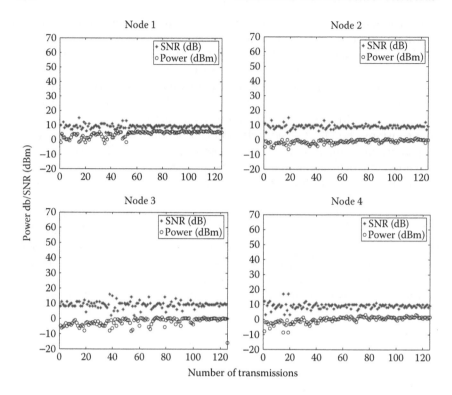

FIGURE 6.39
Network performances of 4 nodes using DAPC.

that two of the nodes (or readers) never achieve their desired SNR whereas others have very unstable SNR. Now, the DAPC is turned on at each of the nodes (or readers). As shown in Figure 6.39, all four nodes attain the desired SNR of 10 dB.

6.12   Conclusions

A novel power control MAC protocol is presented for wireless ad hoc networks. The proposed DPC scheme allows fully DPC and has resulted in better performance in the presence of radio channel uncertainties. The proposed DPC scheme uses significantly less transmitter power per bit compared to 802.11; hence, the energy is saved and the life-span of the wireless nodes is extended. Additionally, the network capacity is maximized in terms of high-aggregate throughput. In conclusion, the power control MAC protocol offers a superior performance in terms of convergence and maximizes the network capacity compared to 802.11 standards. Results justify theoretical conclusions.

The increased throughput is because of the increased channel utilization and not because of higher spatial reuse, because the pulse train to overcome the hidden-terminal problem is sent at the maximum power. Consequently, the spatial reuse factor is not significantly altered. To further increase the throughput and spatial reuse factor, the transmission power has to be changed dynamically for all frames. This amounts to adaptive selection of the first RTS-CTS exchange, as well as power used for the train of pulses. Therefore, potential future work may involve adaptive selection of transmitter powers for the train of pulses to mitigate the hidden-terminal problem, while increasing the spatial reuse in the presence of fading channels.

Next, energy-efficient rate adaptation protocols are introduced that adaptively select the rate based on channel state and queue utilization. Additionally, the algorithms minimize energy consumption. The selection of the rate is performed online by taking into account the congestion, required throughput, and buffer occupancy. The heuristic scheme minimizes buffer overflows by altering back-off intervals based on congestion level. Alternatively, the solution based on the dynamic programming approach is able to precisely control incoming flow, thus preventing any packet losses due to buffer overflows. Moreover, by precisely controlling the congestion level, a 96% higher throughput and 131% higher energy-efficiency is achieved over the RBAR protocol. Simulations confirm that the data can be transmitted faster with fewer dropped packets while consuming less energy. Thus, the network lifetime is extended and QoS improved.

Finally, in this chapter, the implementation of a novel DAPC algorithm for wireless communication systems was discussed. Now, with hardware implementation, the algorithm can be put to test in the real-world wireless channel. A test platform for evaluating wireless MAC protocols was developed with a vision for future expansion and development. Satisfactory performance from the DAPC was observed from the experimental results. The protocol was shown to provide suitable power adjustment to maintain desired SNR even under extreme channel conditions.

Future work will involve testing the protocol for larger networks with more paired connections. Moreover, more complex DAPC MAC protocol can be tested for cellular network, RFID reader network, and wireless ad-hoc or sensor networks.

References

Angel, E. and Bellman, R., *Dynamic Programming and Partial Differential Equations (Mathematics in Science and Engineering)*, Vol. 88, Academic Press, 1972.

Bambos, N. and Kandukuri, S., Power controlled multiple Access (PCMA) in wireless communication networks, *Proceedings of the IEEE INFOCOM*, 2000, pp. 386–395.

Bambos, N., Chen, S., and Pottie, G.J., Channel access algorithms with active link protection for wireless communication networks with power control, *IEEE ACM Transactions on Networking*, 583–597, October 2000.

Bambos, N., Towards power-sensitive network architectures in wireless communications: concepts, issues and design aspects, *IEEE Personal Communications*, June 1998, pp. 50–59.

Bertsekas, D.P., *Dynamic Programming: Deterministic and Stochastic Models*, Prentice Hall, Englewood Cliffs, NJ, 1987.

Canchi, R. and Akaiwa, Y., Performance of adaptive transmit power control in $\pi/4$ DQPSK mobile radio systems in flat Rayleigh fading channels, *Proceedings of the IEEE Vehicular Technology Conference*, Vol. 2, 1999, pp. 1261–1265.

Cha, K., Ramachandran, A., and Jagannathan, S., Adaptive and probabilistic power control schemes and hardware implementation for dense RFID networks, *Proceedings of the IEEE International Conference of Decision and Control*, to appear in 2006.

Dontula, S. and Jagannathan, S., Active link protection for wireless peer-to-peer and cellular networks with power control, *Proceedings of the World Wireless Congress*, May 2004, pp. 612–617.

Fonda, J., Zawodniok, M., Jagannathan, S., and Watkins, S.E., Development and implementation of optimized energy-delay sub-network routing protocol for wireless sensor networks, *Proceedings of the IEEE International Symposium on Intelligent Control*, to appear in 2006.

Gomez, J., Campbell, A.T., Naghshineh, M., and Bisdikian, C., Conserving transmission power in wireless ad hoc networks, *Proceedings of the ICNP'01*, November 2001.

Hashem, B. and Sousa, E., Performance of cellular DS/CDMA systems employing power control under slow Rician/Rayleigh fading channels, *Proceedings of the International Symposium on Spread Spectrum Techniques and Applications*, Vol. 2, September 1998, pp. 425–429.

Holland, G., Vaidya, N., and Bahl, P., A rate-adaptive MAC protocol for multihop wireless networks, *Proceedings of the ACM/IEEE MOBICOM*, July 2001.

Jagannathan, S., Chronopoulos, A.T., and Ponipireddy, S., Distributed power control in wireless communication systems, *Proceedings of the IEEE International Conference on Computer Communications and Networks*, November 2002, pp. 493–496.

Jagannathan, S., Zawodniok, M., and Shang, Q., Distributed power control of cellular networks in the presence of channel uncertainties, *Proceedings of the IEEE INFOCOM*, Vol. 2, March 2004, pp. 1055–1066.

Jagannathan, S., Zawodniok, M., and Shang, Q., Distributed power control of cellular networks in the presence of channel uncertainties, *IEEE Transactions on Wireless Communications*, Vol. 5, No. 3, 540–549, February 2006.

Jantti, R. and Kim, S.L., Second-order power control with asymptotically fast convergence, *IEEE Journal on Selected Areas in Communications*, Vol. 18, No. 3, March 2000.

Jung, E.-S. and Vaidya, N.H., A power control MAC protocol for ad hoc networks, *ACM MOBICOM*, 2002.

Kamerman, A. and Monteban, L., WaveLAN-II: A high-performance wireless LAN for the unlicensed band, *Bell Labs Technical Journal*, 118–113, Summer 1997.

Karn, P., MACA—a new channel access method for packet radio, *Proceedings of 9th ARRL Computer Networking Conference*, 1990.

Lee, G. and Park, S.-C., Distributed power control in fading channel, *Electronics Letters*, Vol. 38, No. 13, 653–654, June 2002.

Maniezzo, D., Cesana, M., and Gerla, M., IA-MAC: Interference Aware MAC for WLANs, UCLA-CSD Technical Report Number 020037, December 2002.

Park, S.J. and Sivakumar, R., Quantitative analysis of transmission power control in wireless ad hoc networks, *Proceedings of the ICPPW'02*, August 2002.

Pursley, M.B., Russell, H.B., and Wysocarski, J.S., Energy efficient transmission and routing protocols for wireless multiple-hop networks and spread-spectrum radios, *Proceedings of the EUROCOMM*, 2000, pp. 1–5.

Rappaport, T.S., *Wireless Communications, Principles and Practices*, Prentice Hall, Upper Saddle River, NJ, 1999.

Schurgers, C., Aberthorne, O., and Srivastava, M.B, Modulation scaling for energy aware communication system, *Proceedings of the International Symposium on Low Power Electronics and Design*, 2001, pp. 96–99.

Singh, S. and Raghavendra, C.S., PAMAS: Power Aware Multi-Access Protocol with Signaling for Ad Hoc Networks, *ACM Computer Communication Review*, Vol. 28, No. 3, July 1998, pp. 5–26.

White, D.J., *Dynamic Programming*, Oliver and Boyd, San Francisco, CA, 1969.

Woo, A. and Culler, D.E., A transmission control scheme for media access in sensor networks, *ACM Sigmobile*, 2001.

Ye, W., Heidermann, J., and Estrin, D., An efficient MAC protocol for wireless sensor networks, *Proceedings of the IEEE INFOCOM*, 2002.

Zawodniok, M. and Jagannathan, S., A distributed power control MAC protocol for wireless ad hoc networks, *Proceedings of the IEEE WCNC*, Vol. 3, March 2004, pp. 1915–1920.

Zawodniok, M. and Jagannathan, S., Energy efficient rate adaptation MAC Protocol for ad hoc wireless networks, *Proceedings of IEEE International Performance Computing and Communications Conference (IPCCC)*, March 2005, pp. 389–394.

Problems

Section 6.5

Problem 6.5.1: The DPC protocol is implemented using the IEEE 802.11 network. Use Equation 6.12 to calculate bit overhead for each feedback implementation case from Problem (6.1). Assume a packet size of 256 bytes (octets), standard 802.11 DATA frame header of 18 bytes (octets), size of RTS frame 24 bytes (octets), CTS frame — 18 bytes (octets), ACK frame — 20 bytes (octets), and back-off adding 5-bytes (octets) duration for each packet transmission.

Section 6.6

Problem 6.6.1: Redo the example in Section 6.6.1 for a random placement of 150 nodes and in the presence of fading and shadowing using CBR data.

Section 6.8

Problem 6.8.1: Simulate the power control system from Table 5.4 for power levels ranging from 1 mW to 1 W. Assume signal attenuation to change linearly first from 50 to 100 dB, and then from 100 to 80 dB. The noise level is equal to 90 dBm.

1. Use exact (real) values for power level transmission.
2. Use rounded (integer) values for power level (in mW).
3. Use rounded (integer) values for power levels (in dBm).
4. Repeat (3) with Equation 6.34 and Equation 6.39 applied to adjust power and readings for discreet levels (NOTE: assume that Pt_0 is the exact power level calculated by DPc algorithm, and Pt_k is the corresponding discreet power level to be used; using Equation 6.34, calculate the matching SNR_k.)

Section 6.9

Problem 6.9.1: Using Equation 6.58, calculate stable state value of parameter G. Assume that, $Q = 0.5$, $R = 2.4$, and $E\{Po\} = 0.1$ W. (NOTE: Assume that stable state is reached if the difference between the consecutive values G_k and G_{k-1} differ less than 0.0001.) See Table 6.9.1.

TABLE 6.9.1

Dynamic Programming Parameter Calculation

k	G_k
N	0.500
N1	
N2	
N3	
...	

TABLE 6.9.2

Selected Burst Size u^*

$Po(k)$ $x(k)$	0.001	0.010	0.050	0.100
5				
4				
3				
2				
1				
0				
1				
2				
3				
4				
5				
6				
7				
8				
9				
10				

Problem 6.9.2: Using Equation 6.60, calculate the burst size u^* for varying power levels, $Po(k)$, and queue utilization error, $x(k)$. Fill in the Table 6.9.2. Assume the outgoing flow, $w(k)$, to be equal to 2 packets per second (Note: 2 means that there will be 2 packets dequeued.)

Section 6.10

Problem 6.10.1: Evaluate the dynamic-programming-based rate adaptation scheme for a random topology consisting of 150 nodes and 1000 nodes. Use the parameters from Section 6.10.1. Plot throughput and energy-efficiency as a function of time.

7

Distributed Fair Scheduling in Wireless Ad Hoc and Sensor Networks

In the previous chapter, the distributed adaptive power control scheme (DAPC) or simply DPC and a medium access control (MAC) protocol were introduced for ad hoc wireless and sensor networks. DAPC is utilized to meet energy efficiency besides other quality of service (QoS) performance parameters such as throughput, end-to-end delay, and loss or drop rate. In many wireless networking applications, there are additional QoS performance metrics such as fairness that need to be met. Under such circumstances, the DPAC alone is not sufficient to meet the target QoS. Because bandwidth is one of the main constraints in wireless ad hoc networks, the key to guarantee QoS in this type of networks would be efficient bandwidth management through which fairness is guaranteed.

7.1   Fair Scheduling and Quality of Service

Fairness is a critical issue (Goyal et al. 1997) when accessing a shared wireless channel. A fair scheduling scheme must then be employed in ad hoc wireless networks and wireless sensor networks (WSNs) to provide proper flow of information. In the literature, many algorithms and protocols regarding QoS metrics are found; however, they do not address hardware constraints.

The notion of fairness must be guaranteed among a set of contending flows. Moreover, the proposed scheme should be computationally distributed in its nature. Thus, any distributed solution to WSN fair scheduling must coordinate local interactions to achieve global performance. This should be achieved within the constraints imposed by the hardware. Therefore, any fair scheduling algorithm proposed for a multihop ad hoc wireless network or WSN must consider the following design criteria:

Centralized vs. distributed approaches: Distributed fair scheduling algorithm for WSN is preferred over a centralized scheme.

Fairness metric: Selection of an appropriate fairness metric is important from a design aspect. It should address the fair allocation of service proportional to weights selected by user-defined QoS metrics.

Scalability: The scheduling scheme should deploy well in WSNs with dynamic topology and link failures.

Efficiency of the protocol: Because a trade-off exists between throughput and fairness, fair scheduling should render reasonable throughput to all flows.

Persistency of quality of service: Fair scheduling should meet QoS of all flows during topology changes and dynamic channel states.

A number of fair scheduling schemes for efficient bandwidth management exist in the literature; in which some are centralized (Golestani 1994, Luo et al. 2001, Demers et al. 2000), and others are distributed (Lee 1995, Jain et al. 1996, Luo et al. 2001, Vaidya et al. 2000). There has been work on achieving fairness using the distributed MAC protocol for wireless networks (Golestani 1994, Bennett and Zhang 1996, Jain et al. 1996). A recent work (Vaidya et al. 2000) proposes a distributed fair scheduling protocol for wireless local area networks (LANs). Distributed fair scheduling (DFS) allocates bandwidth proportional to the weights of the flows. This protocol (Vaidya et al. 2000) performs a fair allocation of bandwidth using a self-clocked fair queuing algorithm (Golestani 1994). However, this protocol may not be suitable for multihop networks with dynamic channel conditions and changing topologies. With node mobility, the network state can change demanding weight updates. Additionally, DFS results in large delay variations, or jitter, in the reception of packets at the destinations. Finally, selection of initial weights is not addressed in DFS. Unless weights are selected appropriately, fairness cannot be guaranteed even for wireless networks with stationary nodes.

In general, these fair scheduling schemes determine appropriate weights to meet QoS criteria. In most schemes, weights are assigned and not updated when dynamic network conditions apply, and thus do not provide the advantage seen in an adaptive and distributed fair scheduling (ADFS) -enabled network (Regatte and Jagannathan 2004). The relevant chapter from Regatte and Jagannathan (2004) presents an ADFS protocol for wireless ad hoc networks (which operates in the CSMA/CA paradigm). The proposed algorithm is fully distributed in nature, and it follows the fairness criterion defined in Goyal et al. (1997) and in Vaidya et al. (2000).

The main contribution of the proposed scheme is the dynamic adaptation of weights as a function of delay experienced, number of packets in the queue, and the previous weight of the packet. The initial weights are selected by employing the user-defined QoS. Moreover, weights are updated for each packet when it reaches the front of the queue. The updated weights are used in deciding which packet to place in the output queue and also in the calculation of the backoff interval if there is a potential collision.

7.2 Weighted Fairness Criterion

In wired and wireless ad hoc networks, the fair scheduling schemes implement the distributed algorithm to achieve certain local criterion, which affects the overall global fairness of the network. Observe a typical node shown in Figure 7.1, where it maintains several input queues (belonging to several flows) for storing incoming packets to be transmitted on the output link. A fair queuing algorithm is used to determine which flow to serve next. Consider several such nodes trying to access a shared wireless (or wired) channel in CSMA/CA paradigm. The node that has to get access to the channel in such a scenario is determined by the backoff interval at each node.

Therefore, a fair scheduling protocol for wired and ad hoc wireless networks should be able to implement the fair queuing algorithm along with a fair backoff algorithm to satisfy the required fairness criterion and to achieve global fairness.

Intuitively, allocation of output link bandwidth is fair if equal bandwidth is allocated at every time interval to all the flows. This concept generalizes to weighted fairness, in which the bandwidth must be allocated in proportion to the weights associated with the flows. Formally, if ϕ_f is the weight of flow f and $W_f(t_1, t_2)$ is the aggregate service (in bits)

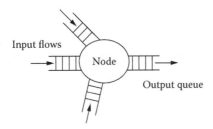

FIGURE 7.1
A node with several contending flows sharing a link.

received by it in the interval $[t_1, t_2]$, then an allocation is fair when both flows f and m are backlogged and satisfies (Goyal et al. 1997)

$$\frac{W_f(t_1, t_2)}{\phi_f} - \frac{W_m(t_1, t_2)}{\phi_m} = 0. \tag{7.1}$$

Clearly, this is an idealized definition of fairness, as it assumes that flows can be served in infinitesimally divisible units. The objective of fair packet scheduling algorithms is to ensure that

$$\left| \frac{W_f(t_1, t_2)}{\phi_f} - \frac{W_m(t_1, t_2)}{\phi_m} \right| \le \varepsilon \tag{7.2}$$

where ε is a small positive number. However, it has been shown in Golestani (1994) that if a packet scheduling algorithm guarantees that

$$\left| \frac{W_f(t_1, t_2)}{\phi_f} - \frac{W_m(t_1, t_2)}{\phi_m} \right| \le H(f, m) \tag{7.3}$$

for all intervals $[t_1, t_2]$ then

$$H(f, m) \ge \frac{1}{2} \left(\frac{l_f^{\max}}{\phi_f} + \frac{l_m^{\max}}{\phi_m} \right) \tag{7.4}$$

where $H(f, m)$ is a function of the properties of flows f and m, while $l_f^{\max}$ and $l_m^{\max}$ denote the maximum lengths of packets of flow f and m, respectively. Several fair, yet centralized, scheduling algorithms that achieve values of $H(f, m)$ close to the lower bound have been proposed in the literature (Golestani 1994, Demers 2000, Bennett and Zhang 1996) first for wired networks, and then for wireless networks. This chapter will discuss the ADFS scheme from Regatte and Jagannathan (2004) for ad hoc networks and its extension to WSN by hardware implementation using UMR motes (Fonda et al. 2006). Besides taking into account channel uncertainties, the proposed analysis is applicable to wired networks.

7.3 Adaptive and Distributed Fair Scheduling (ADFS)

Note that service rate of flow-controlled broadcast medium and wireless links may fluctuate over time. Two service models, fluctuation constrained (FC) and exponential bounded fluctuation (EBF) service model, which are suitable for modeling many variable rate servers have been introduced in Lee (1995).

7.3.1 Fluctuation Constrained and Exponentially Bounded Fluctuation

The variable rate service models for wireless ad hoc networks have to be defined to incorporate the channel- and contention-based protocols. An FC service model for wireless networks in the interval $[t_1, t_2]$ has two parameters, average rate $\lambda(t_1, t_2)$ bps and variations parameter $\psi(\lambda)$ given by $\psi(\lambda) = \chi(\lambda) + \delta(\lambda) + \bar{\omega}(\lambda)$, where $\chi(\lambda)$ is the reduction in wireless channel capacity due to channel uncertainties, $\bar{\omega}(\lambda)$ is the variation due to backoff interval, and $\delta(\lambda)$ is the burstiness in bits.

FC service model (Goyal et al. 1997): A wireless node follows the fluctuation constrained (FC) service model with parameters $(\lambda(t_1, t_2), \psi(\lambda))$, if for all intervals $[t_1, t_2]$ in a busy period of the node, the work done by the node, denoted by $W(t_1, t_2)$, satisfies

$$W(t_1, t_2) \geq \lambda(t_1, t_2)(t_2 - t_1) - \psi(\lambda) \tag{7.5}$$

The EBF service model is a stochastic relaxation of FC service model. Intuitively, the probability of work done by a wireless node following EBF service model deviating from the average rate by more than γ, decreases exponentially with γ.

EBF service model (Goyal et al. 1997): A wireless node follows the EBF service model with parameters $(\lambda(t_1, t_2), B, \omega, \psi(\lambda))$, if for all intervals $[t_1, t_2]$ in a busy period of the node, the work done by the node, denoted by $W(t_1, t_2)$, satisfies

$$P(W(t_1, t_2) < \lambda(t_1, t_2)(t_2 - t_1) - \psi(\lambda) - \gamma) \leq Be^{-\omega\gamma} \tag{7.6}$$

NOTE From now, the weight of a packet of flow f at node l is denoted as $\phi_{f,l}$, and it is given by $\phi_{f,l} = \sigma_f \phi_f$.

The proposed scheme from Regatte and Jagannathan (2004) works well even with variable rate wireless nodes. From now on, we define a variable rate wireless node with the proposed scheme as an ADFS wireless node. The ADFS scheme satisfies the fairness criterion given in Equation 7.3. The description of the scheme and the MAC protocol is given next.

The main goal of the ADFS protocol is to achieve fairness in wireless ad hoc networks. To accomplish this, the protocol has to be implemented both at the queuing algorithm level, for proper scheduling, and at the MAC protocol level, to control the dynamic backoff algorithm for accessing the channel.

7.3.2 Fairness Protocol Development

To achieve fairness at the scheduling level, the proposed ADFS protocol implements the *start-time fair queuing* (SFQ) (Goyal et al. 1997) *scheme*, defined as follows:

1. On arrival, a packet p_f^j of flow f is stamped with start tag $S(p_f^j)$, defined as

$$S\left(p_f^j\right) = \max\left\{ v\left(A\left(p_f^j\right)\right), F\left(p_f^{j-1}\right)\right\} \quad j \geq 1 \qquad (7.7)$$

where $F(p_f^j)$, finish tag of packet p_f^j, is defined as

$$F\left(p_f^j\right) = S\left(p_f^j\right) + \frac{l_f^j}{\phi_f} \quad j \geq 1 \qquad (7.8)$$

where $F(p_f^0)=0$ and ϕ_f is the weight of flow f.

2. Initially, the virtual time of the wireless node is set to zero. During transmission, the node's virtual time at time t, $v(t)$ is defined to be equal to the start tag of the packet being transmitted at time t. At the end of a transmission, $v(t)$ is set to the maximum of finish tag assigned to any packets that have been transmitted by time t.

3. Packets are transmitted in the increasing order of the start tags; ties are broken arbitrarily.

7.3.2.1 Dynamic Weight Adaptation

To account for the changing traffic and channel conditions that affect the fairness and end-to-end delay, the weights for the flows are updated dynamically. The actual weight for the ith flow, jth packet denoted by $\hat{\phi}_{ij}$, is updated as

$$\hat{\phi}_{ij}(k+1) = \alpha.\hat{\phi}_{ij}(k) - \beta.E_{ij} \qquad (7.9)$$

where $\hat{\phi}_{ij}(k)$ is the previous weight of the packet, α and β are design constants, $\{\alpha, \beta\} \in [-1,1]$, and E_{ij} is defined as:

$$E_{ij} = e_{ij,\,queue} + \frac{1}{e_{ij,\,delay}} \qquad (7.10)$$

where $e_{ij,queue}$ is the error between the expected length of the queue and the actual size of the queue, and $e_{ij,delay}$ is the error between the expected delay and the delay experienced by the packet so far. Note that the E_{ij} value is bounded because of finite queue length and delay, as packets experiencing delay greater than the delay error limit will be dropped.

To calculate the backoff interval and to implement the scheduling scheme, the updated weights at each node have to be transmitted in the data frame of the MAC protocol. To enable this, changes are made to the data packet header to accommodate the current weight of the packet. Whenever a packet is received, the current weight is used to update the weights dynamically using Equation 7.9. Then the weight field in the packet header is replaced with the updated weight.

7.3.2.2 MAC Protocol — Dynamic Backoff Intervals

The proposed ADFS protocol follows the CSMA/CA paradigm, similar to the IEEE 802.11 protocol. Because multiple nodes in a wireless network try to transmit simultaneously, when nodes compete to access the shared channel as in Figure 7.2, the selection of the backoff interval plays a critical role in deciding which node gets access to the channel. To achieve global fairness, the nodes must access the channel in a fair manner.

The proposed ADFS is implemented as the MAC protocol to control the access of the shared medium by the wireless nodes, by adjusting the dynamic backoff intervals. ADFS calculates the backoff interval relative to the weight of the packet. The backoff procedure is similar to DFS (Vaidya et al. 2000). However, the weights are updated via Equation 7.9;

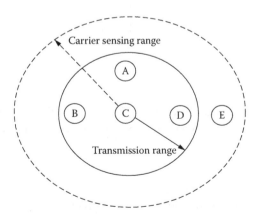

FIGURE 7.2
Nodes contending for shared wireless channel.

and hence, the backoff interval is also updated at each node. Backoff interval, BI_{ij}, for ith flow jth packet with packet length L_{ij} and weight ϕ_{ij} is defined as

$$BI_{ij} = \left\lfloor \rho * SF * \frac{L_{ij}}{\phi_{ij}} \right\rfloor \tag{7.11}$$

where SF is the scaling factor and ρ is a random variable with mean one. The collision handling mechanism is similar to the one in Vaidya et al. (2000). This results in fair allocation of the bandwidth.

7.3.3   Fairness Guarantee

To prove that ADFS is fair, we need to prove a bound on $|\frac{W_f(t1,t2)}{\phi_f} - \frac{W_m(t1,t2)}{\phi_m}|$ for a sufficiently long interval $[t_1, t_2]$ in which both flows, f and m, are backlogged. To proceed, the following assumption is needed.

ASSUMPTION
To arrive at a fair scheduling scheme, we assume that there exists a weight vector ϕ_{ij} for a jth flow, ith packet, at each node l as

$$\phi_{ij} = \begin{bmatrix} \phi_{ijl} \\ \cdot \\ \cdot \\ \cdot \\ \phi_{ijm} \end{bmatrix} \tag{7.12}$$

REMARK 1
In fact, the weight update (Equation 7.9) ensures that the actual weight for the packet at each node converges close to its target value.

REMARK 2
ϕ_{ij} is finite for each flow at each node.
   Let $\tilde{\phi}_{ij}$ be the weight error defined as $\tilde{\phi}_{ij}$, given by

$$\tilde{\phi}_{ij} = \phi_{ij} - \hat{\phi}_{ij} \tag{7.13}$$

LEMMA 7.3.1

If the weights are updated as in Equation 7.9 for a sufficiently long interval $[t_1, t_2]$, *then the weight error* $\tilde{\phi}_{ij}(k+1)$ *is bounded, provided* $|\alpha| < 1$

PROOF Using Equation 7.9 and Equation 7.13, the weight estimation error is expressed as

$$\tilde{\phi}_{ij}(k+1) = \alpha.\tilde{\phi}_{ij}(k) + (1-\alpha)\phi_{ij} + \beta.E_{ij} \qquad (7.14)$$

Choose a Lyapunov function

$$V = \tilde{\phi}_{ij}^2(k) \qquad (7.15)$$

The first difference of the Lyapunov equation can be obtained as

$$\Delta V = V(k+1) - V(k) \qquad (7.16)$$

or

$$\Delta V = \tilde{\phi}_{ij}^2(k+1) - \tilde{\phi}_{ij}^2(k) \qquad (7.17)$$

Substituting Equation 7.14 in Equation 7.17 to get

$$\Delta V = [\alpha.\tilde{\phi}_{ij}(k) + (1-\alpha)\phi_{ij} + \beta.E_{ij}]^2 - \tilde{\phi}_{ij}^2(k) \qquad (7.18)$$

Equation 7.18 can be rewritten as

$$\Delta V = \alpha^2\tilde{\phi}_{ij}^2(k) + (1-\alpha)^2\phi_{ij}^2 + \beta^2 E_{ij}^2 + 2\alpha\,\tilde{\phi}_{ij}(1-\alpha)\phi_{ij}$$
$$+ 2(1-\alpha)\phi_{ij}\beta E_{ij} + 2\alpha\tilde{\phi}_{ij}\beta E_{ij} - \tilde{\phi}_{ij}^2(k) \qquad (7.19)$$

Equation 7.19 can be simplified as

$$\Delta V = -(1-\alpha^2)\tilde{\phi}_{ij}^2 + (1-\alpha)^2\phi_{ij}^2 + \beta^2 E_{ij}^2 + 2\alpha\,\tilde{\phi}_{ij}(1-\alpha)\phi_{ij}$$
$$+ 2(1-\alpha)\phi_{ij}\beta E_{ij} + 2\alpha\,\tilde{\phi}_{ij}\beta E_{ij} \qquad (7.20)$$

This further implies that

$$|\Delta V| \le -(1-\alpha^2)\left|\tilde{\phi}_{ij}\right|^2 + 2\alpha\left|\tilde{\phi}_{ij}\right| a + b, \qquad (7.21)$$

where

$$a = \|[(1-\alpha)\phi_{ij} + \beta E_{ij}]\| \tag{7.22}$$

and

$$b = \left|(1-\alpha)^2 \phi_{ij}^2 + \beta^2 E_{ij}^2 + 2(1-\alpha)\phi_{ij}\beta E_{ij}\right| \tag{7.23}$$

$$|\Delta V| \leq -(1-\alpha^2)\left[\left|\tilde{\phi}_{ij}\right|^2 - \frac{2\alpha}{(1-\alpha^2)}\left|\tilde{\phi}_{ij}\right|a - \frac{b}{(1-\alpha^2)}\right]. \tag{7.24}$$

$|\Delta V| \leq 0$ implies that

$$\left|\tilde{\phi}_{ij}\right| \geq \frac{\alpha + \sqrt{\alpha^2 a^2 + b(1-\alpha^2)}}{(1-\alpha^2)} \tag{7.25}$$

Let $B_{ij,\phi}$ be the bound on the weight estimation error, then

$$B_{ij,\phi} = \frac{\alpha + \sqrt{\alpha^2 + b(1-\alpha^2)}}{(1-\alpha^2)} \tag{7.26}$$

For $\left|\tilde{\phi}_{ij}\right| \geq B_{ij,\phi}$ $\Delta V < 0$. Because $\left|\tilde{\phi}_{ij}\right| \geq B_{ij,\phi}$, from Equation 7.13 we get

$$\hat{\phi}_{ij} \leq \sigma \phi_{ij} \tag{7.27}$$

for some σ.

LEMMA 7.3.2

The actual weights $\hat{\phi}_{ij}$ at each node using Equation 7.9 converge close to their target values in a finite time.

PROOF Because $|\alpha| < 1$, define $\tilde{\phi}_{ij}(k) = x(k)$, then Equation 7.14 can be expressed as

$$x(k+1) = c\,x(k) + d\,u(k) \tag{7.28}$$

where

$$c = \alpha, d = [(1-\alpha) \quad \beta], \quad u(k) = \begin{bmatrix} \phi_{ij} \\ E_{ij} \end{bmatrix} \tag{7.29}$$

This equation is a stable linear system (Brogan 1991), driven by a bounded input $u(k)$ (see Remark 2). According to linear system theory (Brogan 1991), $x(k)$ converges close to its target value in a finite time.

LEMMA 7.3.3

If flow f is backlogged throughout the interval $[t_1, t_2]$, then in an ADFS wireless node

$$\phi_{f,l} \cdot (v_2 - v_1) - l_f^{\max} \le W_f(t_1, t_2) \, , \tag{7.30}$$

where $v_1 = v(t_1)$ and $v_2 = v(t_2)$.

PROOF The steps in the proof follow similar to the method of Goyal et al. (1997).

Because $W_f(t_1, t_2) \ge 0$, if $\phi_{f,l}(v_2 - v_1) - l_f^{\max} \le 0$, Equation 7.30 holds trivially. Hence, consider the case where $\phi_{f,l}(v_2 - v_1) - l_f^{\max} > 0$, i.e., $v_2 > v_1 + \frac{l_f^{\max}}{\phi_{f,l}}$. Let packet p_f^k be the first packet of flow f that receives service in the open interval (v_1, v_2). To observe that such a packet exists, consider the following two cases:

CASE 1 Packet p_f^n such that $S(p_f^n) < v_1$ and $F(p_f^n) > v_1$ exists.

Because flow f is backlogged in $[t_1, t_2]$, we conclude $v(A(p_f^{n+1})) \le v_1$. From Equation 7.7 and Equation 7.8, we get:

$$S\left(p_f^{n+1}\right) = F\left(p_f^n\right) \tag{7.31}$$

Because $F(p_f^n) \le S(p_f^n) + \frac{l_f^{\max}}{\phi_{f,l}}$ and $S(p_f^n) < v_1$, we get:

$$S\left(p_f^{n+1}\right) < v_1 + \frac{l_f^{\max}}{\phi_{f,l}} \tag{7.32}$$

$$< v_2 \tag{7.33}$$

Because $S(p_f^{n+1}) = F(p_f^n) > v_1$, using Equation 7.33, we conclude $S(p_f^{n+1}) \in (v_1, v_2)$.

CASE 2 Packet p_f^n such that $S(p_f^n) = v_1$ exists p_f^n may finish service at time $t < t_1$ or $t \ge t_1$. In either case, because the flow f is backlogged in $[t_1, t_2]$, $v(A(p_f^{n+1})) \le v_1$. Hence, $S(p_f^{n+1}) = F(p_f^n)$.

Because $F(p_f^n) \le S(p_f^n) + \frac{l_f^{\max}}{\phi_{f,l}}$ and $S(p_f^n) < v_1$, we get,

$$S\left(p_f^{n+1}\right) \le v_1 + \frac{l_f^{\max}}{\phi_{f,l}} \tag{7.34}$$

$$< v_2 \tag{7.35}$$

Because $S(p_f^{n+1}) = F(p_f^n) > v_1$, using Equation 7.35, we conclude $S(p_f^{n+1}) \in (v_1, v_2)$.

Because either of the two cases always holds, we conclude that packet p_f^k such that $S(p_f^k) \in (v_1, v_2)$ exists. Furthermore from Equation 7.32 and Equation 7.34, we get

$$S\left(p_f^k\right) \le v_1 + \frac{l_f^{\max}}{\phi_{f,l}} \tag{7.36}$$

Let p_f^{k+m} be the last packet to receive service in the virtual time interval (v_1, v_2). Hence,

$$F\left(p_f^{k+m}\right) \ge v_2 \tag{7.37}$$

From Equation 7.36 and Equation 7.37, we conclude

$$F\left(p_f^{k+m}\right) - S\left(p_f^k\right) \ge (v_2 - v_1) - \frac{l_f^{\max}}{\phi_{f,l}} \tag{7.38}$$

But because flow f is backlogged in the interval (v_1, v_2), from Equation 7.7 and Equation 7.8 we know

$$F\left(p_f^{k+m}\right) = S\left(p_f^k\right) + \sum_{n=0}^{n=m} \frac{l_f^{k+n}}{\phi_{f,l}} \tag{7.39}$$

$$F\left(p_f^{k+m}\right) - S\left(p_f^k\right) = \sum_{n=0}^{n=m} \frac{l_f^{k+n}}{\phi_{f,l}} \tag{7.40}$$

Hence, from Equation 7.38 and Equation 7.40 we get

$$\sum_{n=0}^{n=m} \frac{l_f^{k+n}}{\phi_{f,l}} \ge (v_2 - v_1) - \frac{l_f^{\max}}{\phi_{f,l}} \tag{7.41}$$

$$\sum_{n=0}^{n=m} l_f^{k+n} \ge \phi_{f,l}(v_2 - v_1) - l_f^{\max} \tag{7.42}$$

Because $S(p_f^{n+1}) < v_2$, packet p_f^{k+m} is guaranteed to have been transmitted by t_2.

Hence, $W_f(t_1, t_2) \geq \sum_{n=0}^{n=m} l_f^{k+n}$, and the lemma follows.

LEMMA 7.3.4
In an ADFS-based wireless node, during any interval $[t_1, t_2]$

$$W_f(t_1, t_2) \leq \phi_{f,l}(v_2 - v_1) + l_f^{max} \tag{7.43}$$

where $v_1 = v(t_1)$ and $v_2 = v(t_2)$

PROOF The steps in the proof follow similar to the method of Goyal et al. (1997).

From the definition of ADFS, the set of flow f packets served in the interval $[v_1, v_2]$ have service tag of at least v_1 and at most v_2.

Hence, the set can be partitioned into two sets such as following:

- Set D consisting of packets that have service tag of at least v_1 and finish time at most v_2. Formally,

$$D = \left\{ k | v_1 \leq S\left(p_f^k\right) \leq v_2 \wedge F\left(p_f^k\right) \leq v_2 \right\} \tag{7.44}$$

From Equation 7.7 and Equation 7.8, we conclude

$$\sum_{k \in d} l_f^k \leq \phi_{f,l}(v_2 - v_1) \tag{7.45}$$

- Set E consisting of packets that have service tag at most v_2 and finish time greater than v_2. Formally,

$$E = \left\{ k \Big| v_1 \leq S\left(p_f^k\right) \leq v_2 \wedge F\left(p_f^k\right) > v_2 \right\} \tag{7.46}$$

Clearly, at most one packet can belong to this set. Hence,

$$\sum_{k \in E} l_f^k \leq l_f^{max} \tag{7.47}$$

From Equation 7.45 and Equation 7.47 we conclude that Equation 7.43 holds.

Because unfairness between two flows in any interval is maximum when one flow receives maximum possible service and the other minimum service. Theorem 7.3.1 follows directly from Lemma 7.3.3 and Lemma 7.3.4.

THEOREM 7.3.1

For any interval $[t_1, t_2]$ *in which flows f and m are backlogged during the entire interval, the difference in the service received by two flows at an ADFS wireless node is given as*

$$\left| \frac{W_f(t_1, t_2)}{\phi_{f,l}} - \frac{W_m(t_1, t_2)}{\phi_{m,l}} \right| \leq \frac{l_f^{\max}}{\phi_{f,l}} + \frac{l_m^{\max}}{\phi_{m,l}} \qquad (7.48)$$

REMARK 3

If $E_{ij} = 0$ at each node, then the proposed ADFS will become a DFS scheme (Vaidya et al. 2000).

REMARK 4

In Theorem 7.3.1, no assumption was made about the service rate of the wireless node. Hence, this theorem holds regardless of the service rate of the wireless node. This demonstrates that ADFS achieves fair allocation of bandwidth and thus meets a fundamental requirement of fair scheduling algorithms for integrated services networks.

7.3.4 Throughput Guarantee

Theorem 7.3.2 and Theorem 7.3.3 establish the throughput guaranteed to a flow by an ADFS FC and EBF service model, respectively, when appropriate admission control procedures are used.

THEOREM 7.3.2

If Q is the set of flows served by an ADFS node following the FC service model with parameters $(\lambda(t_1, t_2), \psi(\lambda))$, *and* $\sum_{n \in Q} \phi_{n,l} \leq \lambda(t_1, t_2)$, *then for all intervals* $[t_1, t_2]$ *in which flow f is backlogged throughout the interval,* $W_f(t_1, t_2)$ *is given as*

$$W_f(t_1, t_2) \geq \phi_{f,l}(t_2 - t_1) - \phi_{f,l} \frac{\sum_{n \in Q} l_n^{\max}}{\lambda(t_1, t_2)} - \phi_{f,l} \frac{\psi(\lambda)}{\lambda(t_1, t_2)} - l_f^{\max} \qquad (7.49)$$

PROOF   The steps in the proof follow similar to that of Goyal et al. 1997.

Let $v_1 = v(t_1)$ and let $\hat{L}(v_1, v_2)$ denote the aggregate length of packets served by the wireless node in the virtual time interval $[v_1, v_2]$. Then, from Lemma 7.3.4, we conclude

$$\hat{L}(v_1, v_2) \leq \sum_{n \in Q} \phi_{n,l}(v_2, v_1) + \sum_{n \in Q} l_n^{\max} \tag{7.50}$$

Because $\sum_{n \in Q} \phi_{n,l} \leq \lambda(t_1, t_2)$,

$$\hat{L}(v_1, v_2) \leq \lambda(t_1, t_2)(v_2, v_1) + \sum_{n \in Q} l_n^{\max} \tag{7.51}$$

Define v_2 as

$$v_2 = v_1 + t_2 - t_1 - \frac{\sum_{n \in Q} l_n^{\max}}{\lambda(t_1, t_2)} - \frac{\psi(\lambda)}{\lambda(t_1, t_2)} \tag{7.52}$$

Then, from Equation 7.51, we conclude

$$\hat{L}(v_1, v_2) \leq \lambda(t_1, t_2) \left(v_1 + t_2 - t_1 - \frac{\sum_{n \in Q} l_n^{\max}}{\lambda(t_1, t_2)} - \frac{\psi(\lambda)}{\lambda(t_1, t_2)} - v_1 \right) + \sum_{n \in Q} l_n^{\max} \tag{7.53}$$

$$\leq \lambda(t_1, t_2)(t_2 - t_1) - \psi(\lambda) \tag{7.54}$$

Let $\hat{t}_2$ be such that $v(\hat{t}_2) = v_2$. Also, let $T(w)$ be the time taken by the wireless node to serve packets with aggregate length w in its busy period. Then,

$$\hat{t}_2 \leq t_1 + T(\hat{L}(v_1, v_2)) \tag{7.55}$$

$$\leq t_1 + T(\lambda(t_1, t_2)(t_2 - t_1) - \psi(\lambda)) \tag{7.56}$$

From the definition of FC service model, we get

$$T(w) \leq \frac{w}{\lambda(t_1, t_2)} + \frac{\psi(\lambda)}{\lambda(t_1, t_2)} \tag{7.57}$$

From Equation 7.56 and Equation 7.57 we obtain

$$\hat{t}_2 \le t_1 + \frac{\lambda(t_1,t_2)(t_2-t_1)-\psi(\lambda)}{\lambda(t_1,t_2)} + \frac{\psi(\lambda)}{\lambda(t_1,t_2)} \tag{7.58}$$

$$\le t_2 \tag{7.59}$$

From Lemma 7.3.3, it is clear that

$$W_f(t_1,\hat{t}_2) \ge \phi_{f,l}(v_2-v_1)-l_f^{\max} \tag{7.60}$$

Since $\hat{t}_2 \le t_2$, using Equation 7.52 it follows that

$$W_f(t_1,t_2) \ge \phi_{f,l}(t_2-t_1)-\phi_{f,l}\frac{\sum_{n\in Q}l_n^{\max}}{\lambda(t_1,t_2)}-\phi_{f,l}\frac{\psi(\lambda)}{\lambda(t_1,t_2)}-l_f^{\max} \tag{7.61}$$

THEOREM 7.3.3

If Q is the set of flows served by an ADFS node following EBF service model with parameters $(\lambda(t_1,t_2),B,\omega,\psi(\lambda))$, $\gamma \ge 0$, and $\sum_{n\in Q}\phi_{n,l} \le \lambda(t_1,t_2)$, then for all intervals $[t_1,t_2]$ in which flow f is backlogged throughout the interval, $W_f(t_1,t_2)$ is given as

$$P\left(W_f(t_1,t_2) < \phi_{f,l}(t_2-t_1)-\phi_{f,l}\frac{\sum_{n\in Q}l_n^{\max}}{\lambda(t_1,t_2)}\right.$$
$$\left. -\phi_{f,l}\frac{\psi(\lambda)}{\lambda(t_1,t_2)}-\phi_{f,l}\frac{\gamma}{\lambda(t_1,t_2)}-l_f^{\max}\right) \le Be^{-\omega\gamma} \tag{7.62}$$

7.3.5 Delay Guarantee

Generally, a network can provide a bound on delay only if its capacity is not exceeded. The weight $\phi_{f,l}$ can also mean the rate assigned to a packet of flow f at node l. Let the rate function for flow f at virtual time v, denoted by $R_f(v)$, be defined as the rate assigned to the packet that has start tag less than v and finish tag greater than v. Formally,

$$R_f(v) = \begin{cases} \phi_{f,l} & \text{if} \quad \exists j \ni \left(S(p_f^j) \le v < F(p_f^j)\right) \\ 0 & \text{otherwise} \end{cases} \tag{7.63}$$

Let Q be the set of flows served by the node. Then, an FC or EBF node with average rate $\lambda(t_1, t_2)$, is defined to have exceeded its capacity at virtual time v if $\sum_{n \in Q} R_n(v) > \lambda(t_1, t_2)$. If the capacity of a SFQ–based node is not exceeded, then it guarantees a deadline to a packet based on its expected arrival time. *Expected arrival time* of packet P_f^j, denoted by $T_a(P_f^j, \phi_{f,j})$, is defined as

$$\max \left\{ A\left(P_f^j\right), T_a\left(P_f^{j-1}, \phi_{f,j-1}\right) + \frac{l_f^{j-1}}{\phi_{f,j-1}} \right\} \quad j \geq 1 \qquad (7.64)$$

where $T_a(P_f^0, \phi_{f,0}) = -\infty$. A deadline guarantee based on expected arrival time has been referred to as *delay guarantee*. Theorem 7.3.4 and Theorem 7.3.5 establish the delay guarantee for FC and EBF service models, respectively, and follow the steps for proof similar to that of Goyal et al. 1997.

THEOREM 7.3.4
If Q is the set of flows served by an ADFS node following the FC service model with parameters $(\lambda(t_1, t_2), \psi(\lambda))$, and $\sum_{n \in Q} R_n(v) \leq \lambda(t_1, t_2)$ for all v, then the departure time of packet P_f^j at the node, denoted by $T_d(P_f^j)$, is given by

$$T_d\left(P_f^j\right) \leq T_a\left(P_f^j, \phi_{f,j}\right) + \sum_{n \in Q \wedge n \neq f} \frac{l_n^{\max}}{\lambda(t_1, t_2)} + \frac{l_f^j}{\lambda(t_1, t_2)} + \frac{\psi(\lambda)}{\lambda(t_1, t_2)} \qquad (7.65)$$

PROOF Let H be defined as follows:

$$H = \left\{ m | m > 0 \wedge S\left(P_f^m\right) = v\left(A\left(P_f^m\right)\right) \right\} \qquad (7.66)$$

Let $k \leq j$ be largest integer in H. Also, let $v_1 = v(A(P_f^k))$ and $v_2 = S(P_f^k)$. Observe that as the node virtual time is set to the maximum finish tag assigned to any packet at the end of a busy period, packets P_f^k and P_f^j are served in the same busy period of a wireless node. From the definition of ADFS, the set of flow f packets served in the interval $[v_1, v_2]$ have a start tag at least v_1 and at most v_2. Hence, the set can be partitioned into two sets:

- This set consists of packets that have start tag at least v_1 and finish tag at most v_2. Formally the set of packets of flow n, denoted by D_n, in this set is

$$D_n = \left\{ m | v_1 \leq S\left(p_n^m\right) \leq v_2 \wedge F\left(p_n^m\right) \leq v_2 \right\} \qquad (7.67)$$

Then, from the definition of $R_n(v)$ and $F(P_n^m)$, we know that the cumulative length of such flow n packets served by wireless node in the virtual time interval $[v_1, v_2]$, denoted by $C_n(v_1, v_2)$, is given as

$$C_n(v_1, v_2) \le \int_{v_1}^{v_2} R_n(v)dv \qquad (7.68)$$

Hence, aggregate length of packets in this set, $\sum_{n \in Q} C_n(v_1, v_2)$, is given as

$$\sum_{n \in Q} C_n(v_1, v_2) \le \sum_{n \in Q} \int_{v_1}^{v_2} R_n(v)dv \qquad (7.69)$$

$$\le \int_{v_1}^{v_2} \lambda(t_1, t_2)dv \qquad (7.70)$$

$$\le \lambda(t_1, t_2).(v_2 - v_1) \qquad (7.71)$$

But because $v_2 = S(P_f^k)$, from the definition of k, $v_2 - v_1 = \sum_{n=0}^{n=j-k-1} \dfrac{l_f^{k+n}}{\phi_{f,k+n}}$. Hence,

$$\sum_{n \in Q} C_n(v_1, v_2) \le \lambda(t_1, t_2) \sum_{n=0}^{n=j-k-1} \frac{l_f^{k+n}}{\phi_{f,k+n}} \qquad (7.72)$$

- This set consists of packets that have a start tag at most v_2 and finish tag greater than v_2. Formally, the set of packets of flow n, denoted by E_n, in this set is

$$E_n = \left\{ m | v_1 \le S\left(p_n^m\right) \le v_2 \wedge F\left(p_n^m\right) > v_2 \right\} \qquad (7.73)$$

Clearly, at most one packet of flow n can belong to this set. Furthermore, $E_f = \{j\}$. Hence, the maximum aggregate length of packets in this set is

$$\sum_{n \in Q \wedge n \neq f} l_n^{\max} + l_f^j \qquad (7.74)$$

Hence, the aggregate length of packets served by the wireless node in the interval $[v_1, v_2]$, denoted by $\hat{L}(v_1, v_2)$, is

$$\hat{L}(v_1, v_2) \le \lambda(t_1, t_2) \sum_{n=0}^{n=j-k-1} \frac{l_f^{k+n}}{\phi_{f,k+n}} + \sum_{n \in Q \wedge n \neq f} l_n^{\max} + l_f^j \tag{7.75}$$

Let T(w) be the time taken by the wireless node to serve packets with aggregate length w in its busy period. From the definition of FC service model, we get

$$T(w) \le \frac{w}{\lambda(t_1, t_2)} + \frac{\psi(\lambda)}{\lambda(t_1, t_2)} \tag{7.76}$$

Because packet P_f^j departs at system virtual time v_2 and all the packets served in the time interval $[v_1, v_2]$ are served in the same busy period of the wireless node, we get

$$A\left(p_f^k\right) + T(\hat{L}(v_1, v_2)) \ge T_d\left(P_f^j\right) \tag{7.77}$$

$$A\left(p_f^k\right) + \sum_{n=0}^{n=j-k-1} \frac{l_f^{k+n}}{\phi_{f,k+n}} + \sum_{n \in Q \wedge n \neq f} \frac{l_n^{\max}}{\lambda(t_1, t_2)} + \frac{l_f^j}{\lambda(t_1, t_2)} + \frac{\psi(\lambda)}{\lambda(t_1, t_2)} \ge T_d\left(P_f^j\right) \tag{7.78}$$

From Equation 7.64, we get

$$T_a\left(P_f^j, \phi_{f,j}\right) + \sum_{n \in Q \wedge n \neq f} \frac{l_n^{\max}}{\lambda(t_1, t_2)} + \frac{l_f^j}{\lambda(t_1, t_2)} + \frac{\psi(\lambda)}{\lambda(t_1, t_2)} \ge T_d\left(P_f^j\right) \tag{7.79}$$

THEOREM 7.3.5

If Q is the set of flows served by an ADFS node following EBF service model with parameters $(\lambda(t_1, t_2), B, \omega, \psi(\lambda))$, $\gamma \ge 0$, and $\sum_{n \in Q} R_n(v) \le \lambda(t_1, t_2)$ for all v, then the departure time of packet P_f^j at the node, denoted by $T_d(P_f^j)$, is given by

$$P\left(T_d\left(P_f^j\right) \le T_a\left(P_f^j, \phi_{f,j}\right) + \sum_{n \in Q \wedge n \neq f} \frac{l_n^{\max}}{\lambda(t_1, t_2)} \right.$$

$$\left. + \frac{l_f^j}{\lambda(t_1, t_2)} + \frac{\psi(\lambda)}{\lambda(t_1, t_2)} + \frac{\gamma}{\lambda(t_1, t_2)} \right) \ge 1 - Be^{-\omega\gamma} \tag{7.80}$$

Theorem 7.3.4 and Theorem 7.3.5 can be used to determine delay guarantees even when a node has flows with different priorities and services them in the priority order.

THEOREM 7.3.6
The end-to-end delay denoted by $T_{EED}\left(P_f^j\right)$, *is given by*

$$T_{EED}\left(P_f^j\right) = \sum_{i=1}^{m}\left(T_{d,i}\left(P_f^j\right) - T_{a,i}\left(P_f^j, \phi_{f,j}\right)\right) + T_{prop}\left(P_f^j\right) \qquad (7.81)$$

where $T_{d,i}(P_f^j)$ and $T_{a,i}(P_f^j, \phi_{f,j})$ are the departure time and expected arrival time of packet P_f^j at hop i in the multihop network. T_{prop} is the total propagation delay experienced by the packet, from source to destination.

REMARK 7
As expected the end-to-end delay is a function of packet length, channel uncertainties, and backoff interval of the CSMA/CA protocol.

7.3.6 Overhead Analysis

Analysis is performed to estimate the overhead for data transmission using the proposed ADFS protocol. The additional overhead in ADFS protocol is due to inclusion of the current weight value in the header field of the data packet. Note that the weight information is only transmitted for the data packets but not for the request to send (RTS), clear to send (CTS) and acknowledgment (ACK) packets. For the overhead analysis, we denote the four bytes required for transmitting the weight, separately from the actual data packet. The term T_x is defined as the size of the packet type x. The *efficiency of the protocol* is defined as the ratio of data portion (in bytes) in each data transmission to the total information transmitted including the control message overhead, and it is given by

$$\eta = \frac{T_{data}}{T_{data} + T_{weight} + T_{RTS} + T_{CTS} + T_{ACK}} \qquad (7.82)$$

The packet sizes specified in the IEEE 802.11 MAC protocol are used. The selected packet sizes are 24 bytes for RTS, 18 bytes for CTS, 20 bytes for ACK, 6 backoff slots (each slot is set to be 5 B), 512 bytes for data, and 4 bytes for weight. Then, the efficiency of the proposed ADFS protocol obtained from Equation 7.82 is around 84.2% with the additional overhead due to the addition of weights being less than 16%. This efficiency value is acceptable to meet the performance. The IEEE 802.11 MAC protocol renders 84.8% for the same packet size. It can be observed that the additional overhead due to the weight transmission is only about 0.6%. With very little increase in transmission and computational overheads, the proposed ADFS protocol is able to achieve significant increase in fairness and aggregate throughput, as shown in the next section.

7.4 Performance Evaluation

The performance of the proposed ADFS protocol was evaluated using NS-2 simulator [5] as an extension to wireless networks. ADFS requires modifications at both the MAC protocol level and at the interface queue level. The following values were used for all the simulations, unless otherwise specified: channel bandwidth is taken as 2 Mbps, with $\alpha = 0.9$, $\beta = -0.1$, $SF = 0.02$, sum of initial weights of all flows is equal to unity, and ρ is a random variable uniformly distributed in the interval [0.9, 1.1]. The AODV routing protocol was used. The constant bit rate (CBR) traffic was used with flows always backlogged, and the packet size was 584 bytes.

Example 7.4.1: Star Topology

To evaluate the fairness of the proposed ADFS protocol, consider a star topology with 16 wireless nodes transmitting to a destination and the flows from each source have a weight of 1/16.

Figure 7.3 represents the throughput/weight (normalized weights) ratio vs. the flows for the star topology with 16 wireless nodes. Ideally, the throughput to initial weight curve should be a straight line parallel to the *x*-axis for fair scheduling schemes. It is visible that the ADFS results in a fair allocation of bandwidth compared to the 802.11 MAC protocol. Figure 7.4 presents the delay variations for the star topology. Delay variation is calculated as the difference between the end-to-end delays for successive packets received at the destination. It can be observed that ADFS results in minimal delay variations, whereas those of 802.11 MAC protocol are relatively high. Huge delay variations can degrade the QoS of an ad hoc network.

The performance of the ADFS protocol was also evaluated for networks with varying packet sizes and initial weights. The results shown in Figure 7.5 are for a network with star topology having 16 flows. Packet sizes of 584, 328, 400, and 256 are used for different flows. It can be seen that ADFS achieves fair allocation of bandwidth, even when packets of different sizes were used. Figure 7.6 depicts the throughput/weight ratio for the flows having different initial weights. The initial weights of the

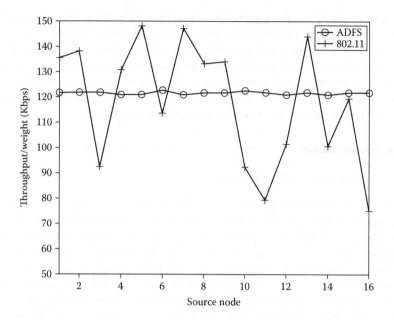

FIGURE 7.3
ADFS performance.

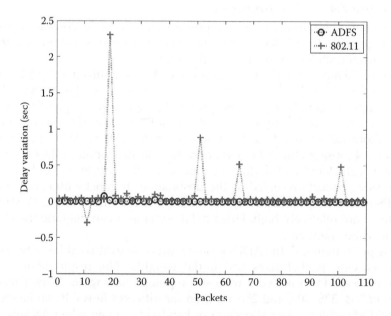

FIGURE 7.4
Delay variations.

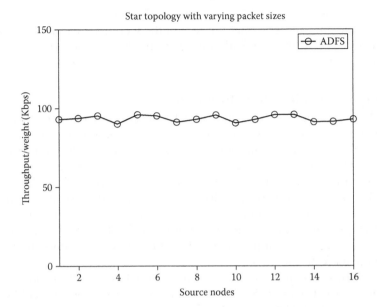

FIGURE 7.5
Star topology with varying packet sizes.

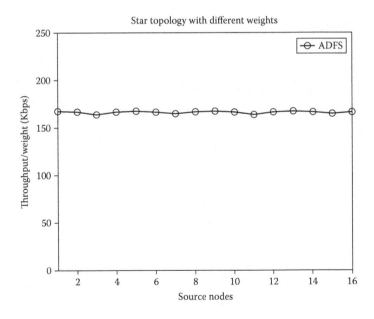

FIGURE 7.6
Star topology with different initial weights.

flows are chosen as 0.1, 0.075, 0.05, and 0.025, with 4 flows selecting each of these weights. Observe that the sum of the weights is equal to one. The figure shows that the ADFS algorithm results in fair allocation of bandwidth even when flows have been assigned different initial weights.

Example 7.4.2: Random Topologies

The ADFS protocol is now simulated for scenarios with n nodes, and $\frac{n}{2}$ flows where the weight of each flow is $\frac{2}{n}$. In general, initial weights are assigned based on user specified QoS. Figure 7.7, Figure 7.8, and Figure 7.9 represent the throughput/weight (normalized weights) ratio vs. the flows for networks with 32, 64, and 128 nodes, respectively. Plots show that ADFS and DFS result in fair allocation of bandwidth compared to the 802.11 MAC protocol. Moreover, it can be observed that, though ADFS and DFS give almost the same degree of fairness, ADFS renders 10 to 20% higher throughput compared to DFS, due to weight adaptation and dynamic backoff intervals.

The results shown in Figure 7.10 are for a network with 32 nodes and 16 flows. Packet sizes of 584, 328, 400, and 256 are used for different flows. It can be seen that ADFS achieves fair allocation of bandwidth, even when packets of different sizes were used. Figure 7.11 depicts the throughput/ weight ratio for the flows having different initial weights. The initial

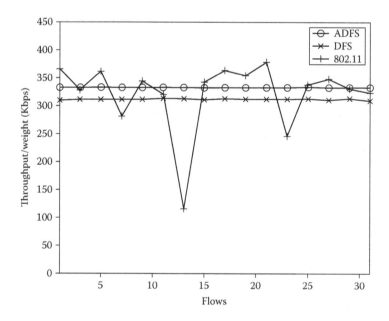

FIGURE 7.7
Performance of ADFS with 32 nodes.

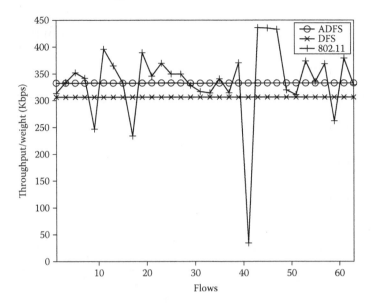

FIGURE 7.8
Performance of ADFS with 64 nodes.

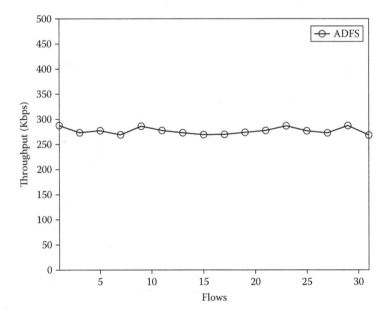

FIGURE 7.9
Performance of ADFS with varying packet sizes.

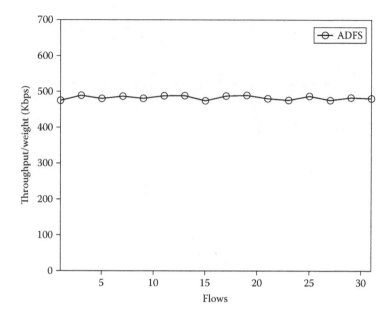

FIGURE 7.10
Performance of ADFS with different initial weights.

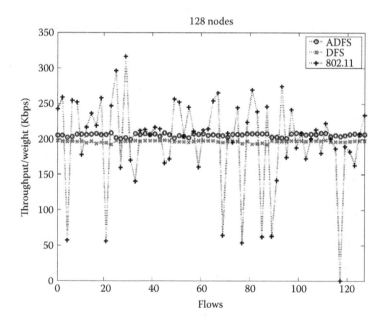

FIGURE 7.11
Performance of ADFS with 128 nodes.

weights of the flows are chosen as 0.1, 0.075, 0.05, and 0.025, with 4 flows selecting each of these weights. Observe that the sum of the weights is equal to one. The figure shows that the ADFS algorithm results in fair allocation of bandwidth even when flows have been assigned different initial weights.

The *fairness index* (FI) (Bennett 1996) is defined as

$$FI = \frac{\left(\sum_f \frac{T_f}{\phi_f} \right)^2}{\eta^* \sum_f \left(\frac{T_f}{\phi_f} \right)^2} \qquad (7.83)$$

where T_f is the throughput of flow f and η is the number of flows. Figure 7.12 displays the FI of networks with different numbers of flows. Note that the fairness indices of ADFS and DFS are close to one, whereas that of IEEE 802.11 is less than one. This confirms that the proposed ADFS and DFS render fair allocation of bandwidth, whereas IEEE 802.11 does not. Delay variations in the network with 64 nodes are presented in

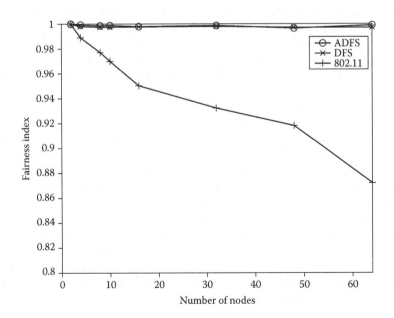

FIGURE 7.12
Fairness index comparison.

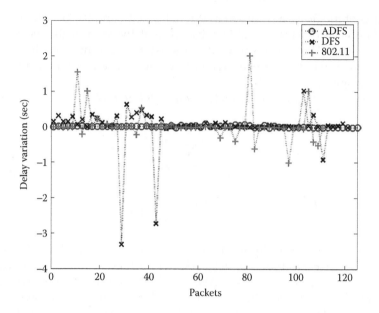

FIGURE 7.13
Delay variations.

Figure 7.13. It can be observed that ADFS results in minimal delay variations, whereas those of DFS and IEEE 802.11 are relatively high. Huge delay variations can adversely affect the QoS of the network.

To evaluate the performance of the proposed algorithm, the network with 32 nodes was simulated with different flow rates, such as 200, 250, 300, 350, 400, and 500 kbps per flow. Figure 7.14 presents the throughput/ weight ratios of 4 randomly selected flows (4, 10, 12, and 26), for the ADFS, DFS, and IEEE 802.11 protocols with different flow rates. The channel gets congested around 300 kbps per flow rate, after which the fairness of the protocols is evaluated. Observe that using ADFS and DFS protocols, the four flows get equal throughput/weight ratios after 300 kbps per flow, whereas the throughput/weight ratios of the flows using IEEE 802.11 keep varying. Although both ADFS and DFS result in fair allocation of bandwidth, notice that the ADFS results in better throughput per flow. The FI of the network is calculated using Equation 7.83. Figure 7.15 presents the fairness indices of the network with different per-flow rates. The fairness indices of the ADFS and DFS protocols are close to one, showing fair allocation of bandwidth, while those of 802.11 keep decreasing with increase in per flow rates.

The contention time for a packet is calculated as the time from when the packet is received by the MAC layer for transmission, to the time when it receives a successful CTS message to transmit the packet.

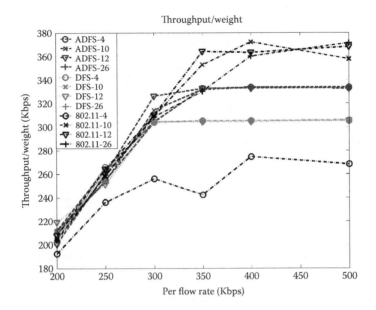

FIGURE 7.14
Performance evaluation with low rates.

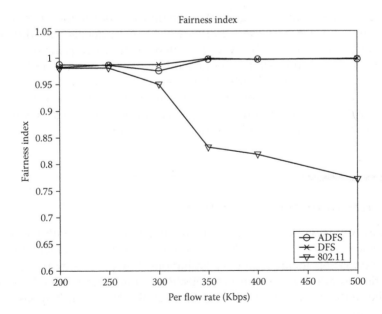

FIGURE 7.15
Fairness index with different flow rates.

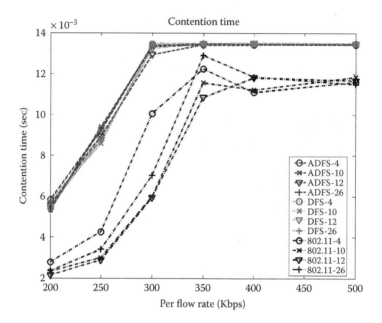

FIGURE 7.16
Contention times with different flow rates.

Figure 7.16 shows the average contention times of four randomly selected flows, using ADFS, DFS, and IEEE 802.11 MAC protocols with different flow rates. Observe that the average contention time keeps increasing till the channel gets congested (around 300 Kbps per flow rate), and almost remains constant after 300 Kbps. It can also be noted that there is small increase in the average contention times using both ADFS and DFS protocols, whereas compared to 802.11 protocol, due to the slight computational complexity involved in using these protocols.

7.5   Hardware Implementation

As part of this work, hardware implementation of WSNs with the ADFS protocol was also shown. Challenges for hardware implementation of ADFS on WSNs include memory limitations, low processing power, and selection of priority sensor flows. Selection of fair bandwidth allocations for the sensor flows, or based on QoS, must also be established based on user requirements.

Introduction of the 802.15 standard has accelerated the application of WSNs in industrial environments. Use of small, low-power, radio-enabled networks provide observability in a cost-effective and deployable platform. Research into WSNs has shown the ability to provide dynamic routing (Ratnaraj et al. 2006), intelligent processing of data, and observability in harsh environments. Because bandwidth is a major constraint in WSNs, one key factor in guaranteeing the QoS is to manage radio resources effectively and fairly. The focus of this work is to address challenges in, and present, a hardware implementation of a fair scheduling network protocol (Regatte and Jagannathan 2004) on a WSN test bed.

Applications for work presented here are based on industrial needs for distributed sensing. Consider the case of distributed sensing of air pressure and volumetric air flow in compressed air systems used for tooling. In this case, sensed parameters must be communicated back to a base station (BS) without aggregation or sensor fusion to provide observability for each parameter. Thus, fair scheduling of sensor flows is needed to provide equal observability to all priority measurands. Moreover, nonaggregated data are required to allow analysis of the independent sensors.

Work presented in this section from Fonda et al. (2006) focuses on implementation of the ADFS scheduling protocol. The ADFS protocol was initially developed at University of Missouri–Rolla (UMR) for ad hoc wireless networks. In this chapter, the protocol is ported to WSN and implemented on a hardware platform developed at UMR. Hardware implementation is based on a platform developed at the UMR. Hardware-testing results are shown to provide a comparison of the performance of the ADFS scheme. Hardware implementation provides comparison of the ADFS in a single hardware cluster. Intercluster scheduling provides useful bandwidth allocation to allow the cluster head (CH) to route the sensor information through the rest of the network using the optimal energy delay subnetwork routing (OEDSR) (Ratnaraj et al. 2006).

In this section, an overview of the hardware implementation of the ADFS scheduling protocol is given. Use of customized hardware for development of sensing, processing, and networking is also presented. A description of capabilities, limitations, and support for networking applications is next presented. Also in this section, an overview of the software architecture is given with respect to the ADFS protocol and its requirements on the hardware.

7.5.1 UMR Mote Description

Hardware for implementation of the ADFS was selected to be energy conservative, performance oriented, and of small form-factor. Use of Silicon Laboratories® 8051 variant hardware was selected for its ability to provide fast 8-bit processing, low power consumption, and ease of interfacing to peripheral hardware components. A Maxstream XBee™ RF module was also employed for this work. The use of external RAM (XRAM),

UART interfacing, and A/D sensing allow the microcontroller to perform the tasks needed for a sensor node platform. Next, a treatment of the hardware capabilities and limitations will be given.

Hardware implementation of any algorithm is constrained by the limitations of the hardware. Use of specific hardware must be weighed against the precision, speed, and criticality of an algorithm's implementation. For this protocol, low power consumption was given the highest priority. In turn, the demand for low power limits the types of processor architectures that can be deployed. The selection of the Silicon Laboratories 8051 variants was based on these criteria. Limitations for the implementation that are incurred through the use of the 8051 variant family are a small memory space and a maximum processing speed. In the next section, a description of the specifications for the hardware implemented nodes will be given.

This section outlines the hardware and software components of the ADFS implementation. A presentation of hardware capabilities is given. The software implementation is also discussed. Software architecture, control-flow, and hardware implications are shown. In this section, the system architecture of the nodes is discussed.

7.5.2   Sensor Node Hardware

The generation-4 smart sensor node (G4-SSN), seen in Figure 7.17, was originally developed at UMR and subsequently updated at St. Louis University (SLU). The G4-SSN has various abilities for sensing and processing.

FIGURE 7.17
G4-SSN.

TABLE 7.1

G4-SSN Capabilities

	Ic @ 3.3V [mA]	Flash Memory [bytes]	RAM [bytes]	ADC Sampling Rate [kHz]	Form-Factor	MIPS
G4-SSN	35	128k	8448	100 @ 10/12-bit	100-pin LQFP	100

The former include strain gauges, accelerometers, thermocouples, and general A/D sensing. The latter include analog filtering, compact flash (CF) memory interfacing, and 8-bit data processing at a maximum of 100 MIPS.

7.5.3 G4-SSN Capabilities

The abilities of the UMR and SLU motes nodes are shown in Table 7.1. As seen in the table, the G4-SSN provides powerful 8-bit processing, a suitable amount of RAM, and a low-power small form-factor. Another strong point of the G4-SSN is the available code space found on the Silicon Laboratories C8051F12x variant.

ADFS requires that the sensor nodes are synchronized. For this reason, a test was performed on the G4-SSN real-time clock (RTC) capabilities. Experimentation consisted of a statistical analysis of the RTC accuracy. Extended use of the RTC without resynchronization can cause drift to occur, and this drift must be quantified to provide a confidence measure of the RTC. A 32.768-kHz quartz crystal is used to feed a timer on the 8051 and is used at the time-base of the RTC. The RTC was allowed to run for 10 min, and the results are tabulated in Table 7.2.

As seen in Table 7.2, the RTC has a drift error of around a ∫ sec over 10 min; this translates to 3.5 sec/h. In the context of this application, this drift is acceptable as the RTC will be synchronized with the BS every 30 sec.

7.5.4 Hardware Results

In this section, hardware implementation results are shown. Using the 802.15.4 standard with 250kbps RF data bandwidth, the scheduling algorithm

TABLE 7.2

G4-SSN RTC Drift Testing Results

Test Time [min]	t [sec]	Variance [μ-sec]	Mean [sec]	STD [sec]	Error [sec]
10	0.05	50.45	0.0504	0.007	0.515

is tested. CBR traffic is generated on the source nodes and is routed to the BS via OEDSR (Ratnaraj et al. 2006). The nodes internally provide 38.4 kbps throughput to the 802.15.4 module. There is no data aggregation, or data fusion, performed yet in the network because the considered application requires data from independent locations, whereas the proposed scenario is a good test for queuing schemes for testing fairness. Due to hardware limitations of the 802.15.4 module, interfacing the back off interval time slots are constrained to a minimum of 15 msec. This limits the overall performance of the implementation; however, the issue can be addressed in future work using the Chipcon CC2420.

Testing of the hardware implementation is now discussed. The results were obtained by use of a star topology with five source nodes and a single CH. CBR traffic is generated at each source node, and the initial weight of each flow is equal to 1/5 and a value of = 0.4, and = 0.6. Other parameters include SF = 0.032 and the packet length of maximum 100 bytes with an 88-byte data payload. During testing, the BS is used to record network activity for analysis. Performance of the ADFS implementation is evaluated using the exponential back off scheme and drop-tail queuing. A comparison of these two methods shows the performance increase in the ADFS enabled network.

The ADFS scheduling scheme takes into account the weight of the packets and proportionally allocates bandwidth to the flows. In contrast, the networks without ADFS functionality lack the ability to differentiate QoS in this manner. Thus, poor fairness is observed in networks without ADFS.

In Table 7.3, the results for throughput and FI are shown. In case of ADFS implementation, the throughput is higher for every flow because it is able to maintain a steady and proportional traffic, thus reducing buffer overflows. The reference scheme is not able to distribute the available bandwidth proportionally to all flows; thus, the buffer overflows occur more often than in case of ADFS protocol. Overall, the ADFS network achieves a 13.3% increase in the throughput over the first-in-first-out (FIFO) queued scheme.

Moreover, a larger FI value is observed for the ADFS implementation than for the reference scheme. The ADFS scheme allocates the resources proportionally to the packet weight, thus adapting to a changing channel and network state.

TABLE 7.3

Throughput and Fairness Comparison Results

	Flow 1 [kB/sec]	Flow 2 [kB/sec]	Flow 3 [kB/sec]	Flow 4 [kB/sec]	Flow 5 [kB/sec]	Overall	FI
ADFS	88.8	85.1	84.7	81.7	81.0	84.3	0. 9989
FIFO	83.8	67.7	76.7	68.7	75.3	74.4	0. 9938

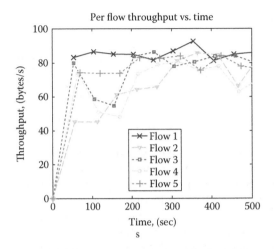

FIGURE 7.18
Throughput of FIFO queued network.

In Figure 7.18 and Figure 7.19, the throughput of each flow for a star-topology using drop-tail queuing and proposed ADFS, respectively, are shown. The non-ADFS protocol experiences high throughput variance, whereas the ADFS system provides more constant performance through establishing fairness of the scheduling. Moreover, ADFS achieves higher overall throughput because it allows all flows to share bandwidth in a more even manner over time when compared to FIFO queued network. This clearly demonstrates the fairness of ADFS protocol.

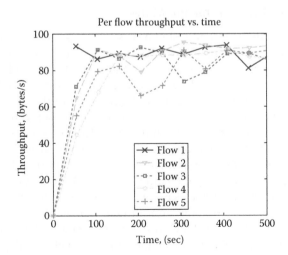

FIGURE 7.19
Throughput of ADFS enabled network.

TABLE 7.4

Delay Comparison

[sec]		Flow 1	Flow 2	Flow 3	Flow 4	Flow 5	Overall
ADFS	mean	11.24	11.41	11.69	11.56	11.30	11.44
	standard	1.011	1.167	1.068	1.015	0.834	0.17
FIFO	Mean	11.11	11.60	11.71	11.61	12.30	11.66
	standard	1.434	1.12	1.459	1.212	1.512	0.38

In Table 7.4, the results for average end-to-end delay and its standard variation (std) are shown. The mean delay values for both protocols are similar with ADFS, achieving smaller delay for four flows out of five flows when compared to the FIFO protocol with drop-tail queue and exponential backoff. Overall, the improvement of 2% in end-to-end delay is observed for the ADFS scheme over the FIFO queue even though there is an additional overhead with the proposed protocol. The advantage of ADFS is due to a fair allocation of radio resources because the ADFS selects a back off interval proportionally to a packet weight. Moreover, the standard deviation is higher in case of the FIFO-queued scheme due to higher variation in the back off.

The results show that the proposed protocol can achieve fair allocation of bandwidth with a 13.3% increase in throughput, slightly lower end-to-end delays, and delay variations reduced by 55% (standard), resulting in a better QoS. This shows the ability of the ADFS scheduler to be effectively applied to WSN systems to enhance network performance.

7.5.5 Future Direction for the Motes

Future work on the UMR and SLU motes includes increased capabilities for RF communications, memory density, and efficient networking implementations. Specifically, a discussion of the current state and direction of the RF communication module is now given. Currently, a Maxstream XBee™ module is used to implement the RF layer of the G4-SSN. Future introduction of the Chipcon CC2420 transceiver chipset to the G4-SSN is now being developed at UMR. With this introduction, several advantages are perceived. The primary perceived advantage is direct access to the physical layer in terms of parameter access and control. For example, direct and real-time sensing of channel access (CA) state will reduce back off interval slots. Next, power savings are expected with the CC2420 on the order of 62%. Additionally, more finely adjustable RF transmission power levels will allow for DPC in the WSN. The CC2420 also provides a smaller physical footprint to enable higher level of integration and miniaturization. Finally, the subsequent frame-processing step on the XBee™ is avoided, and frames are passed directly to the transceiver.

7.6 Energy-Aware MAC Protocol for Wireless Sensor Networks

Sensor networks are an important emerging area of wireless networking. They originated from ad hoc networks; however, the application requirements are different from an ad hoc network perspective. Because of the hardware and application considerations, the sensor networks are resource-limited. The main objective of a sensor network is to collect and forward sampled information about the environment. Normally, a sensor generates data only when there is an unusual event with long idle periods between when no data are collected. However, during transmission of data from the unusual event, a particular sensor node can generate an intense traffic. To minimize redundancy and network congestion, the traffic is aggregated within groups of sensors called *clusters* in which a sensor node acts as a CH. Thus, a cluster topology is normally used. A large amount of data is typically forwarded through the whole sensor network via the CHs toward the BS where it is tagged for further analysis. In certain sensor network applications, for example military and planetary exploration, QoS requirements, in terms of finite end-to-end delay, low loss in information along with high throughput have to be met in the presence of fading channels to make useful decisions with the data. In these applications, energy efficient protocols that can render satisfactory QoS are necessary to extend the life time of the sensor nodes.

Sensor nodes are energy constrained because they run on batteries. Hence, it is desirable to conserve energy even with long idle periods between transmissions. It was shown in Reghunathan et al. 2002 that energy consumption while listening or receiving is significant (similar as in transmission mode). Thus, a large amount of energy can be saved by turning off the RF circuitry during these idle periods. Rigorous work in this area has been in place (Woo and Culler 2001, Ye et al. 2002, Singh and Raghavendra 1998) for asynchronous protocols. PAMAS (Singh and Raghavendra 1998) and S-MAC (Ye et al. 2002) save energy by periodically placing the nodes into a sleep mode. However, they still require nodes to listen to the radio channel for a significant amount of time, and energy conservation is moderate. On the other hand, LEACH protocol (Heinzelman et al. 2002) uses time division multiple access (TDMA) to reduce the time spent on radio communication. The nodes are assigned time slots for the transmission and the reception of data. These nodes sleep for the rest of the TDMA cycle. Hence, an improvement in energy is observed when compared to an asynchronous protocol. However, maintaining synchronization throughout the network is difficult, and TDMA protocols are inflexible and not dynamic in the allocation of radio resources.

Moreover, resources such as bandwidth and buffer space at the sensor nodes are quite limited. Fairness becomes a critical issue in the presence of tight resource constraints and when several sensor nodes access the shared wireless channel in the CSMA/CA paradigm. For example, during a forest fire application, not only sensors within a cluster collect a large amount of data, but also it is desirable to collect the same amount of data from each deployed cluster, so that the temperature gradient can be inferred (Woo and Culler 2001). Therefore, a fair allocation of bandwidth from each node over multihops is required to transmit data to the BS. Thus, the scheduling protocol should deliver data from each sensor node fairly during unusual events, and it also must incorporate the time-varying channel state to meet certain throughput and end-to-end delay requirements.

In this section, we introduce a novel sleep mode for CSMA-based asynchronous networks during idle periods, thus reducing energy spent over the radio communication to a minimum. A cluster topology similar to that of Heinzelman et al. (2002) is utilized with only periodic communication between sensors and its CH. Thus, sensors are in sleep mode for most of the time, saving energy. However, in the case of an event, a measuring circuit will be able to wake up the sensor to process and send the data to the CH. During sleep mode, because the sensors do not listen for an incoming traffic, a query or maintenance packet sent from the CH to the sensor will not be received, which is clearly a drawback. However, the CHs buffer packets for the sensor and delivers them periodically in the sensor wake state. Because sensors usually send data whereas only sporadically receive packets or queries, the proposed sleep mode will greatly increase energy savings while minimizing latency.

A large amount of energy is also consumed during active periods when the nodes transmit data. In the case of an unusual event, the traffic generated by the sensors can be huge. A significant percentage of energy will be spent on transmission; hence, transmitter power control is important to save energy. The DPC algorithm from Jagannathan et al. (2006) and from Zawodniok and Jagannathan (2004), for cellular and ad hoc networks, takes into account different radio channel uncertainties in a wireless network: path loss, shadowing, and Rayleigh fading. The DPC implementation proposed in Zawodniok and Jagannathan (2004) was designed for ad hoc, nonpersistent CSMA/CA wireless networks. In this section, this MAC protocol utilizes a scheme for sleep mode in clustered, CSMA-based sensor networks, thus increasing the lifetime of sensor nodes while meeting the application constraints. Moreover, the proposed protocol uses the DPC scheme from the previous chapter to estimate the time-varying channel and to update the transmitter power.

Finally, a DFS protocol from the previous section is incorporated with the proposed energy-aware scheme so that certain QoS requirements can be met. The net result is an energy-efficient fair MAC protocol for WSNs. The performance of this MAC protocol is demonstrated via simulation, which, in turn, demonstrates that the energy-aware protocol results in energy savings during channel uncertainties for different node densities and traffic patterns, while being fair. Comparison of this protocol with that of the 802.11 is also included in the simulation.

7.6.1 Sleep Mode

The CHs in a WSN remain powered all the time. They can accept an incoming packet from the sensors or data forwarded by other nodes (CHs, BS, etc). Thus, CHs consume energy. At the same time, sensor nodes save energy by turning off their computational and RF circuits. To maximize lifetime of the network, nodes volunteer as CHs.

7.6.1.1 Sensor Data Transmission

When there is any data to be reported to the BS or not, the sensing circuit of a sensor node wakes up the processing and RF circuitry. The sensor node transmits the data to its CH and switches itself to the sleep mode after transmission. At the same time, the wake-up timer is set to the predefined interval. Thus, when the sensor does not wake up during this time instant (e.g., there is no event detected), the timer will expire, and it will wake up the sensor. Next, the sensor will communicate to its CH to check if there is any incoming packet waiting for delivery. Afterwards, the sensor again switches into sleep mode by setting the timer.

7.6.1.2 Delivery of Data to a Sensor Node

The communication between a sensor and its CH is always initiated by the sensor during its wake state. The CH never starts the transmission toward a sensor. Instead, the CH buffers packets to its sensors and waits for the sensors to request packets from the CH. When the sensor sends a MAC frame, the CH checks the buffer for any packet to be delivered to this sensor node. If the packets are found, the CH notifies the sensor and piggybacks the packets to the ACK frame so that the information is received by the sensor. In consequence, sensor nodes do not have to listen for any incoming traffic and can be in sleep mode.

In summary, the sensors wake themselves up and communicate with the CH in two ways:

- *Event occurrence:* A sensing circuit will wake up the sensor to process and send data towards the CH.
- *Periodic wake up:* It is to enable incoming communication during long idle intervals.

The data from the sensor node toward the BS are transmitted whenever an event occurs, without unnecessary delays. However, big delays can occur for the packets sent to sensor nodes as they have to be buffered at the CHs. For example, during long idle intervals, when the sensor node has no data to transmit, the CH will buffer the packets for the entire idle interval (until next communication from the sensor). To minimize the packet latency, the sensors periodically initiate dummy transmissions to its CH. This communication allows the CH to piggyback the packet to the sensor for delivery. Thus, any outstanding queries or packets to a given sensor are received whereas the delay is kept below a certain threshold.

7.6.1.3 Protocol Comparison with SMAC

The analytical evaluation of the sleep mode from the energy-aware protocol and the sleep mode introduced by SMAC (Ye et al. 2002) is presented in the following text. The extra delay caused by the sleep mode implementation is compared. Then, the relative energy savings of the energy-aware sleep mode is presented. For the energy savings evaluation, it is assumed that the same sleep delay is expected in both the cases. Figure 7.20 presents the basic idea of sleep modes in SMAC and in energy-aware MAC protocol.

The sleep cycle in SMAC consists of the sleep and the listening interval. These intervals are equal, while in the proposed protocol the cycle contains the sleep and communication intervals. The probability of a packet arrival at the source node is constant during the whole frame. The average sleep delay in SMAC is given by

$$T_{\text{sleep SMAC}} = \frac{T_{\text{SMAC}}}{8} = \frac{0 \times T_{\text{SMAC}} + 1/4 \times T_{\text{SMAC}}}{2} \qquad (7.84)$$

where *TSMAC* is the sleep cycle duration. The expected sleep delay is equal to arithmetic average of both — delay experienced when packets

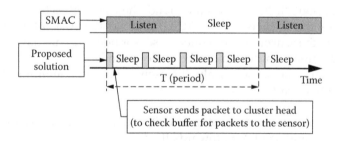

FIGURE 7.20
Sleep mode in SMAC and in the energy-aware protocol.

arrive during listening period – 0 (packet can be sent immediately) –, and average delay when packets arrive during sleep period is equal to half of $T_{sleep} = T_{SMAC}/2$ (average for [T/2, 0>).

In the case of the energy-aware protocol, the average sleep delay for a given sensor is given by

$$D_{sleep\ new} = T_{NEW}/2 \tag{7.85}$$

where T_{NEW} is the time interval between communications. Packets arriving at the CH have to wait for the next communication event. Hence, on an average it will be equal to the half of the communication interval.

The energy consumption can be compared, when other parameters of the protocols are held the same in both the cases. Hence, the delay should to be equal in both cases. From Equation 7.84 and Equation 7.85 this condition implies that

$$T_{NEW} = T_{SMAC}/4 \tag{7.86}$$

Energy expended during communication per sleep cycle for SMAC protocol is equal to

$$E_{SMAC} = E_{LISTEN} * T_{SMAC}/2 \tag{7.87}$$

where E_{LISTEN} is power consumed by node while listening to radio resources, and T_{SMAC} is the duration of the sleep cycle. For the proposed protocol, the energy consumption is equal to

$$E_{NEW} = 4 * E_{TRANS} * T_{TRANS} \tag{7.88}$$

where E_{TRANS} is the power consumed during communication with a CH, and T_{TRANS} is a duration of this communication. Then, the relative saving of Equation 7.88 when compared to Equation 7.87 is equal to

$$Saving = \frac{8 * E_{TRANS} * T_{TRANS}}{E_{LISTEN} * T_{SMAC}} \tag{7.89}$$

Using some typical values: $E_{TRANS} = 2 * E_{LISTEN}$; $T_{SMAC} = 600$ msec; and $T_{TRANS} = 1$ msec, the energy saving ratio is equal to 0.02666. In other words, the energy-aware protocol consumes $1/0.0266 = 37.5$ times less energy than SMAC because of its new sleep-mode option. Similarly, energy savings can be calculated between the energy-aware and other available protocols.

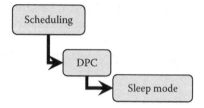

FIGURE 7.21
Overview of energy-aware scheme.

7.6.2   ADFC with Sleep Mode

The proposed algorithm consists of three main items:

1. Fair scheduling algorithm from Section 7.3
2. Distributed power control (DPC) algorithm from Chapter 5
3. Novel sleep mode

Figure 7.21 presents general overview of interactions between these components.

The fair scheduling proposed from Section 7.1 to Section 7.4 utilizes start-time fair queuing (SFQ) with dynamic weight adaptation to sort packets for delivery. A backoff mechanism is used to fairly allocate radio resources between neighboring nodes. The energy conservation is addressed by using the DPC algorithm from Chapter 6 during transmission of packets and the sleep mode during idle intervals. All three elements work collectively to provide reliable and energy-efficient services for WSNs.

7.6.2.1   Scheduling Algorithm

The main goal of the ADFS protocol is to achieve fairness for WSN. To accomplish this, the protocol has to be implemented both at the queuing algorithm level, for proper scheduling and at the MAC protocol level, to control the dynamic backoff algorithm for accessing the channel. The packets are classified according to the traffic flow to which they belong. The nodes in the network store weights assigned to the particular flows. These weights are fixed. Additionally, each user data packet contains its weight. Initially, the value of packet's weight is set to the flow weight. Then, as the packet is forwarded through network, each CH updates the weights. Whenever the packet is going to be transmitted, the MAC protocol calculates backoff interval for a given packet. The packet waits until the backoff time will elapse. Then, the usual four-way handshake exchange is performed in the next subsection. Next, when the packet is received at an intermediate node, the packet's weight is updated and queued accordingly.

7.6.2.2 Protocol Implementation

To achieve fairness at the scheduling level, the proposed ADFS protocol implements the SFQ scheme, defined as follows:

1. On arrival, a packet p_f^j of flow f, is stamped with start tag $S(p_f^j)$, defined as

$$S\left(p_f^j\right) = \max\left\{ v\left(A\left(p_f^j\right)\right), F\left(p_f^{j-1}\right)\right\}, \quad j \geq 1,$$

where $F(p_f^j)$, finish tag of packet p_f^j, is defined as $F(p_f^j) = S(p_f^j) + \frac{l_f^j}{\phi_f}, j \geq 1$,
where $F(p_f^0) = 0$ and ϕ_f is the weight of flow f.

2. Initially, the virtual time of the sensor node is set to zero. During transmission, the node's virtual time at time t, $v(t)$ is defined to be equal to the start tag of the packet being transmitted at time t. At the end of a transmission, $v(t)$ is set to the maximum of finish tag assigned to any packets that have been transmitted by time t.

3. Packets are transmitted in the increasing order of the start tags; ties are broken arbitrarily.

7.6.2.3 Dynamic Weight Adaptation and Backoff Interval Calculation

To account for the changing traffic and channel conditions that affect the fairness and end-to-end delay, the weights for the flows are updated dynamically. The ADFS, which is detailed in Section 7.1 to Section 7.4, calculates the backoff interval by using the packet weight and updated at each node due to weight adaptation.

7.6.2.4 Distributed Power Control and Reset

The basic communication used to send a user packet requires a four-way handshake exchange which consists of four frames:

1. Request-to-send (RTS) — from source to destination
2. Clear-to-send (CTS) — from destination to source
3. DATA frame — from source to destination
4. Acknowledgment (ACK) — from destination to source

All these frames are transmitted over the single radio channel. Hence, the communication between the nodes is carried over shared half duplex medium.

Additionally, handshake between other nodes can exist at the same time instance. As a result, the interference can increase due to multiple nodes communicating with their destinations by accessing the channel. The estimated transmitter power has to overcome such an interference, which is normally not known. In addition, the packet transmission as well as arrival times will vary between frames due to the following reasons. First, the frames within the 4-way handshake have different size, ranging from few bytes (ACK) to over 2500 bytes (DATA) due to the frame type. Second, the delay between two consecutive handshakes can vary because of channel contention. As the result, the estimation error will also vary depending on these time gaps. Therefore, the target signal-to-interference ratio (SIR) has to be chosen to overcome the worst-case scenario due to these uncertainties. Thus in our implementation, the target SIR value is calculated by multiplying the minimum SIR with a safety factor.

In the real scenario, it is possible that channel conditions can change too quickly, which will prevent any algorithm to accurately estimate the power value. Then, the frame losses will occur. Two mechanisms are introduced to overcome or mitigate such problems: increase of transmission power in the case of retransmission and reset of power in the case of a long idle connection.

A retransmission indicates that the received signal has been attenuated. A simple approach would be to retransmit the frame with the same power as the first transmission by hoping that the channel conditions are better. However, the channel attenuation or interferences could also be worse than before. To overcome this problem, an active approach can be used. In this approach, for each retransmission, the transmission power is increased by a certain safety margin. Unfortunately, this will increase interference as well as power consumption. However, experiments indicated that active adaptation of power with the DPC yielded higher throughput and resulted in less retransmissions when compared to the passive method where no safety margin was utilized. As the result, safety margin is used with the DPC scheme.

Additionally, the estimated power value for a given destination can become inaccurate when significant delays are present between any two consecutive frames or delayed feedback. To overcome this, after certain idle interval, the DPC algorithm will reset the transmission power to the maximum value defined for the network. The DPC process described in Chapter 6 is then restarted from the beginning.

7.6.3 Energy-Aware MAC Protocol

For this DPC implementation, the original MAC protocol for 802.11 has been modified. These modifications occur at different layers. Furthermore, an idea, which is presented in Zawodniok and Jagannathan (2004), for ad hoc networks, is used in the energy-aware protocol.

7.6.3.1 Scheduling

To store packet's weight, each user data packet is extended with ADFS header. The source node allocates the initial weight according to the traffic flow the packet belongs to. The MAC layer updates the packet's weight after the packet is received. Next, this weight is used by queuing algorithm to store and dispatch packets. When the packet is released from the queue and passed to the MAC layer, a backoff interval is calculated and the backoff timer is set accordingly. After the timer expires, the packet is transmitted using standard RTS-CTS-DATA-ACK procedure.

7.6.3.2 DPC Protocol

The DPC scheme is used to calculate transmission power for all, except broadcast, messages that are sent over the radio interface. However, simple reduction of a transmission power leads to the degradation of quality of service due to increased hidden-terminal problem, as indicated in Jung and Vaidya (2002) and Zawodniok and Jagannathan (2004). In consequence, more collisions occur reducing throughput and yielding higher energy consumption. By using a train of pulses presented in Chapter 6 with increased power magnitude will reduce the hidden terminal problem, which in turn increased throughput.

In the energy-aware protocol, only the initial RTS-CTS frames during link set up have to be transmitted using maximum power defined by the link. Subsequently, all frames, including RTS-CTS-DATA-ACK frames, will use transmission power calculated according to the DPC. To accommodate the DPC, the MAC header was changed to allow power information to be sent between the communicating nodes. As a result, an increase in overhead is observed, which could cause a decrease in throughput. However, it is found that the increase in throughput, due to better channel utilization, overcomes the penalty introduced by the additional overhead.

7.6.3.3 Sleep Mode Implementation

The packets to the sensor nodes are buffered at the local CHs. The MAC layer allocates memory for such a buffer. Any buffered packet at the CH is delivered after the sensor node wakes up and sends the RTS frame. First, the CH responds with the CTS frame. This frame contains a flag indicating that there is a buffered packet to a particular sensor node. Next, the sensor node sends the DATA frame with most recent collected data. Afterward, the acknowledgment frame is sent from the CH to the sensor node, where the packet to the sensor node is piggybacked. Finally, the sensor node acknowledges the correct reception of the packet by sending a standard ACK message.

In case of a sensor node, the computational and RF circuits are turned off but the MAC frames are exchanged with the CH. To enable periodic wake up of the sensor node, the MAC layer contains a wake-up timer. When the timer expires and there are no data to be sent, a small dummy packet is created and sent using a short DATA-ACK exchange. When the sensor collects the data for delivery, it immediately wakes up the sensor node and starts RTS-CTS-DATA-ACK exchange with the CH. During such communications, the sensor node will accept DATA frame sent instead of ACK frame from the CH, and sends additional ACK frames to acknowledge the reception of data packets. Afterward, the sensor node switches into sleep mode and sets the wake-up timer.

7.6.4 Simulation

The NS-2 simulator was used for evaluating the proposed DPC scheme. Two sensor topologies are used to evaluate the proposed scheme. The topologies were chosen to reflect typical configurations of a sensor network. The 12-cluster sensor network consists of clusters placed within a designated area. Each cluster consists of a CH and eight sensor nodes randomly located around the CH (distance to the CH is less than 100 m). In the 12-cluster topology, the clusters were spread in an area of 550 x 1000 m, forming a multihop clustered network. The BS is located as shown in the Figure 7.22.

The performance of two MAC protocols — DPC without and with sleep mode are evaluated under similar channel conditions, identical node placements, node movement, and data flows (type, rate, start time, source, destination, etc.), and signal-to-noise ratio (SNR) or SIR thresholds, with a deterministic propagation model. The effects of path loss, shadowing, and Rayleigh fading were introduced in the propagation model. The path loss is calculated, based on *propagation/shadowing* object from NS-2 simulator. The shadowing and Rayleigh fading effects have been calculated in advance and stored in a sample file. To create channel variations, path loss exponent of 2.0, shadowing deviation of 5.0 (dB), and reference distance of 1 m, and typical Rayleigh random variable were used. This ensured

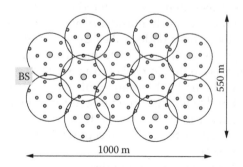

FIGURE 7.22
Twelve-cluster topology.

that the channel uncertainties for the simulations are the same at the respective time instants. Sample attenuation observed for a random receiver node was presented in Chapter 6. The results were repeated for a number of different, randomly generated scenarios, and they were averaged.

The standard AODV routing protocol was used, whereas any routing protocol can be employed because the proposed work is independent of a routing protocol. The following values were used for all the simulations, unless otherwise specified: channel bandwidth is taken as 2 Mbps. The maximum power used for transmission is selected as 0.2818 W; the target SIR for the proposed DPC is equal to 10, the minimum SIR for error-free reception is equal to 4 (~6 db). The target SIR value is 2.5 times higher than minimum SIR to overcome an unpredictable fading. For the proposed DPC, the design parameters are selected as Kv = 0.01, and = 0.01. The safety margin in case of retransmissions is set to 1.5 for the proposed DPC scheme. For the sleep mode, each sensor will initiate periodic dummy communication toward its CH after an idle period of 0.5 sec.

Example 7.6.1: Twelve-Cluster Sensor Network Results

The network consists of 109 nodes: 96 sensor nodes, 12 CHs, and 1 BS. Each sensor generates steady traffic. The flow rates are varied between simulations. Figure 7.23 shows contention time for different per flow rates. The DPC outperforms 802.11 by about 10%. The DPC with the sleep mode has the shortest contention time for all simulated traffic rates due to the inclusion of the sleep mode.

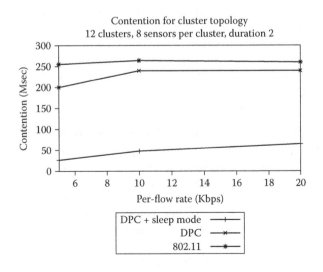

FIGURE 7.23
Contention time for varying per flow rate.

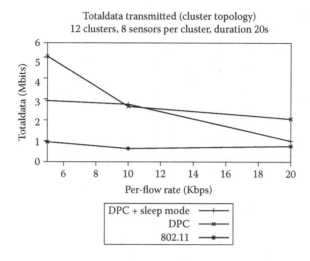

FIGURE 7.24
Total data transmitted for varying per flow rates.

In terms of total data transmitted, the DPC without and with sleep mode achieve better results as shown in Figure 7.24. For both, DPC without and with sleep mode, the amount of transmitted data decreases with an increase in per flow rates. This indicates that the network is congested; hence, the utilization decreases. At low per flow rates, the sleep mode is not as beneficial; hence, low throughput whereas for moderate to high flow rates, sleep mode provides a reasonable throughput compared to other methods. Figure 7.25 presents the energy efficiency of each protocol where the DPC with sleep mode performs better than DPC without sleep mode and 802.11 protocols. This will extend lifetime of the sensors and the network.

Example 7.6.2: Twenty-Cluster Sensor Network Results

To simulate significant traffic, a 20-cluster topology is used. Similar to the 12-cluster case, the DPC with sleep mode achieves lower contention time. Also, from Figure 7.26 and Figure 7.27, with total data transmitted and total data transmitted per joule, the DPC without and with sleep mode outperform 802.11. Additionally, DPC with sleep mode is able to increase throughput and energy efficiency as the traffic increases. However, in general, the 20-cluster network is able to transmit less data than a smaller network due to heavy congestion and presence of a bottleneck link because the entire traffic is directed to one node — BS.

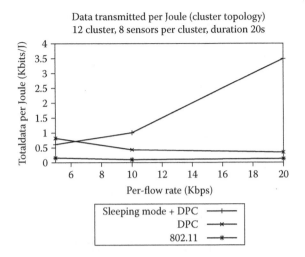

FIGURE 7.25
Total data transmitted per joule.

Example 7.6.3: Fairness Results during Intense Traffic

In this simulation, topology shown in Figure 7.28 is used where the main objective is to test the fairness of the scheduling protocol. Clusters numbered 0 to 5 generate steady traffic flows of 150 kbps toward BS. The cluster 6, where the unusual event occurs, generates higher traffic of 200 kbps.

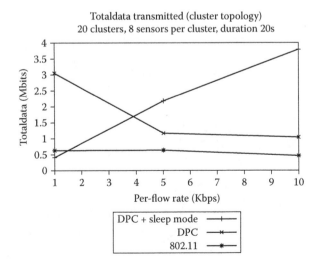

FIGURE 7.26
Total data for varying per flow rates.

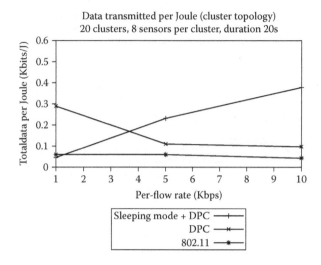

FIGURE 7.27
Total data transmitted per joule.

Each flow coming out of the CH is assigned equal weight, 0.1428, for the sake of convenience the ADFS algorithm parameters: $\alpha = 0.9$, $SF = 0.02$ $\beta = 0.1$, expected delay is 1.0 sec, expected queue length is 10, sum of initial weights assigned to all the flows is equal to unity, ρ is a

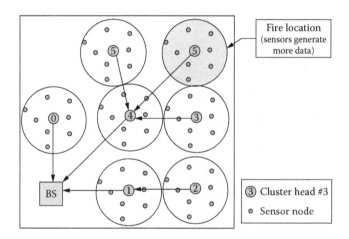

FIGURE 7.28
Contention time for varying per flow rate.

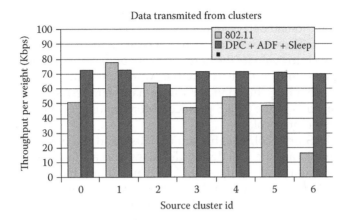

FIGURE 7.29
Throughput over weight ratio with flows.

random variable uniformly distributed in the interval [0.9, 1.1], and with a delay error limit of 0.1 sec. The fairness is tested in the presence of fading channels.

Figure 7.29 shows the throughput over weight ratio achieved for each traffic flow in case of standard 802.11 and with the proposed MAC protocol. The proposed protocol with ADFS scheduling maintains similar throughput per weight ratio (close) for all clusters whereas 802.11 is not. Hence, the proposed protocol is fair to all sources. Assigning different initial weights to the flows at each CH will change the throughputs of the flows.

7.7 Conclusions

In this chapter, a novel adaptive and distributed fair scheduling (ADFS) protocol for wireless ad hoc networks is presented. The objective is to develop a fully distributed fair scheduling algorithm, which meets the overall QoS. The proposed ADFS protocol uses dynamic adaptation of the weights, depending upon the delay experienced, number of packets in the queue, and the previous weight of the packet. The updated weights are used in making the scheduling decisions and also in the calculation of the backoff intervals in the CSMA/CA paradigm.

The effectiveness of our proposed ADFS protocol was evaluated using the NS simulator. The results show that the proposed protocol can achieve fair allocation of bandwidth with a 10 to 20% increase in throughput and minimum end-to-end delays and delay variations, resulting in a better QoS.

Next, the implementation of an ADFS scheme for WSNs is discussed. The objective is to evaluate the hardware capabilities and implementation viability. The results indicate where hardware constraints must be alleviated, thus providing direction for future hardware redesign that meets the requirements of the ADFS protocol. More topologies will be utilized to evaluate the protocol.

In the ADFS protocol, weights are updated, and the updated weights are used in making the scheduling decisions and also in the calculation of the backoff intervals in the CSMA/CA paradigm. Initial weights are assigned to the flows based on the service they expect from the network.

The effectiveness of the proposed ADFS protocol was evaluated through hardware experimentation. The results show that the proposed protocol can achieve fair allocation of bandwidth with a 13.3% increase in throughput, lower end-to-end delays, and delay variations reduced by 55% (standard), resulting in a better QoS. This shows the ability of the ADFS scheduler to be effectively applied to WSN systems to enhance network performance; however, further study is needed to address the current hardware constraints and improve overall performance of the implementation.

This chapter also introduced the energy-aware and fair protocol for WSNs by taking into account the application considerations. This protocol enables higher throughput in the sensor network and consumes less energy per bit when compared to the standard 802.11. It was demonstrated that the scheme allows fully DPC and has rendered better performance in the presence of radio channel uncertainties. Introduction of sleep mode has not decreased throughput of the network. Moreover, the protocol with the sleep-mode solution uses significantly less transmitter power per bit compared to 802.11 or the original DPC implementation; hence, the energy is saved and lifetime of sensors extended. Second, the fair scheduling protocol is incorporated that ensures performance in the presence of unusual events for sensor networks. The performance of the energy-aware scheme consisting of scheduling, DPC with sleep mode is demonstrated in the presence of fading channels. Simulation results justify theoretical conclusions.

References

Bennett, J.C.R. and Zhang, H., WF2Q: worst-case fair weighted fair queueing, *Proceedings of the IEEE INFOCOM'96*, Vol. 1, March 1996, pp. 120–128.

Brogan, W.L., *Modern Control Theory*, 3rd ed., Prentice Hall, NJ, 1991.

Demers, A., Keshav, S., and Shenker, S., Analysis and simulation of a fair queuing algorithm, *Proceedings of the IEEE INFOCOM*, 2000.

Fall, K. and Varadhan, K., ns Notes and Documentation, Technical report UC Berkley LBNL USC/IS Xerox PARC, 2002.

Fonda, J., Zawodniok, M., Jagannathan, S., and Watkins, S.E., Adaptive distributed fair scheduling and its implementation in wireless sensor networks, *Proceedings of the IEEE International Conference on Systems, Man and Cybernetics,* pp. 3382–3387 in 2006.

Golestani, S.J., A self-clocked fair queueing scheme for broadband applications, *Proceedings of the IEEE INFOCOM'94,* Vol. 2, June 1994, pp. 636–646.

Goyal, P., Vin, H.M., and Cheng, H., Start-time fair queueing: a scheduling algorithm for integrated services packet switching networks, *IEEE/ACM Transactions on Networking,* Vol. 5, pp. 690–704, October 1997.

Heinzelman, W.B., Chandrakasan, A.P., Balakrishnan, H., An application-specific protocol architecture for wireless microsensor networks, *IEEE Transactions on Wireless Communications,* 2002.

Jagannathan, S., Zawodniok, M., and Shang, Q., Distributed power control of cellular networks in the presence of channel uncertainties, *Proceedings of the IEEE INFOCOM,* 2004.

Jain, R., Babic, G., Nagendra, B., and Lam, C., Fairness, Call Establishment Latency and Other Performance Metrics, Technical Report ATM_Forum/96–1173, August 1996.

Jung, E.-S. and Vaidya, N.H., A power control MAC protocol for ad hoc networks, *ACM MOBICOM,* 2002.

Lee, K., Performance bounds in communication networks with variable-rate links, *Proceedings of the ACM SIGCOMM,* 1995, pp. 126–136.

Luo, H., Medvedev, P., Cheng, J., and Lu, S., A self-coordinating approach to distributed fair queueing in ad hoc wireless networks, *Proceedings of the IEEE INFOCOM,* 2001, pp. 1370–1379.

Raghunathan, V., Schurgers, C., Park, S., and Srivastava, M.B., Energy-Aware Wireless Microsensor Networks, *IEEE Signal Processing Magazine,* 2002.

Ratnaraj, S., Jagannathan, S., and Rao, V., OEDSR: optimal energy delay subnet routing protocol for wireless sensor networks, *Proceedings of the IEEE Conference on Sensing, Networking, and Control,* April 2006, pp. 787–792.

Regatte, N. and Jagannathan, S., Adaptive and distributed fair scheduling scheme for wireless ad hoc networks, *Proceedings of the World Wireless Congress,* May 2004, pp. 101–106.

Singh, S. and Raghavendra, C.S., PAMAS: Power Aware Multi-Access Protocol with Signaling for Ad Hoc Networks, *ACM Computer Communication Review,* Vol. 28, No. 3, July 1998, pp. 5–26.

Vaidya, N.H., Bahl, P., and Gupta, S., Distributed fair scheduling in a wireless LAN, *Proceedings of the 6th Annual International Conference on Mobile Computing and Networking,* August 2000.

Woo, A. and Culler, A., Transmission control scheme for media access in sensor networks, *ACM Sigmobile,* 2001.

Ye, W., Heidemann, J., and Estrin, D., An energy-efficient MAC protocol for wireless sensor networks, *Proceedings of the IEEE INFOCOM,* 2002.

Zawodniok, M. and Jagannathan, S., A distributed power control MAC protocol for wireless ad hoc networks, *Proceedings of the IEEE WCNC'04,* 2004.

Problems

Section 7.4

Problem 7.4.1: (Fair Scheduling in WSN): Evaluate the performance of ADFS in a WSN consisting of 20 clusters where each cluster will consist of 50 nodes, and data are routed through multihop at the CH level. Vary packet sizes and use the same parameters from Example 7.4.1.

Problem 7.4.2: (Fair Scheduling in WSN): Re-do Problem 7.4.1 now with same weights and by introducing channel uncertainties.

8

Optimized Energy and Delay-Based Routing in Wireless Ad hoc and Sensor Networks

In the last chapter, the development and implementation of the distributed and fair scheduling scheme for wireless ad hoc and sensor networks to meet certain quality of service (QoS) performance requirements was presented. Implementation aspects were also covered. In this chapter, an optimized energy-delay routing (OEDR) protocol for ad hoc wireless networks from Regatte and Jagannathan (2005) is presented first, where the product of the transmission energy and end-to-end (E2E) delay-link costs between any two nodes is utilized to determine a least-cost routing path. The OEDR uses the concept of multipoint relays (MPRs) (Qayyum et al. 2002) similar to the optimized link state routing (OLSR) protocol (Jacquet et al. 2001, Clausen and Jacquet 2003). However, in OEDR, the HELLO control messages are used to determine the transmission energy and delay values between a given node and its neighbors, in addition to performing neighbor sensing. The energy-delay product of these values is considered as the link cost, and this information is relayed by the MPR nodes to other nodes in the network, using topology control (TC) messages.

The OEDR minimizes the energy-delay product which, in turn, minimizes the transmission energy and delay between any two nodes on the route, thus guaranteeing optimal cost, in contrast with OLSR (Clausen and Jacquet 2003). The remaining energy available at the nodes is used to select the MPR nodes and to compute the optimal routes using the minimum-cost spanning tree algorithm. Analytical results are presented to demonstrate the performance of the proposed OEDR protocol in MPR selection and optimal route computation. Network simulator (NS-2) (Fall and Varadhan 2002) results demonstrate that the OEDR protocol leads to a minimum delay, better throughput/delay, and a smaller energy-delay product over OLSR and AODV protocols.

Subsequently, an optimized energy-delay subnetwork routing (OEDSR) protocol for wireless sensor networks (WSN) is presented from Ratnaraj et al. (2006) based on the OEDR protocol. This on-demand routing protocol minimizes a different link cost factor, which is defined using available energy, E2E delay, and distance from a node to the base station (BS), along with clustering, to effectively route information to the BS. Initially, the nodes are either in idle or sleep mode, but once an event is detected, the nodes near the event become active and start forming subnetworks. Formation of the inactive network into a subnetwork saves energy because only a portion of the network is active in response to an event. Later, the subnetworks organize themselves into clusters and elect cluster heads (CHs) in the subnetwork portion whereas relay nodes (RNs) are selected outside the subnetworks.

The data from the CHs are sent to the BS via RNs that are located outside the subnetworks in a multihop manner. This routing protocol improves the lifetime of the network and the scalability. This routing protocol is implemented in the medium access control (MAC) layer using UMR nodes. Simulation and experimental results indicate that the OEDSR protocol results in lower average E2E delay, fewer collisions, and less energy consumed when compared with the DSR, AODV, and Bellman Ford routing protocols.

8.1 Routing in Ad hoc Wireless Networks

Wireless Ad hoc and sensor networks have gained great importance in recent years due to an unprecedented growth in wireless communication technologies and with the advent of the IEEE 802.11 standards. An ad hoc network is a group of wireless mobile nodes dynamically forming a temporary network without any fixed infrastructure or centralized administration.

A routing protocol is used to determine an appropriate path over which data is transmitted in a network. It also specifies how the network nodes share information with each other and report changes in the topology. Moreover, the decisions of the routing protocol have to be dynamic, in response to changes in network topology. Thus, the route selection process greatly affects the overall network performance, defined using E2E delay, throughput, and network energy efficiency. Therefore, routing is a critical issue in wireless ad hoc and sensor networks, where data is transmitted generally over multihop paths.

The challenges involved in developing a routing protocol for mobile ad hoc and sensor networks are very different and more complex than those of static wired networks. The routing protocol should be capable of rapidly adapting to link failures and topology changes caused by node movements. Therefore, the routing protocol should work in a distributed manner with self-organizing capability. The goal of the routing protocol is to compute the optimal path between any source–destination pair with minimal control traffic overhead. This should be achieved within the constraints imposed by the ad hoc networking environment. Therefore, any routing protocol proposed for a multihop ad hoc wireless network must consider the following design issues:

Reactive vs. proactive routing approach: In the reactive routing approach (on-demand), a route is not identified unless it is required. By contrast, proactive protocols periodically exchange control messages and provide the required routes instantly when needed. In the selection of reactive vs. proactive techniques, there is a trade-off between the latency in finding the route to a destination and the control traffic overhead.

Centralized vs. distributed approaches: Because there is no centralized control in wireless ad hoc networks, a distributed routing protocol is preferred.

Optimal route: The definition of an optimal route is very important for the design of a routing protocol. General criteria are that the number of hops on the path or the overall link cost is used as the metric to determine the optimal route.

Scalability: The routing protocol should scale well in large wireless ad hoc and sensor networks, with rapid topology changes and link failures.

Control traffic overhead: The routing protocols must minimize the control traffic overhead required to discover the routes.

Efficiency: The routes selected by the routing protocol affect the performance of the network in terms of delay, throughput, and energy efficiency. Therefore, the routing protocol should aim at improving the overall network efficiency.

In this chapter, the available routing protocols (Clausen and Jacquet 2003, Perkins et al. 2003, Johnson et al. 2003, Park and Corson 1997, Sivakumar et al. 1999, Perkins and Bhagwat 1994, Aceves and Spohn 1999) are reviewed, and an optimal energy-delay routing scheme is introduced. Reactive protocols like AODV (Perkins et al. 2003), DSR (Johnson et al. 2003), TORA (Park and Corson 1997), and CEDAR (Sivakumar et al. 1999) compute the routes on-demand, which reduces the control overhead at the cost of increased latency. The rapid changes to the topology and link failures present in ad hoc networks, and the need to react quickly to the routing demands, make proactive protocols more suitable for such networks. A few examples of proactive routing protocols are DSDV (Perkins and Bhagwat 1994), STAR (Aceves and Spohn 1999), and OLSR (Jacquet et al. 2001, Clausen and Jacquet 2003). However, these routing protocols attempt to find the minimal hop path from the source to the destination, an approach that may not provide the optimal path in terms of delay and energy efficiency.

To accommodate quality of service (QoS) issues in routing, a new routing scheme using the OLSR algorithm is presented in Ying et al. (2003), which routes packets based on the path with maximum bandwidth bottleneck. However, in an attempt to find the maximum bandwidth path, the protocol can result in longer paths with an increase in the E2E delay from the source to the destination, especially in large dense networks. By contrast, energy efficiency is more important than minimizing the number of hops in wireless ad hoc networks because minimal transmission energy consumption implies a good wireless channel, and further minimizing E2E delay renders the best possible throughput. It is important to notice that two nodes close to each other may still require significant transmission energy if the wireless channel between them is highly undesirable.

The first few sections of this chapter present an OEDR proactive protocol (Regatte and Jagannathan 2005) for ad hoc wireless networks. The proposed algorithm is fully distributed in nature. The control packets (HELLO and TC) are generated similar to the OLSR protocol (Jacquet et al. 2001, Clausen and Jacquet 2003). However, the HELLO packets in OEDR are used to calculate the energy consumed and delay experienced during transmission from the neighbors and the energy levels of the neighbors, along with the neighbor sensing. Using this information, the multipoint relay nodes (MPRs) (Qayyum et al. 2002) are selected among the one-hop neighbors to reach all the two-hop neighbors with the minimum energy-delay metric as the link cost. MPR nodes, in turn, transmit the TC messages with link cost information from their respective MPR selectors to all the nodes in the network. Once available, the link cost information in the network is used by the minimum-cost spanning tree algorithm, to compute optimal routes between a source and the destination pair.

Analytical results are included to prove that the MPR selection algorithm of the proposed OEDR protocol will result in an optimal route both between two-hop neighbors and source–destination pairs. The NS-2 simulation results indicate that the proposed scheme renders minimum delay, a better throughput/delay ratio, and a smaller energy-delay product over the OLSR and AODV protocols. Section 8.2 provides the description of the OLSR protocol. Section 8.3 describes the novel OEDR protocol. In Section 8.4 the optimality of the OEDR is demonstrated analytically. Section 8.5 presents the NS simulator results to evaluate the performance of the OEDR algorithm.

8.2 Optimized Link State Routing (OLSR) Protocol

Before presenting the OEDR protocol, it is important to understand the OLSR protocol. The OLSR protocol (Jacquet et al. 2001, Clausen and Jacquet 2003) inherits the stability of a link state algorithm and has the advantage of reduced latency in route discovery due to its proactive nature. OLSR is an optimization over the classical link state algorithm that is tailored for mobile ad hoc networks. The important concept introduced in the protocol is the multipoint relays (Qayyum et al. 2002) where only selected nodes, referred to as RNs or MPRs, forward the broadcast messages during the flooding process. The purpose of MPR nodes is to minimize the overhead of flooding messages in the network by reducing the number of duplicate retransmissions while forwarding a broadcast packet. According to MPR selection criteria presented in Qayyum et al. (2003) and Clausen and Jacquet (2003), each node in the network selects a set of nodes from its one-hop neighbors as MPRs. These nodes are chosen such that they reach the maximum number of uncovered two-hop neighbors until all of them are covered.

The OLSR protocol uses HELLO messages for neighbor sensing and TC messages to declare the MPR information. MPR nodes periodically announce information about the neighbors that have selected it as an MPR, by broadcasting a TC message. In contrast to the classic link state algorithm, only a small subset of links between neighbors (only the MPR selectors) is declared and, therefore, the overhead due to the control messages is reduced. Upon receiving the TC message, each node in the network stores the information in the topology table, and only the MPR nodes forward the TC messages to the next-hop neighbors until all the nodes in the network receive the message. Whenever the topology table changes, routing table entries are calculated at each node, using a shortest path algorithm (number of hops is the metric used in OLSR) to determine the routes to all the destinations.

An example of the MPR selection algorithm used by the OLSR protocol is shown in Figure 8.1. Consider a node s with one-hop neighbors:

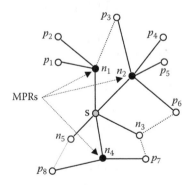

FIGURE 8.1
MPR selection using OLSR protocol.

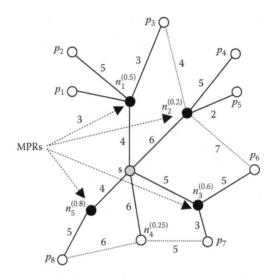

FIGURE 8.2
MPR selection using OEDR protocol.

n_1, n_2, n_3, n_4, n_5 and two-hop neighbors: $p_1, p_2, p_3, p_4, p_5, p_6, p_7, p_8$ reachable as shown in the figure. The one-hop neighbors that are selected as MPRs are n_1, n_2, and n_4, because these one-hop neighbors reach the maximum number of uncovered two-hop neighbors (Qayyum et al. 2002). In the next section, the proposed OEDR (see Figure 8.2) protocol is discussed in detail.

8.3   Optimized Energy-Delay Routing (OEDR) Protocol

The main steps in the proposed OEDR protocol operation are: neighbor sensing, energy delay metric calculation, MPR selection, declaration of energy-delay information, and routing table calculation. To proceed further, the following notations are used in the rest of the chapter. Define:

N — Set of nodes in the network

s — Source node

$N(s)$ — One-hop neighbors of node s

$N^2(s)$ — Two-hop neighbors of node s

$MPR(s)$ — Selected MPR set of nodes s

$RT(s)$ — Routing table of node s, containing the route entries

$C_{x,y}$ — (Energy $[x \ -> \ y]$ * Delay $[x \ -> \ y]$); energy-delay (cost) of the direct link between nodes x and y

E_x — Energy level or total available energy of node x.

C_{s,n_1,n_2}^{MPR} — Cost for MPR selection of node s, to reach the two-hop neighbor n_2, with one-hop neighbor n_1 as the intermediate node $(s \to n_1 \to n_2)$; it is given by

$$C_{s,n_1,n_2}^{MPR} = C_{s,n_1} + C_{n_1,n_2} + (1/E_{n_1}) \tag{8.1}$$

where $n_1 \in N(s)$ and $n_2 \in N^2(s)$.

$Cost_{s,d}$ — Energy-delay (cost) of the entire path between a source s and the destination d, given by:

$$Cost_{s,d} = \sum (C_{s,n1}, C_{n1,n2}, \ldots, C_{nk-1,nk}, C_{nk,d}) \tag{8.2}$$

where $n_1, n_2 \ldots n_k$ are intermediate nodes on the path.

Optimal-route — Optimal route, between any source–destination pair s and d, is the route with minimum energy-delay cost ($Cost_{s,d}$).

8.3.1 Neighbor Sensing and Energy-Delay Metrics Calculation

Each node in the network periodically generates HELLO messages and transmits to all the one-hop neighbors, similar to the OLSR implementation given in (Clausen and Jacquet 2003). However, changes are made to the HELLO message header format to include various fields, like the transmission time, transmission energy, and the energy level (available energy) of the source node, for calculating the energy-delay metrics. Moreover, the HELLO messages also contain information about the list of the one-hop neighbors and the link costs ($C_{x,y}$) between the source node and its neighbors. This information is obtained from the node's "neighbor table," which is used to maintain the list of one-hop and two-hop neighbors, and their associated link costs.

When a HELLO packet is received by a node, it can calculate the transmission delay of the packet as the difference between the transmission time stamped in the packet at the source and the received time at the destination. However, to calculate the exact delay, the clocks of the neighboring nodes have to be synchronized. To achieve this, a mechanism like the one presented in Mock et al. (2000) can be used to synchronize the clocks of all the nodes with high precision in ad hoc wireless networks. Similarly, the energy used can be calculated as the difference between the transmission energy stamped in the packet and the received energy. By using the energy consumed and delay, an energy-delay product for the link is computed by each node. This information is treated as the link cost and, along with the energy level of the neighbors, it is recorded in the neighbor table for the one-hop neighbors. Moreover, the neighbors of the HELLO message originator node (one-hop neighbor) are recorded as two-hop neighbors along with the costs of their links.

Therefore, after receiving the HELLO messages from all the neighbors, each node will have the following information about the one-hop and two-hop topology in the neighbor table: link costs to all the one-hop neighbors, energy levels of the one-hop neighbors, lists of the two-hop neighbors that can be reached from each of the one-hop neighbors, and their link costs.

8.3.2 Multipoint Relay (MPR) Selection

The criteria for MPR selection in the OEDR protocol are to minimize the energy-delay cost to reach the two-hop neighbors and to consider the energy level of the one-hop nodes to increase their lifetime. The proposed OEDR protocol uses the following algorithm to select the MPR nodes (flowchart for the algorithm is presented in Figure 8.3).

MPR Selection algorithm can be detailed as follows:

1. Start with an empty MPR set $MPR(s)$ of node s.
2. First identify those two-hop neighbor nodes in $N^2(s)$ which have only one neighbor in the one-hop neighbor set $N(s)$. Add these nodes of $N(s)$ to the MPR set $MPR(s)$ if they are not already in $MPR(s)$.
3. If there exists a node in $N^2(s)$ for which MPR node is not selected, do the following:
   - For each node in $N^2(s)$, with multiple neighbors from $N(s)$, select a neighbor from $N(s)$ as MPR node which results in minimum cost from s to the node in $N^2(s)$, $C^{MPR}_{s,N(s),N^2(s)}$, according to Equation (8.1).
   - Add that node of $N(s)$ in $MPR(s)$ if it is not already in $MPR(s)$.

The above MPR selection algorithm aims at increasing the lifetime of the one-hop neighbors and reducing the delay, besides improving the energy-efficiency. The lifetime of the nodes is very critical in certain applications like sensor networks (Yee and Kumar 2003), which have severe energy limitations. The subsequent example illustrates the MPR selection algorithm.

Example 8.3.1: MPR Selection

Consider the neighbor topology of node s shown in Figure 8.2. The link costs and the energy levels of the one-hop neighbors are also shown in

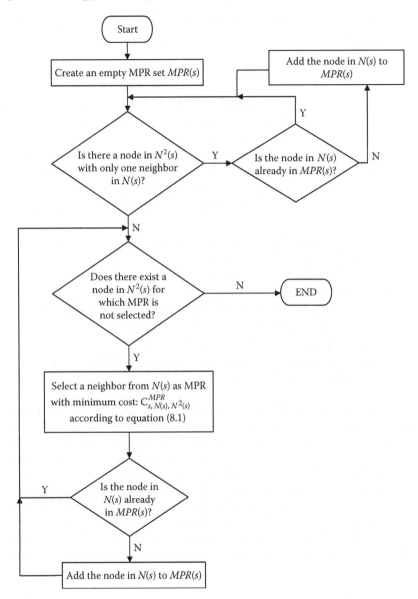

FIGURE 8.3
Flowchart for the MPR selection algorithm.

the figure. The following steps are followed according to the MPR selection algorithm of OEDR:

1. Start with an empty MPR set $MPR(s)$ of node s: $MPR(s) = \{\}$
2. Nodes p_1, p_2, p_4, and p_5 each have only one neighbor in the one-hop neighbor set $N(s)$. Add n_1 and n_2 to $MPR(s)$: $MPR(s) = \{n_1, n_2\}$
3. For each node in $N^2(s)$ for which MPR node is not selected (p_3, p_6, p_7, and p_8), do the following:

   • For node p_3, there are two possible paths from s to reach p_3. For each of the paths, calculate the MPR costs: $C_{s,n_1,p_3}^{MPR} = 4+3+1/0.5 = 9$; $C_{s,n_2,p_3}^{MPR} = 6+4+1/0.2 = 15$. Because C_{s,n_1,p_3}^{MPR} is smaller, select n_1 as MPR. However n_1 is already in the MPR set: $MPR(s) = \{n_1, n_2\}$

   • For node p_6 : $C_{s,n_2,p_6}^{MPR} = 6+7+1/0.2 = 18$; $C_{s,n_3,p_6}^{MPR} = 5+5+1/0.6 = 11.67$. Because C_{s,n_3,p_6}^{MPR} is smaller, select n_3 as MPR and add to the MPR set: $MPR(s) = \{n_1, n_2, n_3\}$

   • For node p_7 : $C_{s,n_3,p_7}^{MPR} = 5+3+1/0.6 = 9.67$; $C_{s,n_4,p_7}^{MPR} = 6+5+1/0.25 = 15$. Because C_{s,n_3,p_7}^{MPR} is smaller, select n_3 as MPR. However, n_3 is already in the MPR set: $MPR(s) = \{n_1, n_2, n_3\}$

   • For node p_8 : $C_{s,n_4,p_8}^{MPR} = 6+6+1/0.25 = 16$; $C_{s,n_5,p_8}^{MPR} = 4+5+1/0.8 = 10.25$. Because C_{s,n_5,p_8}^{MPR} is smaller, select n_5 as MPR and add to the MPR set: $MPR(s) = \{n_1, n_2, n_3, n_5\}$

By following the MPR selection algorithm of the OEDR, the nodes selected as MPRs are n_1, n_2, n_3, and n_5 . In contrast, it can be seen from Figure 8.1 that the nodes selected as MPRs using the OLSR were n_1, n_2, and n_4. The following Table 8.1 shows the link costs, calculated according to equation (8.1), to reach the two-hop neighbors from s, when the MPR selection is performed, using the OLSR and OEDR protocols:

It can be observed from this table that the proposed OEDR protocol significantly reduces the cost to reach the two-hop neighbors, when compared to OLSR. This is achieved by using an efficient MPR selection algorithm that considers the energy-delay costs, to select the MPR nodes.

TABLE 8.1

MPR Selection Algorithm Comparison of OLSR and OEDR Protocols

Protocol	Cost to Reach the Two-Hop Neighbors (Via the MPR node) According to (1)							
	p_1	p_2	p_3	p_4	p_5	p_6	p_7	p_8
OLSR	9 (n_1)	11 (n_1)	15 (n_2)	16 (n_2)	13 (n_2)	18 (n_2)	15 (n_4)	16 (n_4)
OEDR	9 (n_1)	11 (n_1)	9 (n_1)	16 (n_2)	13 (n_2)	11.67 (n_3)	9.67 (n_3)	10.25 (n_5)

8.3.3 MPR and Energy-Delay Information Declaration

Each node in the network that is selected as MPR, by at least one of its neighbors, transmits a TC message periodically. The TC message contains information about the MPR node's selector set, which is a subset of the one-hop neighbors that have selected the sender node as a MPR. Moreover, the TC message format is modified to include the link costs (energy delay) between the MPR node and its selectors. TC messages are forwarded throughout the network like usual broadcast messages, except that only the MPR nodes forward the message to the next hop.

The proposed OEDR protocol significantly reduces the overhead in transmitting the cost information, in contrast to a normal link-state algorithm, which uses flooding to broadcast control messages with cost information. In the OEDR protocol, only the MPR nodes (which are a small subset of the one-hop neighbors) forward the broadcast control messages, instead of all the neighbor nodes. Moreover, only the nodes that are selected as MPRs generate the TC messages that contain cost information, instead of all the nodes in the network. Also, the size of the TC control messages is smaller because it contains the costs of only the links between the source node and its MPR selectors.

Each node in the network maintains a "topology table," in which it records the information about the topology of the network obtained from the TC messages, along with the link costs. An entry in the topology table consists of the address of a destination (an MPR selector in the received TC message), address of the last-hop node to that destination (originator of the TC message), and the cost of the link between the destination and its last-hop. It implies that the destination node can be reached in the last hop through this last-hop node at the given cost.

Whenever a node receives a TC message, it records the information as entries into the topology table with the addresses in the MPR selector set as the destinations, the originator as the last-hop, and the corresponding link costs. Based on this information, the routing table is calculated.

8.3.4 Routing Table Calculation

Each node maintains a "routing table," which enables it to route packets for other destinations in the network. The routing table entries consist of the destination address, next-hop address, estimated distance to destination, and cost of the path from the source to the destination ($Cost_{s,d}$). Each destination has an entry in the routing table, for which a route is known from the given node to the destination. The proposed OEDR protocol determines the routes to destinations, by using a least-cost spanning tree method with the energy-delay product of the links as costs of the edges. By contrast, the OLSR protocol computes the routes, based on a shortest path algorithm, using number of hops as the metric. OLSR's method may not result in optimal routes, in most cases, in terms of delay and energy consumption for ad hoc wireless networks. The following algorithm is used by the OEDR protocol to calculate the routing table, based on the information contained in neighbor and topology tables (flowchart for the algorithm is presented in Figure 8.4). Algorithm:

1. Clear all entries in the routing table, $RT(s)$, of node s .
2. Record the new entries in the $RT(s)$, starting with one-hop neighbors in $N(s)$ as destination nodes. For each neighbor entry in the neighbor table, record a new route entry in the routing table, where destination and next-hop addresses are both set to the address of the neighbor; distance is set to 1, and the cost of the route is set to the cost of the link from the neighbor table.
3. Then record the new entries for destination nodes $i + 1$ hops away in the routing table. The following procedure is executed for each value of i , starting with $i = 1$ and incrementing it by 1 each time. The execution will be stopped if no new entry is recorded in an iteration.

    • For each topology entry in the topology table, if the last-hop address corresponds to the destination address of a route entry with distance equal to i , then check to see if a route entry already exists for this destination address.

        a. If the destination address of the topology entry does not correspond to the destination address of any route entry in the routing table, then a new route entry is recorded in the routing table where:

            • Destination is set to destination address in topology table.

            • Next-hop is set to next-hop of the route entry whose destination is equal to previously mentioned last-hop address.

            • Distance is set to $i + 1$.

            • Cost of the route is set to the sum: $Cost_{s, last-hop}$ $+ C_{last-hop, destination}$ (cost of the route entry in the $RT(s)$ with the last-hop as its destination address + cost of the link between the destination and its last-hop from the topology table).

        b. Else, if there exists a route entry in $RT(s)$ whose destination address corresponds to the destination address of the topology entry, then compute the cost of the new route as the

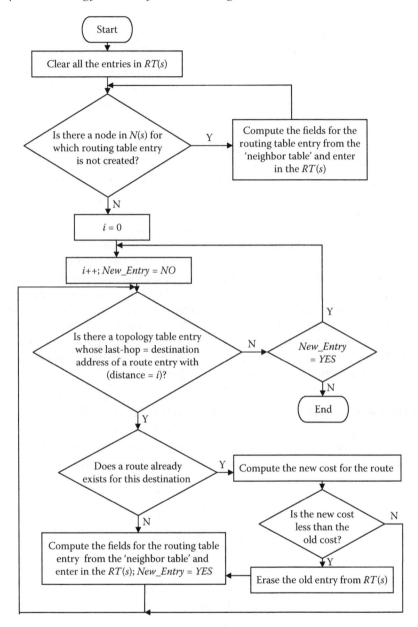

FIGURE 8.4
Flowchart for the routing table calculation algorithm.

sum: $Cost_{s,\,last-hop} + C_{last-hop,\,destination}$, and compare with the cost of the old route in the routing table, corresponding to the same destination address.

- If the new cost is less than the old cost, then erase the old entry in the $RT(s)$, and record a new route entry in the $RT(s)$, with fields for the entry computed similar to step a.

According to the above algorithm, initially start with an empty routing table for the source node step 1. In step 2, enter routing table entries for all the one-hop neighbors, present in the "neighbor table." Next (step 3) iterate through the "topology table" to find the entries (destinations) that are reachable in one hop from a node for which the route has been entered in the previous iteration (this becomes the new route for the destination). Then check to see if the destination already has an entry in the routing table. If a routing table entry does not exist for that destination (step 3a), enter the new route in the routing table. Otherwise, if a route already exists to that destination (step 3b), calculate the cost of the new route and compare with the existing route. If the new cost is less than the old cost to reach the destination, then replace the old routing table entry with the new route. Repeat this process (step 3), until the optimal routes are found to all the destinations in the network topology. An example is presented next to illustrate the routing table calculation.

Example 8.3.2: Routing Table Algorithm Illustration

To illustrate the routing table calculation algorithm, consider the network topology shown in Figure 8.1 and Figure 8.2. Figure 8.5 shows the result of routing table calculations using the shortest hops method of the OLSR

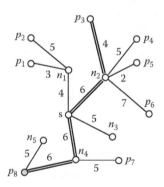

FIGURE 8.5
"Shortest hops path" using the OLSR protocol.

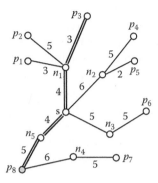

FIGURE 8.6
Least-cost spanning tree using the OEDR protocol.

protocol for node p_8. On the other hand, Figure 8.6 shows the least-cost spanning tree for node p_8 using the OEDR protocol.

To compare the route efficiency of the two protocols, observe the path from the source p_8 to the destination p_3 for both the OLSR and OEDR protocols. The path using the OLSR protocol is given by $p_8 -> n_4 -> s -> n_2 -> p_3$ and the cost of the path is: $Cost_{p_8,p_3} = 6+6+6+4 = 22$, according to equation (8.2). On the contrary, the path using the OEDR protocol is $p_8 -> n_5 -> s -> n_1 -> p_3$ and the cost of the path is: $Cost_{p_8,p_3} = 5 + 4 + 4 + 3 = 16$, which is much smaller than the cost using the OLSR protocol. This clearly shows that the proposed OEDR protocol results in a least cost or optimal cost path from the source to the destination.

8.3.5 The OEDR Protocol Summary

The basic functioning of the OEDR protocol is briefly described in the following steps:

1. *Neighbor sensing:* Each node in the network transmits HELLO packets to its neighbors. When HELLO packets are received, each node updates the information of delay and energy of the links from their neighbors in the neighbor table, along with the energy level (available energy) of the neighbors.

2. *Multipoint relay selection*: For each node in the network, its neighbor table is used to select the MPR nodes from the one-hop neighbors to reach all the two-hop neighbors with minimum costs, using the MPR selection algorithm given in Section 8.3.

3. *Topology information declaration:* All the nodes in the network that are selected as MPR nodes transmit TC messages, with energy-delay

information of the links from its MPR selectors, and are transmitted to all the nodes in the network through broadcast. Upon receiving the TC messages, each node in the network records all the information in the topology table.

4. *Routing table calculation:* Each node in the network proactively computes the routes to all the destination nodes in the network by using the neighbor table and the topology table of the node. OEDR protocol uses the least-cost spanning tree algorithm to calculate the optimal routes described in Section 8.3.4, to all the nodes in the network and records the entries in the routing table.

Next, the analytical results are given to demonstrate the performance of the routing scheme.

8.4   Optimality Analysis for OEDR

To prove that the proposed OEDR protocol is optimal in all cases, it is essential to analyze the optimality of the MPR selection algorithm and the optimal route computation algorithm.

ASSUMPTION 8.4.1

If the one-hop neighbor of a node s has no direct link to at least one of the two-hop neighbors of s , then it is not on the optimal path from s to its two-hop neighbors. However, to reach a two-hop neighbor from s through such a node, the path has to go through another one-hop neighbor, which has a direct link to the two-hop neighbor. This would usually result in more delay and energy being consumed than a direct path through the one-hop neighbor, which has a direct link to this two-hop neighbor.

THEOREM 8.4.1

The MPR selection based on the energy-delay metric and the available energy of the RNs will result in an optimal route between any two-hop neighbors.

PROOF   Consider the following two cases:

CASE I   When the node in $N^2(s)$ has only one neighbor from $N(s)$, then that node in $N(s)$ is selected as an MPR node. In this case, there is only one path from the node s to the node in $N^2(s)$. Hence, the OEDR MPR selection algorithm will select this optimal route between s and the two-hop neighbor in $N^2(s)$.

CASE II   When the node in $N^2(s)$ has more than one neighbor in $N(s)$, the MPR nodes are selected, based on the OEDR MPR selection criteria.

Consider a node s whose one-hop neighbors are given by $N(s)$, and a particular node d in $N^2(s)$, with multiple nodes $n_1, n_2, \ldots, n_k$ ($k > 1$) belonging to $N(s)$ as its neighbors. Let the cost to reach these one-hop neighbors from s be C_{s,n_i}, and the cost to reach d from n_i be given by $C_{n_i,d}$. According to the MPR selection criteria of OEDR, the MPR node to cover d from s is selected as the node n_i, with cost of $Min\{(C_{s,n_1} + C_{n_1,d}), (C_{s,n_2} + C_{n_2,d}), \ldots, (C_{s,n_k} + C_{n_k,d})\}$. Hence, the MPR selection criteria of OEDR will result in an optimal route from s to its two-hop neighbors in $N^2(s)$.

LEMMA 8.4.1
The intermediate nodes on the optimal path are selected as MPRs by the previous nodes on the path.

PROOF   The proof follows on similar lines to that of Ying et al. (2003). A node in the route may not be selected as the MPR by the previous node, if the node does not provide a connection to that node's two-hop neighbors or the node does not meet the MPR selection criteria.

CASE I   The node in $N(s)$ of the previous node s, does not provide a connection to any node in $N^2(s)$. Consider the graph shown in Figure 8.7. Node n_2 only connects to node s's one-hop neighbor n_1. The two possible paths from s to d are $s \,-> \, n_1 \,-> \, d$ and $s \,-> \, n_2 \,-> \, n_1 \,-> \, d$. According to Assumption 1, n_2 is not on the optimal path from s to d.

CASE II   There is an optimal path from source to destination such that all the intermediate nodes on the path are selected as MPRs by their previous nodes on the same path. Without loss of generality, we suppose that in an optimal path, $s \,-> \, n_1 \,-> \, n_2 \,-> \, \ldots \,-> \, n_k \,-> \, n_{k+1} \,-> \, \ldots \,-> \, d$, there are nodes in the route which are not selected as MPRs by their previous nodes. Also, based on the result of Case I, we can assume that for each node on the path, its next node on the path is its one-hop neighbor, and the node two hops away from it is its two-hop neighbor (see Figure 8.8).

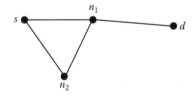

FIGURE 8.7
Case I scenario.

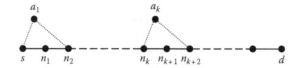

FIGURE 8.8
Case II scenario.

For example, n_1 is s's one-hop neighbor and n_{k+2} is n_k's two-hop neighbor. Consider the following two situations:

1. Suppose that on the optimal route, the first intermediate node n_1 is not selected as MPR by source s. However, n_2 is the two-hop neighbor of s. Based on the basic idea of MPR selection that all the two-hop neighbors of s must be covered by its MPR set, s must have another neighbor a_1, which is selected as its MPR, and is connected to n_2. According to the MPR selection criteria of OEDR, s selects a_1 instead of n_1 as its MPR because the cost to reach n_2 using a_1 is less than or equal to the cost to reach n_2 using n_1. Because route $s \to n_1 \to n_2 \to \cdots \to d$ is an optimal path, $\Rightarrow s \to a_1 \to n_2 \to \cdots \to d$ is also an optimal path. This implies that the source's MPR are on the optimal path.

2. Assume that on the optimal route $s \to n_1 \to n_2 \to \cdots \to n_k \to n_{k+1} \to \cdots \to d$, all the nodes on segment $n_1 \to \cdots \to n_k$ are selected as MPR by their previous node, we now prove that the next hop node of n_k on the optimal route is n_k's MPR.

   Suppose that n_{k+1} is not n_k's MPR. Same as in I, n_{k+2} is the two-hop neighbor of n_k, so n_k must have another neighbor a_k, which is the MPR of n_k and has a connection to n_{k+2}. Again, n_k selects a_k instead of n_{k+1} as its MPR because the cost to reach n_{k+2}, using a_k, is less than or equal to the cost to reach n_{k+2}, using n_{k+1}.

   Because, route $s \to \cdots \to n_k \to n_{k+1} \to n_{k+2} \to \cdots \to d$ is an optimal path,

   $\Rightarrow s \to \cdots \to n_k \to a_k \to n_{k+2} \to \cdots \to d$ is also an optimal path.

   This implies that in an optimal route, the $(k+1)$ th intermediate node is the MPR of the (k) th intermediate node.

Based on I and II all the intermediate nodes of an optimal path are MPRs of the previous node.

LEMMA 8.4.2
A node can correctly compute the optimal path for the whole network topology.

PROOF This statement means that, using a minimum-cost spanning tree, a node can compute optimal routes to all the destinations in the network. In OEDR, each node knows the links between the MPR nodes and their selectors, along with the costs of the links between them, from the TC messages. According to Lemma 1, the intermediate nodes on the optimal path are selected as MPRs by the previous nodes on the path. As a result, for a given node, the optimal paths to all the destinations are covered by the network topology known through the MPR nodes. Therefore, the node can correctly compute the optimal path for the whole network topology. Theorem 8.4.2 directly follows from Lemma 8.4.1 and Lemma 8.4.2.

THEOREM 8.4.2
The OEDR protocol results in an optimal-route (the path with the minimum energy-delay cost) between any source–destination pair.

THEOREM 8.4.3
For all pairs of nodes s and d, s generating and transmitting a broadcast packet P, d receives a copy of P.

PROOF The proof follows on similar lines to (Jacquet et al. 2001). Let k be the number of hops to d from which a copy of packet P has retransmitted. We shall prove that there exists a minimum $k = 1$, that is, a one-hop neighbor of d eventually forwards the packet.

Let n_k be the first forwarder at distance k $(k \geq 2)$ from d, which has retransmitted P. There exists a MPR n'_{k-1} of n_k which is at distance $k-1$ from d. To be convinced, let us imagine a path of length k from n_k to $d : n_k -> n_{k-1} -> n_{k-2} -> \cdots -> n_1 -> d$ and consider for n'_{k-1} as the MPR of n_k, which covers n_{k-2} (on the optimal route calculated using OEDR n_{k-1} will be the MPR n'_{k-1}).

Because n'_{k-1} received a copy of P the first time from n_k (the prior transmitters are necessarily two hops away from n'_{k-1}), n'_{k-1} will automatically forward P : packet P will be retransmitted at distance $k-1$ from d and will reach n_{k-2}. Similarly, packet P will be retransmitted at distance $k-2$, $k-3$, ..., 2, 1 until the packet reaches d. Hence, the theorem is proved.

8.5 Performance Evaluation

The proposed OEDR algorithm is implemented in the NS-2 simulator as a new routing protocol. Simulations were performed by varying node mobility and with variable numbers of nodes in the network. The OEDR, OLSR, and AODV routing protocols and IEEE 802.11 MAC protocol are employed.

The average delay of the packets (E2E) is chosen as one of the metrics to evaluate the performance of the protocols. Moreover, the network throughput is affected by the decisions of the routing protocol, and there exists a trade-off between the average delay and the network throughput. Therefore, in an attempt to minimize the average delay, the OEDR protocol might result in lower network throughput. However, the ratio of through-put to delay (throughput/average delay) is proven to be a more concise metric that can be used to compare different protocols, according to Jain (1991).

The third metric is the energy-delay product, and it is calculated as (total energy used/number of packets received at the destinations) times the average delay. Because the OEDR protocol aims at finding the optimal path with respect to the energy-delay cost of the links, instead of the shortest hops path (like in OLSR), in some cases, it can result in routes with higher number of hops compared to other routing protocols. This could slightly increase the energy consumption for packet delivery though minimizing the overall delay. Therefore, the energy-delay product would serve as a more precise metric to compare the performance of different routing protocols. The average contention time for a packet at an ad hoc node is calculated as the time interval elapsed between the packet being ready for transmission at the MAC layer to the time when it receives a successful CTS message to transmit the packet. The two simulation scenarios are presented next.

Example 8.5.1: Varying Node Mobility

A network of 100 nodes, randomly distributed in an area of 2000 × 2000 m is simulated by varying the node mobility between 20 and 100 km/h, along with the following parameters: simulation time is 100 sec, maximum number of flows is 50, location of the nodes and the flows are generated randomly, channel bandwidth is 1 Mbps, "two-ray ground" propagation model with a "path-loss exponent" of 4.0 is used, initial energy of each node is 10 J, flow rate is 41 kbps, packet size is 512 B, and the queue limit is 50 packets.

According to Figure 8.9, the average E2E delays using the OEDR protocol are much smaller than the delays using the OLSR and AODV protocols, because the proposed OEDR protocol considers the transmission delay of the links in the cost function while computing the routes between the source and destination. Also, observe that the average packet delays tend to increase with an increase in node mobility. With the movement of nodes, links get broken (and created) more frequently as the nodes enter and leave the transmission range of other nodes. Therefore, the routes have to be recomputed dynamically to reflect the changes in network topology.

According to Figure 8.10, the throughput/delay values of the OEDR protocol are always higher than those of OLSR and AODV protocols,

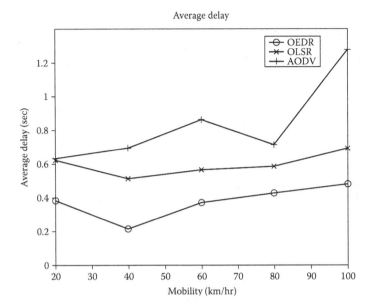

FIGURE 8.9
Average delay and mobility.

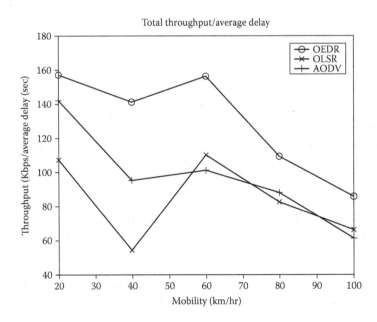

FIGURE 8.10
Throughput/delay and mobility.

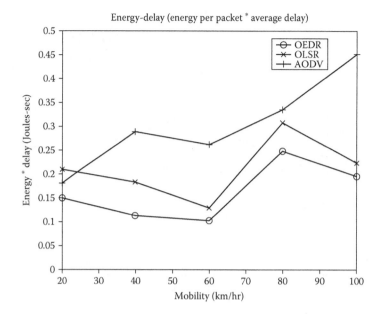

FIGURE 8.11
Energy and delay product with mobility.

because of a significant reduction in the average delays achieved by the OEDR protocol, compared to the reduction in the throughput due to delay optimization. Also, observe that these values decrease with the increase in the mobility, because of the increase in delays and reduction in throughput caused by frequent topological changes.

The energy-delay product values shown in Figure 8.11 are smaller for the OEDR protocol compared to the OLSR and AODV protocols, indicating that the proposed OEDR protocol results in the optimization of the energy-delay function. Similarly, contention time illustrated in Figure 8.12 is smaller for OEDSR and OEDR when compared to AODV.

Example 8.5.2: Variable Number of Nodes

Simulations are performed with network sizes varying from 10 to 200 nodes and the area of the network is selected depending upon the number of nodes. For networks with 10 to 20 nodes, the area of the network is selected as 500 × 500 m and for 50 nodes the area is 1000 × 1000 m. However, for larger networks of 100 to 200 nodes, the nodes are distributed in an area of 2000 × 2000 m. The maximum number of flows is selected as half the number of nodes in the network, and the rest of the simulation parameters are same as those of scenario 1.

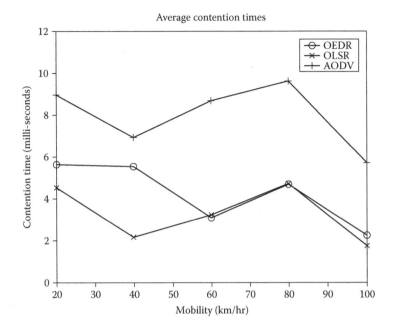

FIGURE 8.12
Contention time and mobility.

The average packet delay is illustrated in Figure 8.13, which indicates that the E2E delays using the OEDR protocol are always smaller than the delays incurred using the OLSR and AODV protocols, owing to the fact that delay information is taken into account in OEDR when calculating the routes. Moreover, the average packet delays tend to increase with an increase in the number of nodes because of higher interferences and channel contentions resulting from the density of the nodes and their traffic.

According to Figure 8.14 the throughput/delay values of the OEDR protocol are higher than those of OLSR and AODV protocols, because of significant reduction in the average delays achieved by the OEDR protocol. As the density of the nodes and the traffic increases, the throughput/delay ratio decreases as shown in the figure, because of an increase in delays caused by the retransmissions.

Energy delay product as a cost function and contention times are shown in Figure 8.15 and Figure 8.16, respectively. The contention times for the OEDR protocol are almost equal to those of the OLSR protocol and the differences are negligible (in milliseconds). This indicates that the OEDR protocol does not result in additional increase in the contention times. The energy and delay product values are smaller for the OEDR protocol

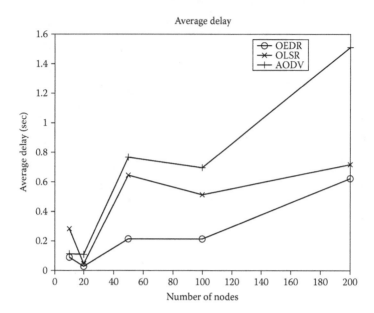

FIGURE 8.13
Average delay with number of nodes.

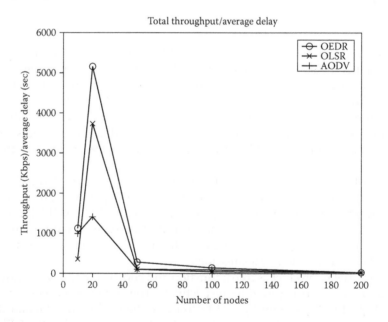

FIGURE 8.14
Throughput/delay with number of nodes.

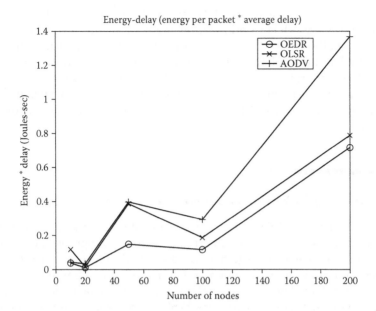

FIGURE 8.15
Energy and delay product with number of nodes.

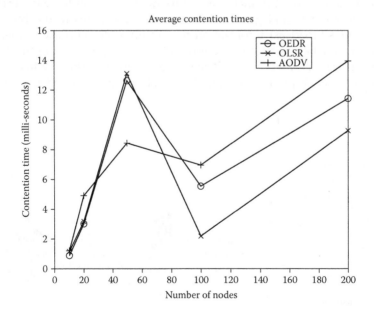

FIGURE 8.16
Contention time with number of nodes.

compared to the OLSR and AODV protocols, according to Figure 8.15, due to the energy-delay optimization methods of the proposed OEDR protocol. On the whole, the NS-2 simulation results suggest that the proposed OEDR protocol finds the minimum delay path from source to destination and optimizes the energy-delay product. OEDR protocol can also be applied to a wireless sensor network. Next, we show an extension of OEDR, known as optimal energy-delay subnet routing protocol (OEDSR), applied to a WSN.

8.6   Routing in Wireless Sensor Networks

Recent advances in WSNs have made it possible for small, inexpensive, low-power sensor nodes to be distributed across a geographical location. The information can then be gathered and sent to the end user or BS through wireless communication, as shown in Figure 8.17. These tiny sensor nodes have sufficient intelligence for signal processing and data broadcasting, but compared to other wireless networks, WSNs have resource constraints such as limited battery power, bandwidth, and memory.

Additionally, because of fading wireless channels, WSNs are prone to failures or intermittent connectivity. Consequently, WSNs have to periodically self-organize and generate routes from the nodes to the BS. The sensor nodes broadcast information to communicate with each other in the network, whereas ad hoc networks use peer-to-peer communication.

The performance metrics usually considered when working with a WSN are power consumption, connectivity, scalability, and limited resources.

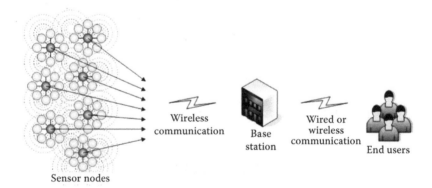

FIGURE 8.17
Wireless sensor network.

Sensor nodes play the dual role of data collection and routing and therefore need energy to perform. Malfunctioning of a few nodes, due to hardware or lack of energy, in the network will cause significant topology changes, leading to greater consumption of energy and rerouting of packets. Therefore, energy efficient schemes and communication protocols are being designed for WSNs.

One of the requirements for a protocol is scalability, meaning that the addition of more nodes to the existing network should not affect the functionality of the network with the protocol. Therefore, the communication protocols for the WSNs should take into consideration their limited resources such as battery power, memory, and bandwidth, and be able to function in the face of topology changes due to node mobility as well as guaranteeing scalability.

Available routing protocols for sensor networks are classified as data-centric, location based, quality of service (QoS) aware, and hierarchical protocols. Data-centric protocols (Esler et al. 1999) such as SPIN (Heizelman et al. 1999), directed diffusion (Intanagonwiwat et al. 2003), and GRAB (Ye et al. 2003, 2005) consolidate redundant data when routing from source to destination. Location-based routing protocols such as GPSR (Hill et al. 2000), GEAR (Zhang et al. 2004), and TTDD (Luo et al. 2002) require GPS signals to determine an optimal path so that the flooding of routing-related control packets is not necessary. From Chapter 1, sensor nodes are not supposed to have GPS devices and therefore such protocols have limited use. On the other hand, QoS aware protocols such as SPEED (He et al. 2003) address various requirements such as energy efficiency, reliability, and real-time requirements. Finally, the hierarchical protocols such as LEACH (Heinzelman et al. 2002), TEEN (Manjeshwar et al. 2001), APTEEN (Manjeshwar et al. 2002), and PEGASIS (Lindsey and Raghavendra 2002) form clusters and CHs to minimize the energy consumption both for processing and transmission of data.

In LEACH, the CH function is rotated among the nodes to reduce the energy consumption because only the CHs communicate to the BS even though the placements of CHs are not uniform. Nodes in TEEN and APTEEN are designed to respond to sudden changes in the sensed attribute when the attribute exceeds a user-defined threshold. On the other hand, PEGASIS is a chain-based protocol in which rather than multiple CH sending data, only one node in the chain is selected to transmit to the BS. LEACH, TEEN, APTEEN, and PEGASIS assume that the position of the BS is fixed and every node in the sensor network can directly communicate to it. This assumption is not always valid in larger networks, because the BS could be out of range with some of the CHs making a multihop routing protocol a necessity.

AODV is an on-demand routing protocol that discovers routes on an "as-needed" basis for wireless ad hoc networks. It uses traditional routing tables where a single entry per destination is entered. AODV relies on the routing table entries to send a route reply (RREP) back to the source and therefore routes data to the destination. To prevent routing loops, AODV uses sequence numbers to determine the originality of the routing information.

By contrast, DSR uses the concept of source routing, wherein the sender knows the complete route to the destination because the routes are stored in a route cache. When a node attempts to send data packets to a destination whose route is unknown, it uses the route discovery process to dynamically determine the route. The network is flooded with a route request (RREQ) packet. Each node that receives this RREQ packet keeps rebroadcasting it until it reaches a node that knows the route to the destination. This node then replies with a RREP that is routed back to the source node. Then using this route, the data is transmitted from the source to the destination.

On the other hand, the Bellman Ford algorithm computes the shortest paths from the source to the destination, based on the distance vector (DV) routing algorithm. The DV requires that each of the RNs informs its neighbors to the routing table. Using the lowest cost assigned for the minimum distance, the best node is selected and added to the routing table. This procedure is followed till the destination is reached.

The OEDR scheme is a routing protocol that generates paths based on a routing metric, defined in terms of energy efficiency, delay, and the state of the radio channel again for wireless ad hoc networks. It uses a concept of MPRs to minimize overhead. The MPR nodes are selected, based on a link cost factor which is a product of energy and delay. Using the MPR nodes and the minimum-cost spanning tree algorithm, the optimum route to the destination is computed. Hierarchical routing protocols such as LEACH are typically employed for wireless sensor networks.

This chapter now presents an OEDSR protocol (Ratnaraj et al. 2006) for WSN, where the information from the CHs to the BS is routed in a multihop fashion by maximizing a routing metric, which in turn, ensures the lifetime of the network. OEDSR uses a link cost factor, which is defined using available energy, E2E delay, and distance from a node to the BS, along with clustering, to effectively route information to the BS. The *link_cost_ factor* is defined at a given node as the ratio of energy available over the E2E delay and shortest distance to the BS. This identifies an optimal route by maximizing the *link_cost_ factor* while guaranteeing the performance analytically.

This routing protocol uses routing metrics similar to OEDR, but it is an on-demand protocol compared to the proactive OEDR protocol. Additionally, clusters are formed only in the subnetworks in the case of OEDSR, and the rest of the network outside the subnetwork is treated as an ad hoc wireless network. The performance of the proposed routing protocol is demonstrated in simulation by comparing it with other routing protocols such as AODV, DSR, Bellman Ford (Ford and Fulkerson 1962), and OEDR (Regatte and Jagannathan 2005). These routing protocols are selected for comparison because they are multihop routing protocols.

As an important first step toward node mobility, the WSN must self-organize. Therefore, Section 8.7 briefly describes the self-organization, using the subnetwork (SOS) protocol (Ratnaraj et al. 2006), which organizes the nodes in the network into a subnetwork; Section 8.8 discusses the novel OEDSR protocol in detail; in Section 8.8, the optimality of the algorithm is demonstrated analytically; Section 8.9 presents the GloMo-Sim simulator results to evaluate the performance of the routing algorithm; Section 8.10 presents hardware implementation results and Section 8.11 concludes the chapter.

8.7 Self-Organization Using Subnetwork Protocol

The OEDSR routing protocol runs in conjunction with the self-organization of the sensor nodes and, therefore, the self-organization protocol is initially dealt with in a brief manner. Initially, all nodes in the network are in sleep mode to save battery power. When an event is detected in the network, nodes around the event wake up and measure the sensed attribute. If the sensed attribute value is greater than a predefined threshold, then the nodes will join the subnetwork. Otherwise, the node goes back to sleep mode. By forming a subnetwork, the size of the active part of a network is limited to the nodes that have significant information. The activated nodes send HELLO messages to all neighbor nodes, which are in the communication range. The HELLO messages contain various fields such as the node ID, energy available in the node, and sensed attribute detected from the event.

Next, all the nodes in the subnetwork are grouped into clusters so that the sensor information can effectively be aggregated before sending it to the BS. First, all the nodes in the subnetwork send out HELLO packets to each other with their node IDs, the amount of energy available, and the sensed attribute. Then the node with the highest energy available elects itself as the temporary subnetwork head (TH), and the other nodes in the subnetwork will become idle as shown in Figure 8.18. The function of the TH is to calculate the required number of CHs and select the nodes that will become CHs.

An appropriate percentage of nodes in the subnetwork are selected as CHs by the TH. The percentage can vary depending on the density of the network, with a dense network requiring more CHs. Once the number of

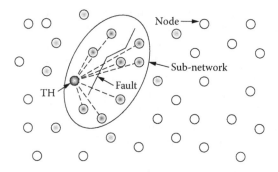

FIGURE 8.18
Selection of temporary subnetwork head.

CHs is calculated, the suitable nodes have to be identified, based on the *CH_selection_ factor*. This factor is calculated as the product of the energy available and the value of the sensed attribute from the event. For every node in the subnetwork, the TH calculates the *CH_selection_ factor* as

$$CH\_selection\_factor = Er_n \times S(n),\qquad(8.3)$$

where Er_n is the energy available per node obtained from the battery monitor, and $S(n)$ is the sensed magnitude attribute from the event.

An ideal CH should have both maximum energy available and the sensed attribute value from the event. However, the nodes near the faulty event are more prone to failure because they will exhaust their energy more quickly by detecting the events subsequent to its occurrence. Also, as a node gets farther away from the event, the sensed attribute from the event decreases, and the nodes will not have accurate information. Therefore, the CHs are selected such that the *CH_selection_ factor* is closer to the signal strength value, which is the median of the collected signal strength values, so that the CHs are distributed evenly. In other words, the TH arranges all the *CH_selection_factors* in ascending order and chooses the node whose *CH_selection_ factor* is closer to the median of the collection as the CHs. The median is considered instead of taking the average because it works better where extreme sensed attributes are present. It also ensures that the selected CHs are at an optimum distance from the event and also have maximum energy available. Once the CHs are selected, the TH broadcasts this information to all the nodes in the subnetwork by sending a cluster head selection (CH_SELECT) packet. Subsequently, the TH becomes a regular node.

All the sensors use the received signal strength indicator (RSSI) to identify the radio frequency (RF) signal strength used for communication from other nodes in the network. RSSI is a widely used parameter for wireless networks in general. It is important to notice that the RSSI provides an indirect assessment of the channel state, which has to be taken into account both during self-organization and routing. Channel state is totally ignored in the literature in most protocols. Once the CHs are selected, these CHs broadcast a beacon to all the nodes in the subnetwork. The nodes in the subnetwork measure the RSSI for these beacons, and depending on the strength of each signal, the nodes join a CH with highest signal strength. Once a node selects a CH, it sends a JOIN packet to the CH to indicate that it is joining a particular cluster. Sensor nodes in each cluster will now relay information directly to the corresponding CHs. By doing this, they avoid flooding the network with data and allow the CHs to perform data aggregation so that the redundant data are compressed. After the network self-organizes itself, then the data is routed through the network as follows.

8.8 Optimized Energy-Delay Subnetwork Routing Protocol

Once the CHs are identified and the nodes are clustered within the subnetwork, the CHs initiate routing towards the BS by checking if the BS is within the communication range so that the data can be sent directly. Otherwise, the data from the CHs in the subnetwork have to be sent over a multihop route to the BS. The proposed on-demand routing algorithm is fully distributed and adaptive because it requires local information for building routes and adapting to the topological changes. A high-power beacon from the BS is sent periodically to the entire network such that all the nodes in the network have knowledge of the distance to the BS. It is important to note that the BS has a sufficient power source and therefore it is not energy constrained. Though the OEDSR protocol borrows the idea of using an energy-delay type metric from the OEDR, the selection of RNs is not based on maximizing the number of two-hop neighbors and available energy. Instead, the link cost factor is calculated by using a totally different metric as given in Equation 8.4. Here, the selection of a relay node depends on maximizing the *link_cost_factor*, which includes the distance from the relay node to the BS. Moreover, the route selection is performed differently between the OEDR and the OEDSR because the OEDR is a proactive routing protocol whereas the OEDSR is an on-demand routing protocol. Within the subnetworks, CHs are involved in the routing process, whereas RNs are selected outside the subnetworks for the purpose of routing. These RNs are viewed as CHs. Next, the selection of optimal RNs is discussed.

8.8.1   Optimal Relay Node Selection

The *link_cost_factor*, which is defined as the ratio of the energy available over the product of the average E2E delay and the distance from the node to the BS, is given by

$$Link\_cost\_factor = \frac{Er_n}{Delay \times Dist} \tag{8.4}$$

where Er_n is the energy available in the given node obtained from the battery monitor, *Delay* is the average E2E delay between any two CHs, and *Dist* is the distance from the given node to the BS. In Equation 8.2, the energy available at a node has to be higher whereas the average E2E delay and the distance to the BS have to be at a minimum to select an optimal route that renders a long network lifetime. Therefore, a relay node on the optimal route should maximize the *link_cost_factor*. Ensuring that the delay and the distance to the BS are minimal will help reduce the number of hops and the overall delay in routing the information from a CH to the BS. When multiple RNs are available for routing, an optimal RN is selected such that the *link_cost_factor* is maximized. The nodes that are common between the CH are taken into consideration to become candidate RNs. Figure 8.19 shows the relay-node selection for the proposed OEDSR in contrast with OEDR wherein RNs have to satisfy certain stringent requirements such as maximizing the number of two-hop neighbors. The details of the relay-node selection in OEDSR are discussed next.

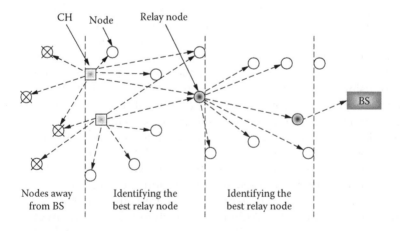

FIGURE 8.19
Relay node selection.

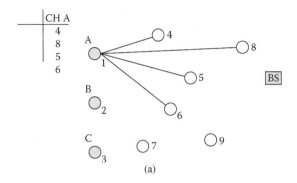

(a)

FIGURE 8.20 (a)

Steps to select a relay node: (a) A, B, C are the CHs and 1, 2, 3, ..., and 9 are the node IDs. In the first step, CH A sends out HELLO packets to all the nodes that are in range and receives RESPONSE packets. CH A first checks if its distance from the BS is greater than node 4, 8, 5, and 6. If it is, then a list is created with the information of all nodes that are in range with CH A.

In the OEDSR protocol, routes are selected, based on the best *link_cost_factor* of all nodes in range to the CH or a relay node. Initially, a CH broadcasts HELLO packets to all the nodes that are in range and receives RESPONSE packets from all the relay candidates that are in the communication range as shown in Figure 8.20a. The RESPONSE packets contain information such as the node ID, energy available, average E2E delay, and distance to the BS. After receiving RESPONSE packets, CH neglects those nodes whose distances to the BS are greater than the distance from the CH to the BS. This ensures that the route does not take a longer or cyclic path to the BS. If the RESPONSE packet is received from the BS, then it is selected as a next hop node, thus ending the route discovery procedure. Otherwise, the CH broadcasts a HELLO packet to the next CH in the subnetwork with the list of candidate RNs.

Once the next in-sequence CH receives the HELLO packet, it checks whether it is in communication range with any of the listed nodes or not. Then it creates a new list of candidate RNs. The same procedure is carried out for all the CH. Finally, if more than two nodes are selected as candidate RNs, the node with the highest *link_cost_factor* is selected as the RN. Selection of RNs is illustrated in Figure 8.20b and Figure 8.20c. If a CH does not have a common RN, then it sets up a separate route using the *link_cost_factor* to the BS such that it is optimal from the CH. Figure 8.20d shows the route for the remaining CHs, determined in the same way as explained previously.

The advantage of using this routing method is that it reduces the number of RNs that have to forward data in the network, and hence the scheme reduces overhead, number of hops, and communication among nodes to

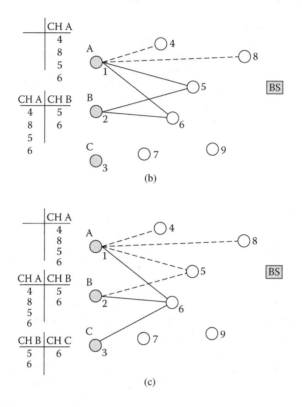

(b)

(c)

FIGURE 8.20 (b c)
(b) In the second step, CH A sends the information it received, to CH B. CH B sends out HELLO packets to all the nodes that are in range with CH A, and checks if it is in range with any of the nodes. In this case, CH B is in range with node 6 and node 5, which are the common relay nodes for both CH B and CH A. Now, a new list with this information is available to CH B. (c) The list CH B created is sent to CH C. The same procedure of finding common relay node is carried out. Here, node 6 is selected as the relay node. By doing this, minimum numbers of nodes are selected as relay nodes and fewer paths are taken to the BS.

overcome flooding. Additionally, clusters are not formed outside the sub-networks. The next section presents the relay-node selection algorithm.

8.8.2 Relay-Node Selection Algorithm

1. The candidate RNs from a CH are identified by sending HELLO packets and receiving RESPONSE packets, as shown in Figure 8.21.

2. The distances from all the nodes to the BS are compared. If the distance from the node to the BS is greater than the distance from the CH to the BS, then that node is not selected as a candidate relay

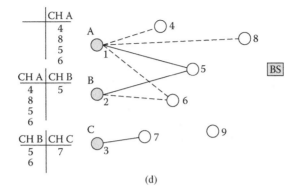

(d)

FIGURE 8.20 (d)
(d) In this case, CH B sends its information about node 5 and node 6 to CH C. CH C checks if it is in range with these nodes. Because CH C is not in range with both the nodes, it creates its own optimal path to the BS. Now CH B chooses the best relay node between node 5 and node 6 based on the *link_cost_factor*. Here, node 6 is chosen as the optimal relay node for CH A and B. Therefore, two separate optimal routes using the *link_cost_factor* are created from the CHs, if no common relay node is identified.

node. By doing this step, only nodes that are closer to the BS are filtered out as candidate RNs, thereby avoiding longer routes.

3. A list containing candidate RNs is sent to the next CH (or relay node). The CH, in turn, checks whether it is in range with any of the nodes in the list or not. The CH sends a HELLO packet to each of the nodes in the list, and if it receives a response packet from any of the nodes, then those nodes are considered to be in range with the CH.

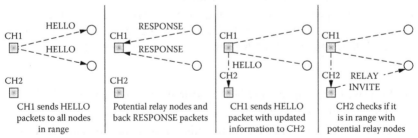

CH with list of potential nodes ID, energy available, delay and distance to BS
Node that receive HELLO message from CH

FIGURE 8.21
Steps to route information to the BS.

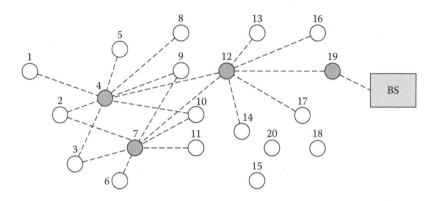

FIGURE 8.22
Relay node selection.

4. If it is in range with any of the candidate RNs, then it becomes a part of the new list of candidate RNs. The same procedure is carried out for all the CHs in the network.

5. Finally, if only one node is identified as the candidate relay node in the list, then that node is assigned as the relay node.

6. On the other hand, if more than one node is identified as a candidate relay node, then the *link_cost_ factor* is calculated, and the node with the highest *link_cost_ factor* is selected as the relay node.

7. The *link_cost_ factor* for each of the nodes is calculated, and the node with the maximum *link_cost_ factor* is selected as the optimal relay node.

Example 8.8.1: Relay-Node Selection

Consider the topology shown in Figure 8.22. The link_cost_factors are calculated to route data to the BS. The following steps are implemented to route data using the OEDSR protocol and the flowchart is depicted in Figure 8.23.

1. Start with an empty relay list for source node n: Relay(n)={ }. Here node n_4 and n_7 are CHs.

2. First, for CH n_4, check which nodes it is in range with. In this case, CH n_4 is in range with nodes n_1, n_2, n_3, n_5, n_8, n_9, n_{12}, and n_{10}.

3. The nodes n_1, n_2, and n_3 are not considered as potential candidate RNs because the distance from them to the BS is greater than the distance from CH n_4 to the BS.

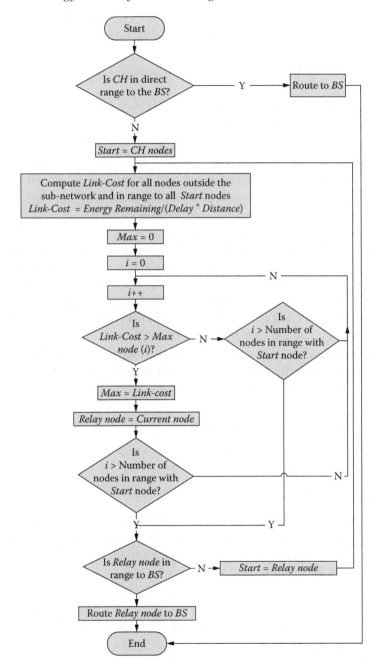

FIGURE 8.23
Flowchart for the OEDSR protocol.

4. Now, all the nodes that are in range with CH n_4 return RESPONSE packets. Then CH n_4 makes a list of candidate RNs n_5, n_8, n_9, n_{12}, and n_{10}.

5. CH n_4 sends the list to CH n_7. CH n_7 checks if it is in range with any of the nodes in the list.

6. Nodes n_9, n_{10}, and n_{12} are in range with both CH n_4 and n_7. They are selected as the common RNs.

7. The *link_cost_factors* for n_9, n_{10}, and n_{12} are calculated.

8. The node with the maximum *link_cost_factor* value is selected as the relay node and assigned to *Relay(n)*. In this case, *Relay(n)*={n_{12}}.

9. Now node n_{12} checks if it is in direct range with the BS, and if it is, then it directly routes the information to the BS.

10. Otherwise, n_{12} is assigned as the relay node, and all the nodes that are in range with node n_{12} and whose distances to the BS are less than their distances to the BS are taken into consideration. Therefore, nodes n_{13}, n_{16}, n_{19}, and n_{17} are taken into consideration.

11. The *link_cost_ factor* is calculated for n_{13}, n_{16}, n_{19}, n_{14}, and n_{17}. The node with the maximum *link_cost_ factor* is selected as the next relay node. In this case *Relay(n)* = {n_{19}}.

12. Next, the relay node n_{19} checks if it is in range with the BS. If it is, then it directly routes the information to the BS. In this case, n_{19} is in direct range, so the information is sent to the BS directly.

8.8.3   Optimality Analysis for OEDSR

To prove that the proposed route created by OEDSR protocol is optimal in all cases, it is essential to show the performance analytically.

ASSUMPTION 8.8.1
All nodes know the position of the BS irrespective of whether they are mobile or not. When the BS is mobile, it sends out its location information periodically to the network. This information is sent using a high-power beacon from the BS to all the nodes in the network.

THEOREM 8.8.1
The link_cost_factor based routing generates viable RNs to the BS.

PROOF   Consider the following two cases:

CASE I   When the CHs are one hop away from the BS, the CH selects the BS directly. In this case, there is only one path from the CH to the BS. Hence, the OEDSR algorithm does not need to be used.

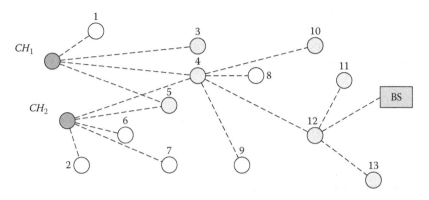

FIGURE 8.24
Link cost calculation.

CASE II When the CHs have more than one node to relay information, the OEDSR algorithm selection criteria are taken into account. In Figure 8.24, there are two CHs, CH_1 and CH_2. Each CH sends signals to all the other nodes in the network that are in range. Here, CH_1 first sends out signals to n_1, n_3, n_4, and n_5 and makes a list on the potential RNs. The list is then forwarded to CH_2. CH_2 checks if it is in range with any of the nodes in the list. Here, n_4 and n_5 are selected as potential common RNs. A single node must be selected from both n_4 and n_5, based on the highest *link_cost_ factor*. The cost to reach n from CH is given by Equation 8.4. So, based on the OEDSR *link_cost_ factor*, n_4 is selected as the relay node for the first hop. Next, n_4 sends signals to all the nodes it is in range of, and selects a node as the relay node using the *link_cost_ factor*. The same procedure is carried on until the data is sent to the BS.

LEMMA 8.8.1
The intermediate nodes on the optimal path are selected as RNs by the previous nodes on the path.

PROOF A node is selected as a relay node only if it has the highest *link_cost_ factor* and is in range with the previous node on the path. Because OEDSR maximizes the *link_cost_ factor*, intermediate nodes that satisfy the metric on the optimal path are automatically selected as RNs.

LEMMA 8.8.2
A node can correctly compute the optimal path (lower E2E delay, minimum distance to the BS and maximum available energy) for the entire network topology.

PROOF When selecting the candidate RNs to the CHs, it is ensured that the distance from the candidate relay node to the BS is less than the distance from the CH to the BS. When calculating the *link_cost_ factor*, available energy is divided by distance and average E2E delay, to ensure that the selected nodes are in range with the CHs and close to the BS. This helps minimize the number of MPR nodes in the network.

THEOREM 8.8.2

OEDSR protocol results in an optimal route (the path with the maximum energy, minimum average E2E delay, and minimum distance from the BS) between the CHs and any source destination.

8.9   Performance Evaluation

The OEDSR algorithm is implemented in GloMoSim as a new routing protocol. Simulations are performed by varying node mobility and number of network nodes. Additionally, the nodes are made mobile. It is important to notice that the OEDSR cannot be directly compared with other hierarchical routing protocols such as LEACH, TEEN, APTEEN, and PEGASIS as these protocols assume that every node is in direct range with the BS and, if that is not the case, each node sends a high-power signal to the BS so that information can be transmitted using one hop. Therefore, OEDR, AODV, DSR, and Bellman Ford's routing protocol are used for comparison because they use multiple RNs to route information when the BS is not in direct range to a given network node.

The total energy consumed and the average E2E delay from a CH to the BS is minimized by OEDSR and, therefore, those two metrics are considered for the analysis. Additionally, the energy available times the average E2E delay metric served as a more precise overall metric to compare the performance of different routing protocols. In Figure 8.25, different network sizes are simulated for the same CH (shown in green) and the same BS (shown in red). It can be seen that when the number of nodes in the network increases, the OEDSR protocol ensures that the optimum path is selected for routing the information by selecting the appropriate RNs, which are shown as blue stars. First, the case where nodes are stationary is discussed.

Example 8.9.1: Stationary Nodes

It is assumed that the network has stationary nodes that are randomly placed on the surface of a structure to monitor incipient faults and other faulty events. The simulation was run for networks with 40, 50, 70, 100, and 150 nodes, randomly distributed in an area of size 2000 × 2000 m. Parameters that were used for this simulation include a packet size of

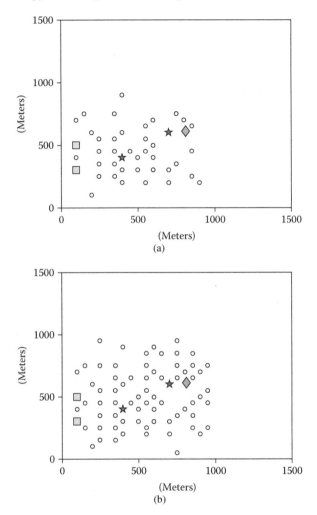

FIGURE 8.25
Network size: (a) Layout of a 40-node random network. (b) Layout of a 70-node random network.

256 B, a simulation time of 1 h, two-ray ground propagation model with a path-loss exponent of 4.0. In addition, the transmission powers of all active nodes are taken as 25 dBm, whereas for the sleep nodes it was set to 5 dBm.

In Figure 8.26, the energy consumed for different routing protocols can be observed. Because the Bellman Ford protocol consumes a radically higher amount of energy than other protocols, the graph is magnified to compare the performance of OEDSR with the other protocols. On magnifying the

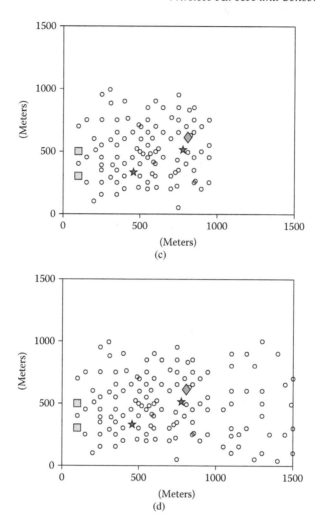

FIGURE 8.25 (Continued)
Network size: (c) Layout of 100-node random network. (d) Layout of a 150-node random network.

plot as shown in Figure 8.27, it can be seen that OEDSR protocol performs better than AODV and DSR. This is because OEDSR uses an optimal route to the BS by selecting nodes based on the available energy over average E2E delay times distance metric. Moreover the distance metric will ensure that the numbers of RNs are minimized. In consequence, the energy used to transmit packets is minimized in case of the proposed scheme as only a few of the nodes are selected to transmit the data to the BS.

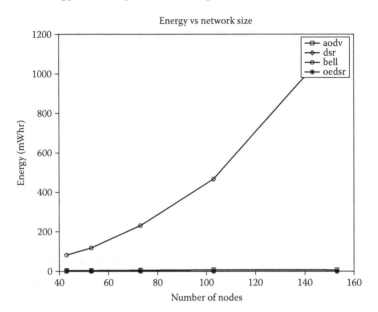

FIGURE 8.26
Energy consumed with network size.

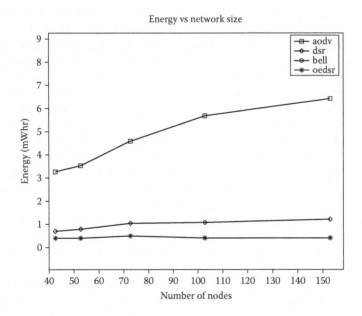

FIGURE 8.27
Energy consumed with network size (magnified).

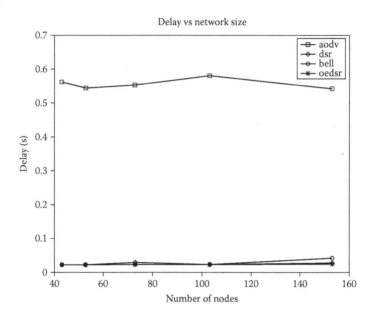

FIGURE 8.28
Average end-to-end delay with network size.

According to Figure 8.28 and Figure 8.29, the average E2E delays observed using the OEDSR protocol are much smaller than the delays incurred using AODV, DSR, and Bellman Ford protocols, because, in the proposed protocol, the E2E delay is considered explicitly in the *link_cost_factor* calculation. OEDSR ensures that the route renders smaller E2E delays, whereas the Bellman Ford and the AODV create a route with higher E2E delay. By contrast, the DSR has much lower E2E delay than the Bellman Ford and the AODV protocols. This is because the DSR uses source routing to determine the RNs. Initially, the entire route has to be determined prior to routing. However, the OEDSR performs better than the DSR, because unlike the DSR, which maintains the same route, the OEDSR changes its route dynamically to lower the E2E delay. Therefore, OEDSR is an on-demand routing protocol.

As observed in Figure 8.30 and Figure 8.31, Bellman Ford consumes more energy, whereas AODV has a greater E2E delay. Therefore, the appropriate overall metric to compare the actual performance of the protocols will be to take the product of the energy consumed and the average E2E delay. Both the energy consumed and the average E2E delay have to be minimized for an optimal routing path. Taking the product of these two metrics helps in identifying the optimal route, and therefore it is considered as the overall metric in OEDR. By doing so, it is observed that the Bellman Ford routing protocol has the highest product. This is because

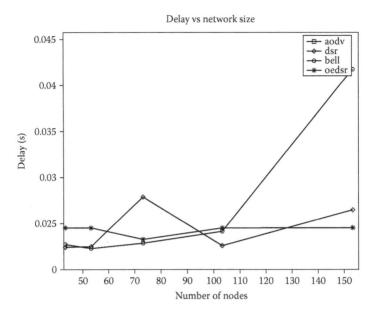

FIGURE 8.29
Average end-to-end delay with network size (magnified).

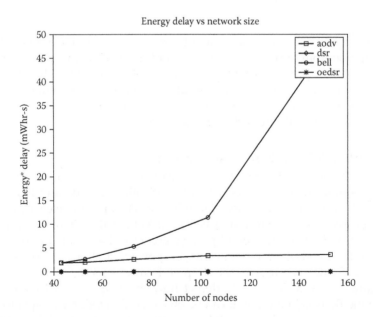

FIGURE 8.30
Energy times average end-to-end delay with network size.

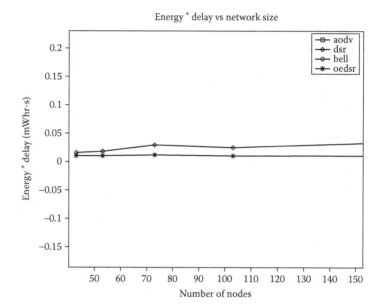

FIGURE 8.31
Energy times average end-to-end delay with network size (magnified).

Bellman Ford consumes high quantities of energy while transmitting information from the CHs to the BS. By contrast, from Figure 8.31, OEDSR has the lowest product. OEDSR performs better than the other routing protocols as it takes into account a route that has maximum energy available and minimum E2E delay while minimizing the number of RNs to effectively route data from the CHs to the BS.

Example 8.9.2: Mobile Base Station

For the networks used in Scenario 1, simulations were run again for a mobile BS. The simulation was run for networks with 40, 50, 70, 100, and 150 nodes randomly distributed in an area of size 2000 × 2000 m. Parameters that were used for this simulation include a packet size of 256 B, a simulation time of one hour, and a two-ray ground propagation model with a path-loss exponent of 4.0. In addition, the transmission powers of all active nodes were taken as 25 dBm, whereas for the sleep nodes it was set to 5 dBm.

Figure 8.32 and Figure 8.33 depict how the RNs were selected even when the BS was in motion. It can be observed that OEDSR adapts to the changing BS position by identifying an efficient route to the BS, irrespective of its position.

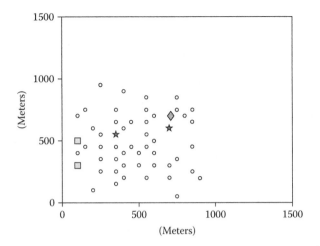

FIGURE 8.32
Mobile base station at 700 m, 700 m.

Figure 8.34 shows how the CHs take two paths to the BS using OEDSR, when they are placed far apart and have no common relay node to route their information.

Figure 8.35 and Figure 8.36 depict the comparison of energy consumed by different routing protocols when the BS is mobile. The OEDSR performs better than the other routing protocols for all network sizes. As the

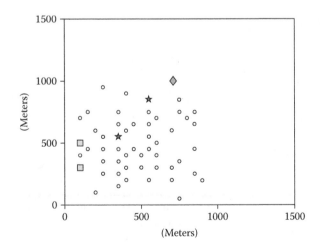

FIGURE 8.33
Mobile base station at 700 m, 1000 m.

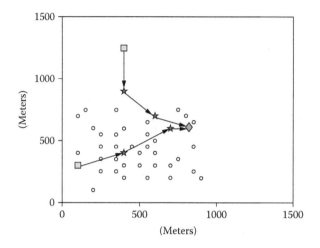

FIGURE 8.34
CHs taking two paths to the BS.

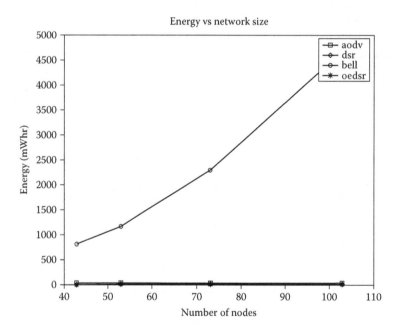

FIGURE 8.35
Energy consumed with network size.

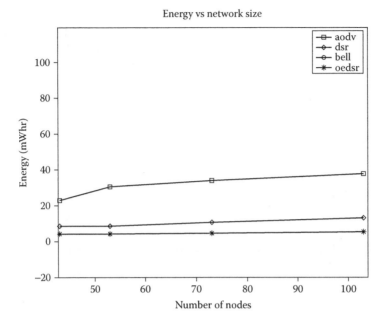

FIGURE 8.36
Energy consumed with network size (magnified).

network size increases, the energy consumed for other protocols increases. At the same time, the nodes in the network using OEDSR consume almost a fixed amount of energy, because only a limited number of nodes are used to route information while the rest are in sleep state. Moreover, subnetwork formation minimizes the energy consumption. By contrast, with an increase in network size, Bellman Ford uses more number of RNs to route information, thus consuming more energy.

Next, the average E2E delay in the case of a mobile BS is computed. In this scenario, the AODV incurs higher E2E delay, compared to the OEDSR as illustrated in Figure 8.37 and Figure 8.38. The E2E delay is calculated as the time taken to complete data transfer from the source to the destination. This time includes the transmission time, which is the time it takes for the data to reach the destination node from the source node, and the wake-up time, which is defined as the time taken for the destination node to wake up and receive this stream of data. Moreover, when multiple packets arrive at the same time, then the queuing time is also considered for calculating the E2E delay. Therefore, when there are fewer RNs, as is the case of OEDSR, fewer nodes have to be kept awake. This reduces the average energy consumed.

Figure 8.39 and Figure 8.40 display the number of collisions observed when using different routing protocols. As was the case with stationary

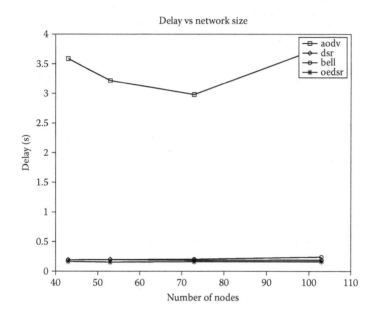

FIGURE 8.37
Average end-to-end delay with network size.

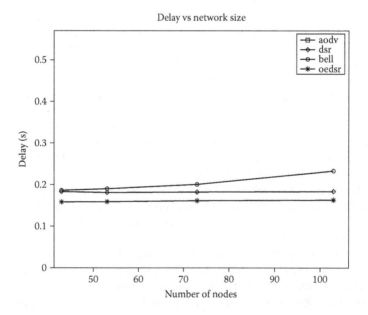

FIGURE 8.38
Average end-to-end delay with network size (magnified).

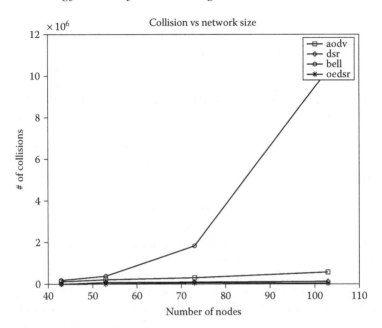

FIGURE 8.39
Number of collisions with network size.

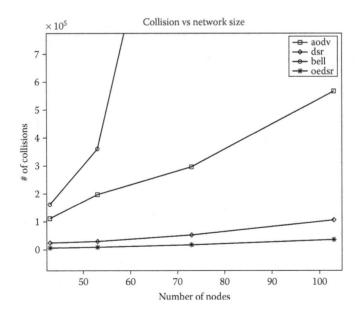

FIGURE 8.40
Number of collisions with network size (magnified).

nodes, the number of collisions observed in the OEDSR is lower because of the fewer RNs used to route information. Congestion is mitigated because of fewer numbers of nodes accessing the channel. For smaller networks, OEDSR and DSR have almost the same number of collisions. This is because both these protocols use almost the same number of RNs to route information. But with an increase in network size, the DSR uses more RNs to route information, while the OEDSR uses the same number of RNs. This is because the OEDSR is an on-demand routing protocol and it uses E2E delay and distance to the BS from each node as the routing metrics to minimize the number of RNs used. Consequently, the traffic in the network is minimized and the numbers of collisions are minimized.

Example 8.9.3: Comparison of OEDSR with OEDR

Static Case: Additional simulations were run for networks with 40, 50, 70, and 100 nodes randomly distributed. Parameters used for this scenario are identical to the previous case.

Figure 8.41 shows the comparison of energy consumed by the OEDSR and the OEDR, wherein the energy consumed is lower for the OEDSR than the OEDR, because the OEDSR considers the distance of a node from the BS apart from the energy-delay metrics for calculating the link cost factor, as this decreases the number of RNs. On the other hand, the OEDR protocol uses more RNs because the link cost factor is calculated based

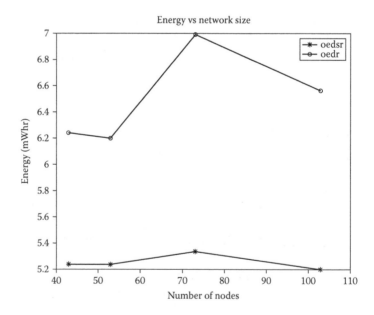

FIGURE 8.41
Energy consumed with network size.

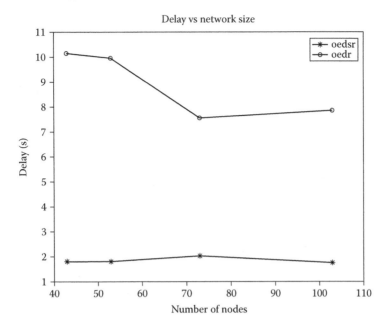

FIGURE 8.42
Average end-to-end delay with network size.

on just the available energy and the delay. It can be noticed that when there are 70 nodes in the network, the energy consumed using OEDR and OEDSR protocols increases because of the nature of the topology. However, the increase in energy using OEDSR is less compared to OEDR because a different and more efficient route was selected.

The average E2E delay is also lower for OEDSR when compared to OEDR as observed in Figure 8.42. This is because there are fewer nodes that are selected as RNs in the OEDSR protocol, and this decreases processing and queuing time. In the case of OEDR, it is assumed that every node knows the route to the BS, based on the MPR sets. Because of this, a longer route was selected to route information as OEDR is not an on-demand routing protocol. The E2E delay includes the wake-up time for a node apart from the transmission time. Therefore, with more RNs, wake-up time goes up, increasing the E2E delay.

It can be observed from Figure 8.43 that the number of collisions in the network is lower for the OEDSR when compared with the OEDR. OEDR uses more RNs to transmit data from the CHs to the BS. Moreover, when selecting the multipoint-relay (MPR) nodes, the nodes send information to their one-hop and two-hop neighbors, and this increases the number of signals being sent in the network. Because of this, the traffic in the network goes up, thereby increasing the number of collisions in the network.

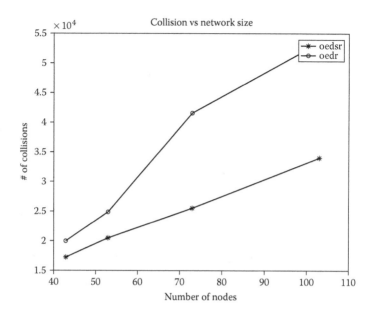

FIGURE 8.43
Number of collisions with network size.

Figure 8.44 shows the overhead comparison between the OEDSR and the OEDR. It is observed that the overhead is significantly less for the OEDSR. The overhead for the OEDSR is mainly because of the HELLO packets being sent from a node to its one-hop neighbors and the RESPONSE packets that the node receives from these one-hop neighbors. Initially, each source node sends out a HELLO packet consisting of the node ID and the distance to the BS to all the one-hop nodes. This HELLO message is 2 bytes in size. Upon receiving this HELLO message, all the potential RNs send back a RESPONSE packet consisting of the node ID, energy available, delay, and the distance to the BS.

The delay is derived from the HELLO packet exchange. Then the source node sends out packets containing the information of the potential RNs to the next source node. This is continued till an optimum relay node is determined, which subsequently informs all the source nodes of this relay node. All these messages sent constitute for the overhead in the OEDSR. But in the case of the OEDR, apart from sending HELLO packets to all the one-hop and two-hop neighbors and receiving the RESPONSE packets, once the MPR nodes are selected, a TC message containing all link cost and selector set information is transmitted periodically to all the nodes in the network in a multihop manner. This significantly increases the overhead, and thus the OEDSR performs better than the OEDR.

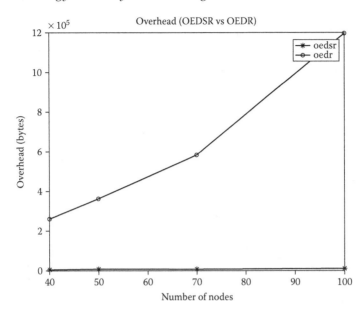

FIGURE 8.44
Overhead with network size.

Even though the distance to the BS from each node is included in the cost factor calculation, the total overhead for the OEDSR is still lower than the OEDR, due to the way the RNs are selected.

To understand the effect of channel fading during routing, another simulation run was carried out but with fading channels. Channel fading was introduced during routing for the OEDSR in a random manner. The channel fading causes the noise level in the terrain to go up, which in turn, increases the percentage of packets lost. With channel fading, the signal-to-noise ratio (SNR) has to be increased because of an increase in the threshold for successful packet reception. Otherwise, the packet is dropped. It can be observed from Figure 8.45 that the number of packets lost increases as the received SNR threshold is increased. When the threshold is set greater than 20 dB, the percentage of packets lost is greater than 20%, that is, one out of every five packets transmitted are lost and have to be retransmitted. This also increases the amount of energy consumed because more packets have to be transmitted from the node. These results clearly show that protocols for ad hoc network cannot be directly deployed for WSNs.

Example 8.9.4: Mobile Network

The simulation was run for networks with 40, 50, 70, and 100 nodes randomly distributed in an area of size 2000 × 2000 m with node mobility.

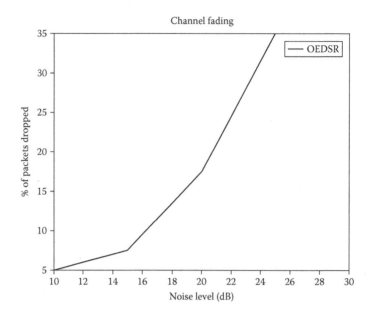

FIGURE 8.45
Packets dropped with channel fading.

Node mobility is generated in a random manner for each node but was kept the same when testing protocols.

Figure 8.46 shows the comparison of energy consumed by the OEDSR and the OEDR for a mobile network. As expected and observed in previous scenarios, the energy consumed using the OEDSR increases significantly as the number of nodes increases in the network, because the nodes in the network move in a random fashion. Due to this, the different routes having varying number of RNs are selected at different times. However, the energy consumed in the OEDSR is far less compared to the OEDR due to the reasons mentioned in the static node case.

The average E2E delay is still lower for the OEDSR when compared to the OEDR as observed in Figure 8.47. Even with node mobility, fewer nodes are selected as RNs in the OEDSR protocol, whereas the numbers of RNs are significantly more with the OEDR, which in turn, increase the E2E delay. The E2E delay includes the wake-up time for a node, apart from the transmission time. Therefore, with more RNs, waking, processing, and queuing times go up increasing the E2E delay.

It can be observed from Figure 8.48 that the number of collisions in the network is fewer for the OEDSR when compared with the OEDR. The OEDR uses more RNs to transmit data from the CHs to the BS. Moreover, when selecting the MPR nodes, the nodes send information to their one-hop and two-hop neighbors, and this increases the number of signals

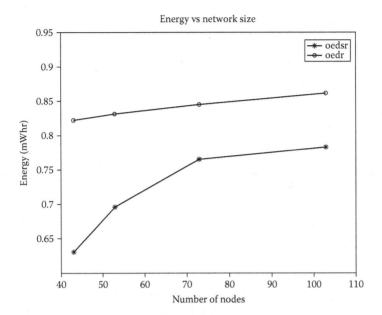

FIGURE 8.46
Energy consumed with network size.

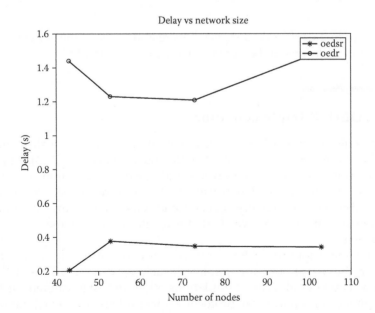

FIGURE 8.47
Average end-to-end delay with network size.

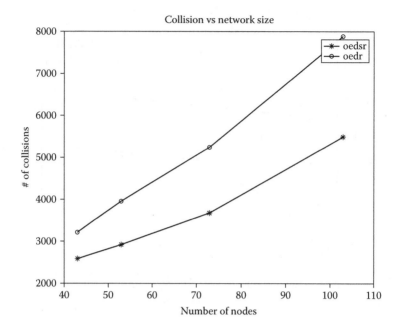

FIGURE 8.48
Number of collisions with network size.

being sent in the network. Because of increased traffic in the network, the number of collisions in the network increases significantly.

8.10   OEDSR Implementation

Energy-efficient network protocols are an integral part of constructing a practical WSN for deployment (Regatte and Jagannathan 2005). Implementation issues that are not always addressed in simulation constrain the type of protocols and hardware that can be deployed. Processing capabilities, on-board battery capacity, and sensor interfacing all become constraints that must be weighed during the design of the hardware components.

Implementations of WSN protocols are traditionally evaluated through the use of network simulators such as NS2, OPNET, and Glo-MoSim (Fonda et al. 2006). Simulations allow for establishment of the performance of a particular protocol against others. However, simulations lack the ability to evaluate the protocol against hardware constraints. In this work, hardware implementation is shown for the OEDSR protocol (Ratnaraj et al. 2006). The OEDSR is used in WSN to provide optimal routing calculations in energy and delay dependent environments. The use of hardware developed at the University of Missouri–Rolla (UMR) is shown as a development platform for this implementation.

Available routing protocols such as data-centric, location-based, QoS aware, and hierarchical for WSNs have been evaluated in simulation. However, there are few, or no, experimental results reported as to how they perform on hardware. In this section (Fonda et al. 2006), the focus is on hardware implementation of OEDSR protocol and its assessment.

This section from Fonda et al. (2006) will present the performance evaluation of OEDSR through hardware implementation. An 8-bit 8051 variant microcontroller-based implementation platform utilizing 802.15.4 RF communication units is shown. The use of this platform provides high-speed processing, interconnectivity with sensors, and a capable RF communications unit to facilitate a development platform for WSN. The hardware description includes considerations and limitations that the algorithm and hardware incur on one another. A description of the software implementation is next described.

8.10.1 Hardware Implementation Description

In this section, an overview of the hardware implementation of the OEDSR protocol is given. Use of customized hardware for development of sensing, processing, and networking will be presented. A description of capabilities, limitations, and support for networking applications are given next. Also, in this section, an overview of the software architectures is given with respect to the routing protocol and its memory requirements on the hardware.

Hardware for implementation of the OEDSR was selected to be energy conservative, performance-oriented, and of small form-factor. Use of Silicon Laboratories® 8051 variant hardware was selected for its ability to provide fast 8-bit processing, low power consumption, and ease of interfacing to peripheral hardware components. Next, a treatment of the hardware capabilities and limitations will be given.

Hardware implementation of any algorithm is constrained by the limitations of the hardware. Use of specific hardware must be weighed against the precision, speed, and criticality of an algorithm's implementation. Constraints addressed for the implementation of the OEDSR were the use of low-power, small form-factor, and fast processing hardware. For this protocol, low power consumption was given the highest priority. In turn, the demand for low power limits the types of processor architectures that can be deployed. The selection of the Silicon Laboratories® 8051 variants was based on these criteria. Limitations for the implementation that are incurred through the use of the 8051 variant family are a small memory space and a maximum processing speed. In the next section, a description of the specifications for the hardware-implemented nodes will be given.

FIGURE 8.49
(a) Instrumentation Sensor Node (ISN) and (b) Generation-4 Smart Sensor Node (G4-SSN).

8.10.1.1 Architecture of the Hardware System and Software

Now, a discussion of hardware and software resources employed for implementation of the OEDSR is given. A hardware performance comparison of sensor node platforms used at UMR is shown. Software implementation in terms of architecture, control-flow, and hardware limitations is shown.

8.10.1.2 Sensor Node: Instrumentation Sensor Node

The UMR Instrumentation Sensor Node (ISN), as seen in Figure 8.49a, is used for interfacing sensors to CHs. The ISN allows a sensor to be monitored by a small and low-power device that can be controlled by CHs which is also another node. In this application, the ISN is used as the source of sensor traffic. The ISN is capable of being interfaced with several sensor types and can be instructed by control packets to transmit data in raw or preprocessed form.

8.10.1.3 Cluster Head and Relay Nodes

The Generation-4 Smart Sensor Node (G4-SSN), seen in Figure 8.49b, originally developed at UMR and subsequently updated at St. Louis University, was chosen as the CH. The G4-SSN has various abilities for sensing and processing. The former include strain gauges, accelerometers, thermocouples, and general A/D sensing. The latter include analog filtering, CF memory interfacing, and 8-bit data processing at a maximum of 100 MIPS. The G4-SSN provides memory and speed advantages over the ISN that make it a suitable choice for implementation as a CH or an RN. Future

TABLE 8.2

Comparison of G4-SSN and ISN Capabilities

	Ic @ 3.3 V [mA]	Flash Memory [bytes]	RAM [bytes]	ADC Sampling Rate [kHz]	Form-Factor	MIPS
G4	35	128k	8448	100 @ 10/12-bit	100-pin LQFP	100
ISN	7	16k	1280	200 @ 10-bit	32-pin LQFP	25
X-Bow	8	128k	4096	15 @ 10-bit	64-pin TQFP	8

work is being undertaken to develop a better CH that is more powerful than an ISN and smaller in size than a G4-SSN.

8.10.1.4 Comparison of ISN and G4-SSN Capabilities

The abilities of the G4-SSN and the ISN sensor nodes are compared in this section. The ISN was designed to be a simplified sensor node with the abilities to sample, process, and transmit data. The ISN has a limited ability to process data relative to the G4-SSN. The abilities of the two nodes are shown in Table 8.2, with a comparison to other commercially available hardware. As seen in the table, the G4-SSN has approximately four times the processing speed available relative to the ISN. Memory constraints are also shown between the two sensor nodes, with the G4-SSN having more available code space and RAM. This translates to the design criteria for the ISN to be a "simple sample-and-send sensor node." In comparison, the G4-SSN is used for networking functionality and other tasks that require more memory and processing ability. In the next section, an overview of the software architecture is given for the OEDSR implementation.

8.10.2 Software Architecture

The software architecture for 8051 platform is presented in this section. The network stack is presented, and the layers are discussed in detail. The software architecture utilized to implement OEDSR protocol on 8051 platform is presented in Figure 8.50.

The three-tier structure is used to provide flexibility to the radio and application design. The wireless radio dependent components are interfaced with networking layers through the message abstraction layer. This layer provides generic access to the physical and link-level parameters and information, for example, transmission power level and RSSI indicator. Consequently, cross-layer protocols such as the OEDSR can be easily implemented.

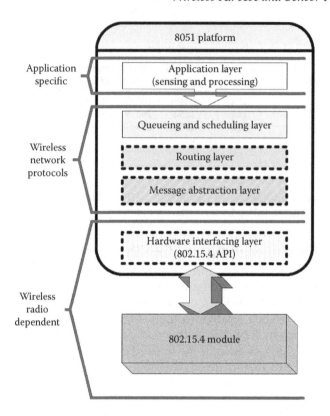

FIGURE 8.50
Software architecture of OEDSR implementation.

The main components of the software architecture consist of the following:

- Physical interface between 8051 and 802.15.4 module — in the used setup a standard serial interface connects processor with radio module
- Abstraction layer — provides generic access to the physical and link layers
- Routing layer — contains the OEDSR implementation
- Queuing — employs a simple drop-tail queuing policy
- Sensing application — application dependent measurement and processing of sensor data

8.10.2.1   Routing Implementation

In this section, the implementation of the routing protocol is described. Packets used by the routing protocol, handling of traffic cases by a node, and memory handling are presented.

8.10.2.2 Routing Packets

The routing aspects of the OEDSR protocol have been implemented on the 8051 platform with an 802.15.4 radio module. Five types of messages have been considered:

- BEAM packet: The BS broadcasts BEAM packets to the whole network to wake-up nodes and initiates data transmission. RSSI is retrieved by the receiving nodes and used to estimate the distance to the BS.
- HELLO packet: The node while searching for a route to the BS, broadcasts HELLO packets to neighbors periodically until ACK is received, or until timeout. The distance to BS is included so that the receiving node can determine the closest node to the BS.
- Acknowledgment (ACK) packet: ACK is sent as a response to the HELLO packet when the node's distance to BS is smaller than the requesting node's distance. Also, ACK contains the node's remaining energy and distance to the BS. The HELLO source node receives ACK packet and calculates a transmission delay. The link cost is calculated and temporarily stored to compare it with later responses.
- SELECT packet: When HELLO/ACK timeout elapses, the node selects the route based on the link costs from the stored ACK information. Subsequently, the SELECT packet is sent to the selected node to indicate route selection. The receiving node starts route discovery toward BS by sending a HELLO packet.
- DATA packet: The DATA packet conveys application specific data to the BS.

8.10.2.3 Traffic Cases

Figure 8.51 presents a block diagram of the routing implementation. The handling of the received message starts at the RX block, where the type of the packet is determined. Next, the processing proceeds depending on the packet type.

8.10.2.4 Memory Limitations

Memory limitations are incurred by the hardware. The routing protocol requires a particular amount of memory to store the routing table and temporary information from ACK. The number of routing table entries depends on expected number of active CHs. Moreover, the routing tables store only a link cost value, calculated from HELLO-ACK exchange. Furthermore, to reduce memory requirements, periodically inactive sources are purged from the routing table.

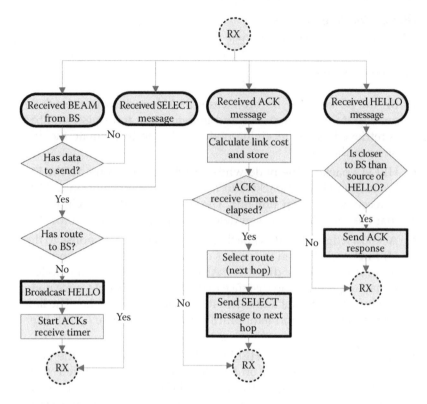

FIGURE 8.51
Control flow scheme for OEDSR routing implementation.

8.11   Performance Evaluation

Experiments for OEDSR were performed using a network of UMR ISNs and the G4-SSNs. Experimental results are compared to static routing to demonstrate the dynamic routing of the OEDSR. Use of static routing provides an initial assessment, whereas future work will provide comparison to existing protocols.

The nodes use 802.15.4 modules transmitting at a 250-kbps RF data rate. The ISN is used to generate CBR traffic and provide data source functionality. CHs and RNs are implemented using the G4-SSN. The CH provides the OEDSR routing capabilities by choosing the RN for routing of traffic toward the BS. The node's processor interfaces to the 802.15.4 module at 38.4 kbps, the maximum supported data rate. The ISN, CH, and RNs are equipped with low-power 1-mW 802.15.4 modules; whereas the BS is

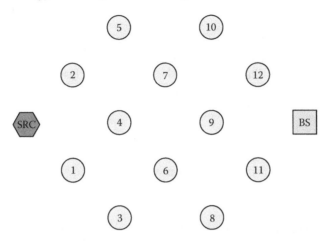

FIGURE 8.52
Network schematic.

equipped with a high transmission power, 100-mW, 802.15.4 module to increase the BS range for beam signals.

8.11.1 Description of the Experimental Scenario

Experimental scenarios were performed with 12 nodes placed in the topology illustrated in Figure 8.52. The topology was then modified by the amount of energy available in each node to perform testing of the protocol's ability to provide dynamic optimal routing based on energy, delay, and distance. Testing demonstrates the ability of the OEDSR protocol to evenly balance the energy consumed in the entire network besides providing suitable delay in the transmission of packets.

8.11.2 Experiment Results

The network performance is measured in terms of throughput, E2E delay, drop rate, and number of total dropped packets. Experiments were repeated for varying energy levels at each node, thus enforcing route changes. In Table 8.3, the performance measurements are shown for the six experimental cases. Each test was run for 3 min and an average result is shown. The experimental scenarios were prepared to generate four-hop routes thus providing comparable data sets. Throughput and E2E delay are consistent across all six cases, because the routing algorithm selects an optimal route regardless of energy distribution in the network. Variance in the number of dropped packets and in the drop rate is attributed to the distribution of packet collisions. In Table 8.4, a comparison of

TABLE 8.3

OEDSR Performance for Differing Topologies

Test	Throughput [bps]	E2E Delay [sec]	Dropped [packets]	Drop Rate [packets/sec]
T1	1152.0	0.7030	181	1.9472
T2	970.6	0.7030	3	0.0321
T3	972.0	0.7030	6	0.0643
T4	1035.5	0.7020	73	0.7811
T5	1048.0	0.7020	83	0.8862
T6	1047.3	0.7030	84	0.8968
AVG	1037.6	0.7027	72	0.7680

OEDSR network performance with varied packet size is shown. The network performance degrades as the packet size reduces and the number of generated packets increases. Because the amount of bandwidth used to transmit overhead bits increases at the expense of user data throughput, decreasing packet size increases overhead.

Figure 8.53 illustrates throughput when an active RN is removed from the network and OEDSR reestablishes communication.

At packet index 25, there is a drop in throughput when the RN is removed. Subsequent reestablishment of an alternate route by OEDSR is demonstrated because the throughput is restored. In comparison, static routing is not able to recover and would require manual intervention causing continued network downtime.

Static routing was compared to the OEDSR protocol. The route was manually configured to mimic a desired route. Experimental results show a similar throughput, E2E delay, and drop rate for the static routing and OEDSR. However, lack of a dynamic network discovery period is observed during network initialization with static routing. In the case of OEDSR, the setup time is dependent on the number of hops and the query time for each hop. In contrast, static routing requires manual setup for each topology change, which can take long periods of time. It is important

TABLE 8.4

OEDSR Performance for Differing Packet Size

Packet Size [bytes]	Throughput [bps]	E2E Delay [sec]	Dropped [packets]	Drop Rate [packets/sec]
30	1075.7	0.2500	188	2.0181
50	1197.0	0.3440	167	2.7991
70	1096.1	0.5620	156	1.6715
90	1047.3	0.7030	84	0.8968

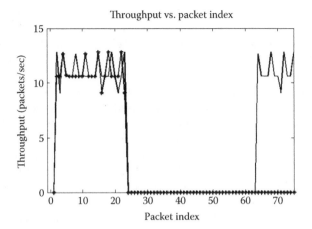

FIGURE 8.53
Throughput for data rate of 1kbps and 90 bytes per data packet.

to note that static routing is not normally preferred because of node mobility and channel fading.

8.11.3 Future Work

Future work will involve evaluating the proposed protocol in the presence of node mobility, and channel fading. Additionally, a performance comparison with other protocols such as LEACH is planned. Preliminary results of the OEDSR hardware implementation, as compared to a static routing, show promise. Future work will include implementation of protocols such as AODV and DSR on UMR hardware. Comparisons of OEDSR to other standard protocols can be shown. Other considerations include larger topologies, differing traffic loads and patterns, and vehicular mobile nodes.

8.12 Conclusions

In this chapter, an OEDR protocol for wireless ad hoc networks is presented. The objective is to develop a fully distributed routing protocol, which finds the optimal routes based on energy-delay costs of the links. The proposed OEDR protocol computes the energy and delay values of the links and shares this information with other nodes using the HELLO and TC control messages. Minimizing this product will render the best possible route because minimizing energy indicates best possible channel available and minimizing delay shows the route selected has maximum available bandwidth. Moreover, it also uses the concept of MPR nodes to minimize the overhead of flooding messages in the network and the number of links to be declared. The MPR selection and the routing table calculation are performed based on the energy-delay costs of the links using a minimum-cost spanning tree algorithm.

The performance of the proposed OEDR protocol is presented and analyzed. The effectiveness of our proposed OEDR protocol was evaluated, using the NS2. The results show that the proposed protocol can achieve smaller delay and better throughput/delay metric. Moreover, the protocol also results in smaller energy-per-packet and delay product, compared to the OLSR and AODV protocols. Therefore, the proposed OEDR protocol reduces the delay and improves the overall energy-delay efficiency of the network.

Next, the OEDSR protocol for WSN is presented. The objective was to develop a fully distributed routing protocol that finds the optimal routes based on energy-delay and distance metric, which is used as the *link_cost_factor*. It could be observed from the results that the path taken from the CH to the BS using the OEDSR is free from loops; it also ensures that the selected route is both energy efficient and has the least E2E delay. Additionally, the lifetime of the network is maximized because the available energy is taken into account while selecting nodes to form a route. Moreover, when a node loses more energy, a different path is computed. Clusters are formed only in the subnetwork portion of the network, whereas the rest of the network is treated as an ad hoc network. Because CH is usually used to route data in a sensor network, the RNs are viewed as CHs in the case of an ad hoc network.

The performance of the OEDSR protocol is presented and discussed. The effectiveness of the proposed OEDSR protocol was evaluated using the GloMoSim network simulator. The results show that the proposed protocol achieves smaller average E2E delay, and lower energy consumption when compared with Bellman Ford, DSR, and AODV protocols. The OEDSR protocol also performs better than OEDR even with node mobility for all the performance metrics mentioned earlier.

In this chapter, the hardware implementation is also shown for the OEDSR WSN protocol. The objective was to develop a fully distributed routing protocol that provides optimal routing. The route selection is based on a metric given by the ratio of energy available and delay multiplied with distance, which is used as the link cost factor.

The OEDSR protocol computes the energy available and average E2E delay values of the links, and this information along with the distance from the BS, determines the best RN. While ensuring that the path from the CH to the BS is free from loops, it also ensures that the selected route is both energy efficient and has the least E2E delay. Additionally, the lifetime of the network is maximized because the energy is taken into account while selecting nodes from a route. Due to the energy level being considered in the routing protocol, there is also a balancing of energy consumption across the network.

Implementation of the OEDSR protocol was shown using the G4-SSN and ISN hardware at UMR. The protocol was shown to provide suitable traffic rates and short E2E delays. Drop rate and E2E delay are dependent on the packet size that is being transmitted. Drop rate increases and E2E delay decreases as the packet size decreases. A decrease in E2E delay is expected due to the larger number of packets required to send the same information; however, higher traffic volume also increases the probability of packet collisions on the channel and increases overhead.

A series of tests taking a nominal of four hops were performed to show the capabilities of the OEDSR routing protocol to provide the needed throughput on the network with dynamic routing capabilities. An average throughput of approximately 1 kbps and an E2E delay of 0.7 sec are observed for a nominal route.

In reference to implementation, several issues were confronted. First, the issue of hardware capabilities is of concern. Selection of hardware must consider the complexity and memory footprint of an algorithm. The constraints of the 8-bit hardware became known during implementation of the OEDSR protocol. For example, the ISN nodes were designed to minimize the physical size of the node and reduce energy consumption. However, the selected processor does not have enough RAM to support the OEDSR routing. Therefore, minimum hardware requirements in terms of memory size, processing power, energy consumption, physical size, and the corresponding tradeoffs have to be explored before the particular protocol is targeted and implemented. Additionally, the limitations of the off-the-shelf radio modules are limiting current capabilities of the proposed solution. In particular, the 38.4-kbps limit on the interface to the 802.15.4 module reduces the overall throughput and increases delay at each hop, when compared to theoretical 802.15.4 capabilities.

References

Chee-Yee, C. and Kumar, S.P., Sensor networks: evolution, opportunities, and challenges, *Proceedings of the IEEE,* Vol. 91, No. 8, August 2003, pp. 1247–1256.

Clausen, T. and Jacquet, P., Optimized link state routing protocol, *IETF MANET Working Group, Internet Draft, draft-ietf-manet-olsr-11.txt,* July 2003.

Esler, M., Hightower, J., Anderson, T., and Borriello, G., Next century challenges: data-centric networking for invisible computing: The Portolano Project at the University of Washington, *MobiCom '99,* Seattle, WA, August 1999, pp. 15–20.

Fall, K. and Varadhan, K., ns Notes and Documentation, Technical report UC Berkley LBNL USC/IS Xerox PARC, 2002.

Fonda, J., Zawodniok, M., Jagannathan, S., and Watkins, S.E., Development and implementation of optimal energy delay routing protocol in wireless sensor networks, *Proceedings of the IEEE Symposium on Intelligent Control*, to appear in October 2006.

Ford, L. and Fulkerson, D., *Flows in Networks*, Princeton University Press, Princeton, NJ, 1962.

Garcia-Luna Aceves, J.J. and Spohn, M., Source-tree routing in wireless networks, *Proceedings of the 7th International Conference on Network Protocols, ICNP '99*, November 1999, pp. 273–282.

Global Mobile Information Systems Simulation Library, http://pcl.cs.ucla.edu/projects/glomosim/.

He, T., Stankovic, J.A., Lu, C., and Abdulzaher, T., SPEED: a stateless protocol for real-time communication in sensor networks, *Proceedings of International Conference on Distributed Computing Systems*, May 2003, pp. 46–55.

Heinzelman, W.R., Chandrakasan, A., and Balakrishnan, H., Energy-efficient communication protocol for wireless microsensor networks, *IEEE Transactions on Wireless Communication*, 660–670, October 2002.

Heinzelman, W.R., Kulik, J., and Balakrishnan, H., Adaptive protocol for information dissemination in wireless sensor networks, *Proceedings of the Fifth Annual ACM/IEEE International Conference on Mobile Computing and Networking (MobiCom)*, August 1999, pp. 174–185.

Hill, J., Szewczyk, R., Woo, A., Hollar, S., Culler, D., and Pister, K., System architecture directions for networked sensors, *Proceedings ASPLOS*, 2000, pp. 93–104.

Intanagonwiwat, C., Govindan, R., Estrin, D., Heidemann, J., and Silva, F., Directed diffusion for wireless sensor networking, *IEEE/ACM Transactions on Networking*, 2–16, February 2003.

Jacquet, P., Muhlethaler, P., Clausen, T., Laouiti, A., Qayyum, A., and Viennot, L., Optimized link state routing protocol for ad hoc networks, *Proceedings of the IEEE International Multi Topic Conference on Technology for the 21st Century*, December 2001, pp. 62–68.

Jain, R., *The Art Of Computer Systems Performance Analysis: Techniques for Experimental Design, Measurement, Simulation, and Modeling*, John Wiley & Sons, New York, April 1991.

Johnson, D., Maltz, D., and Hu, Y., The dynamic source routing protocol for mobile ad hoc networks (DSR), *IETF MANET Working Group, Internet Draft, draft-ietf-manet-dsr-09.txt*, April 2003.

Lindsey, S. and Raghavendra, C., PEGASIS: power-efficient gathering in sensor information systems, *Proceedings of the IEEE Aerospace Conference*, 2002, pp. 1125–1130.

Luo, H., Ye, F., Cheng, J., Lu, S., and Zhang, L., TTDD: a two-tier data dissemination model for large-scale wireless sensor networks, *Proceedings of International Conference on Mobile Computing and Networking (MobiCom)*, September 2002, pp. 148–159.

Manjeshwar, A. and Agrawal, D.P., APTEEN: A hybrid protocol for efficient routing and comprehensive information retrieval in wireless sensor networks, *Proceedings of International Parallel and Distributed Processing Symposium (IPDPS 2002)*, April 2002, pp. 195–202.

Manjeshwar, A. and Agrawal, D.P., TEEN: a routing protocol for enhanced efficiency in wireless sensor networks, *Proceedings of the 15th Parallel Distributed Processing Symposium*, 2001, pp. 2009–2015.

Mock, M., Frings, R., Nett, E., and Trikaliotis, S., Clock synchronization for wireless local area networks, *Proceedings of the 12th Euromicro Conference on Real-Time Systems*, June 2000, pp. 183–189.

Park, V.D. and Corson, M.S., A highly adaptive distributed routing algorithm for mobile wireless networks, *Proceedings of the IEEE INFOCOM '97*, Vol. 3, April 1997, pp. 1405–1413.

Perkins, C.E. and Bhagwat, P., Highly dynamic destination-sequenced distance-vector routing (DSDV) for mobile computers, *ACM SIGCOMM'94*, pp. 234–244, 1994.

Perkins, C.E., Belding, E., and Das, S., Ad hoc on-demand distance vector (AODV) routing *IETF MANET Working Group, Internet Draft, draft-ietf-manet-aodv-13.txt*, February 2003.

Qayyum, A., Viennot, L., and Laouiti, A., Multipoint relaying for flooding broadcast messages in mobile wireless networks, *Proceedings of the 35th Annual Hawaii International Conference on System Sciences*, January 2002, pp. 3866–3875.

Ratnaraj, S., Jagannathan, S., and Rao, V., Optimal energy-delay subnetwork routing protocol for wireless sensor networks, *Proceedings of IEEE Conference on Networking, Sensing and Control*, April 2006, pp. 787–792.

Ratnaraj, S., Jagannathan, S., and Rao, V., SOS: self-organization using subnetwork for wireless sensor network, to appear in 2006 SPIE Conference, San Diego, CA.

Regatte, N. and Jagannathan, S., Optimized energy-delay routing in ad hoc wireless networks, *Proceedings of the World Wireless Congress*, May 2005.

Sivakumar, R., Sinha, P., and Bharghavan, V., CEDAR: a core-extraction distributed ad hoc routing algorithm, *IEEE Journal on Selected Areas in Communications*, Vol. 17, No. 8, 1454–1465, August 1999.

Ye, F., Zhong, G., Lu, S., and Zhang, L., GRAdient broadcast: a robust, long-lived large sensor network, *ACM Wireless Networks*, Vol. 11, No.2, March 2005.

Ye, F., Zhong, G., Lu, S., and Zhang, L., PEAS: a robust energy conserving protocol for long-lived sensor networks, *Proceedings of 23rd International Conference on Distributed Computing Systems (ICDCS)*, May 2003, pp. 28–37.

Ying, G., Kunz, T., and Lamont, L., Quality of service routing in ad-hoc networks using OLSR, *Proceedings of the 36th Annual Hawaii International Conference on System Sciences*, January 2003, pp. 300–308.

Zhang, J., Yang, Z., Cheng, B.H., and McKinley, P.K., Adding safeness to dynamic adaptive techniques, *Proceedings of ICSE 2004 Workshop on Architecting Dependable Systems*, May 2004.

Problems

Section 8.5

Problem 8.5.1: Redo Example 8.5.1 for a 300-node network, using the same parameters defined in the example. Compare the results obtained from OEDR, OLSR and AODV.

Problem 8.5.2: Redo Example 8.5.2 by varying network size consisting of 300 nodes to 3000, with the node increment value of 150. Discuss results using average delay, throughput over delay, energy and delay product, and contention time with number of nodes.

Section 8.8

Problem 8.8.1: For topology presented in Figure 8.20, calculate the number of routing messages transmitted during route discovery for OEDSR protocol.

Problem 8.8.2: For topology presented in Figure 8.20, calculate the number of routing messages transmitted by the AODV protocol (the source floods the whole network with route request message while assuring only single transmission by each node; then the route replay message is passed from destination to source along the same route as for OEDSR in Problem 8.8.1).

Problem 8.8.3: Derive a general equation for the number of routing messages transmitted for the AODV routing scheme. Assume that all n nodes in the network will be involved in routing. Consider the number of hops in the final route as a parameter of the equation.

Problem 8.8.4: Derive general equation for number of routing messages transmitted for OEDSR scheme. Assume that for n nodes in the network there will be only fraction, $0 < \alpha < 1$, of the nodes involved in the routing. This will correspond to a source being inside the network area. Consider the number of hops in the final route as a parameter of the equation.

Section 8.9

Problem 8.9.1: Redo the Example 8.9.1, with higher node density networks consisting of 300 and above.

Problem 8.9.2: Repeat Problem 8.9.1, with a mobile BS. Compare the results with those of Problem 8.9.1.

Problem 8.9.3: Redo Example 8.9.4, with node density starting at 300, and with an increment of 100 nodes until 3000. Introduce channel fading and plot packets dropped, throughput, energy consumed, and E2E delay.

9

Predictive Congestion Control for Wireless Sensor Networks

Previous chapters have covered congestion and admission control schemes for wired networks and power control, scheduling, and routing schemes for wireless ad hoc and sensor networks. As indicated in Chapter 1, quality of service (QoS) guarantee includes throughput, end-to-end delay, and packet drop rate even during congestion and in the presence of unpredictable wireless channel. This chapter will introduce a congestion control scheme for wireless networks that takes into account energy efficiency and fairness. This scheme will be implemented through feedback obtained from one hop.

Available congestion control schemes, for example, the transport control protocol (TCP), when applied to wireless networks, result in a large number of packet drops, unfair scenarios, and low throughputs with a significant amount of wasted energy due to retransmissions. To fully utilize the hop-by-hop feedback information, a decentralized, predictive congestion control method consisting of an adaptive flow and adaptive backoff interval selection scheme is introduced for wireless sensor networks (WSN) (Zawodniok and Jagannathan 2006) in concert with the distributed power control (DPC) (Zawodniok and Jagannathan 2004). Besides providing an energy-efficient solution, the embedded channel estimator algorithm in DPC predicts the channel quality in the subsequent time interval. By using this information together with queue utilization, the onset of network congestion is assessed and a suitable transmission rate is determined by an adaptive flow control scheme. Then, an adaptive backoff interval selection scheme enables a node to transmit packets at the rate determined by the flow control scheme through one-hop feedback, thus preventing congestion. Additionally, because of the recursive application of the proposed congestion control at each node and through piggyback acknowledgments, the onset of congestion is propagated backward toward the source nodes, so that they too reduce their transmission rates. An optional adaptive scheduling scheme from Chapter 7 at each node updates the packet weights to guarantee the weighted fairness during congestion. DPC was discussed in Chapter 6.

Closed-loop stability of the proposed hop-by-hop congestion control is demonstrated by using the Lyapunov-based approach. Simulation results show that this scheme results in fewer dropped packets, better fairness index, higher network efficiency and aggregate throughput, and smaller end-to-end delays over the other available schemes such as Congestion Detection and Avoidance (CODA) (Wan et al., 2003) and IEEE 802.11 protocols.

9.1   Introduction

Network congestion, which is quite common in wireless networks, occurs when the offered load exceeds available capacity or the link bandwidth is reduced because of fading channels. Network congestion causes channel quality to degrade and loss rates to rise. It leads to packet drops at the buffers, increased delays, and wasted energy, and requires retransmissions. Moreover, traffic flow will be unfair for nodes whose data have to traverse a significant number of hops. This considerably reduces the performance and lifetime of the network. Additionally, WSN have constraints imposed on energy, memory, and bandwidth. Therefore, energy-efficient data transmission protocols are required to mitigate congestion resulting from fading channels and excess load. In particular, a congestion control mechanism is needed to balance the load, to prevent packet drops, and to avoid network deadlock.

Rigorous work has been done in wired networks on end-to-end congestion control (Jagannathan 2002, Peng et al. 2006). In spite of several advantages in end-to-end control schemes, the need to propagate the onset of congestion between end-systems makes the approach slow. In general, a hop-by-hop congestion control scheme reacts to congestion faster and is normally preferred to minimize packet losses in wireless networks. Therefore, the scheme from Zawodniok and Jagannathan (2005) uses a novel hop-by-hop flow control algorithm that is capable of predicting the onset of congestion and then gradually reducing the incoming traffic by means of a backpressure signal.

In comparison, the CODA protocol proposed by Wan et al. (2003) uses both a hop-by-hop and an end-to-end congestion control scheme to react to an already present congestion by simply dropping packets at the node preceding the congestion area. Thus, CODA partially minimizes the effects of congestion, and as a result retransmissions still occur. Similar to CODA, Fusion (Hull et al., 2004) uses a static threshold value for detecting the onset of congestion even though it is normally difficult to determine a suitable threshold value that works in dynamic channel environments.

In both CODA and Fusion protocols, nodes use a broadcast message to inform their neighboring nodes of the onset of congestion. Although this is quite interesting, the onset of congestion message occurring inside the network is not guaranteed to reach the sources. Moreover, available protocols (Hull et al. 2004, Wan et al. 2003, Yi and Shakkotai 2004) do not predict the onset of congestion in dynamic environments, for example, due to fading channel conditions. Finally, very few analytical results are presented in the literature in terms of guaranteeing the performance of available congestion control protocols. In contrast to these protocols, the method of Zawodniok and Jagannathan (2005) can predict and mitigate the onset of congestion by gradually reducing the traffic flow defined by using the queue availability and channel state.

Besides predicting the onset of congestion, this scheme guarantees convergence to the calculated target outgoing rate by using a novel, adaptive backoff interval selection algorithm. In CSMA/CA-based wireless networks, a backoff selection mechanism is used to provide simultaneous access to a common transmission medium and to vary transmission rates. Many researchers (Vaidya et al. 2000, Wang et al. 2005, Kuo and Kuo 2003) have focused on the performance analysis of backoff selection schemes for static environments. However, these schemes lack the ability to adapt to a changing channel state, congestion level, and size of the network. For example, in (Vaidya et al. 2000) the backoff intervals for the nodes are selected proportional to a flow weight, thus providing weighted fairness. However, the solution assumes a uniform density of transmitting nodes because the achieved throughput is determined by both a number of competing nodes and their backoff intervals. Consequently, distributed fair scheduling (DFS) will yield an unfair throughput for nodes with different number of neighbors. In contrast, the proposed algorithm dynamically alters backoff intervals according to current network conditions, for instance, the varying number of neighbor nodes and fading channels.

Additionally, the protocol (Zawodniok and Jagannathan 2004) uses weights associated with flows to fairly allocate resources during congestion. By adding an optional dynamic weight adaptation algorithm, weighted fairness can be guaranteed in dynamic environments as shown in the proposed work. Finally, using a Lyapunov-based approach, the stability and convergence of the three algorithms, for buffer control, backoff interval selection and dynamic weight adaptation, are proved.

This chapter is organized as follows. Section 9.2 presents an overview of the proposed methodology of predicting the onset of congestion, mitigation, DPC and weighted fairness. In Section 9.3, the flow control and backoff interval selection schemes, their performance and weighted fairness guarantees are presented. Mathematical analysis through the Lyapunov method is discussed in this section. Section 9.4 details the simulation results of the proposed scheme in comparison with the available congestion control scheme for sensor networks such as CODA and 802.11. Finally, conclusions are drawn.

9.2 Overview of Predictive Congestion Control

The network congestion, shown in Figure 9.1, occurs when either the incoming traffic (received and generated) exceeds the capacity of the outgoing link or link bandwidth drops because of channel fading caused by path loss, shadowing, and Rayleigh fading. The latter one is common to wireless networks. Therefore, the overall objective of this paper is to develop a novel way of utilizing the channel state in rate adaptation and a new MAC protocol using the mathematical framework, capturing channel state, backoff intervals, delay, transmitted power, and throughput.

Reserving resources for multimedia traffic in wired networks is well understood. On the other hand, the impact of reserving resources in wireless networks on meeting QoS is not clear. Although most of the available scheduling schemes (Bennett and Zhang 1996, Vaidya et al. 2000, Wang et al. 2005, Kuo and Kuo 2003) tend to vary the rate based on traffic type, they all assume that the channel is time invariant, which is a strong assumption. In this chapter (Zawodniok and Jagannathan 2006), we consider dynamic channel and network state. To accommodate these changes we will vary the backoff intervals of all the nodes based on the network and channel conditions to transmit data. This idea is utilized even during congestion due to channel fading.

In the next subsection, the predictive congestion strategy is discussed. Subsequently, an overview of the congestion control scheme is presented.

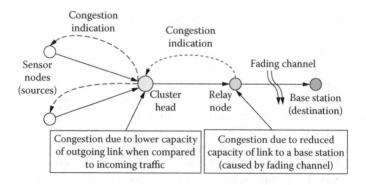

FIGURE 9.1
Congestion in wireless sensor networks.

Then, the DPC protocol is briefly described, and the used metrics are highlighted.

9.2.1 Congestion Prediction and Mitigation

To predict the onset of congestion, the proposed scheme uses both queue utilization and the transmission power under the current channel state at each node. When nodes in a network become congested, the traffic will accumulate at the nodes because there will be an excess amount of incoming traffic over the outgoing one. Hence, the queue utilization has been selected as an indicator of the onset of congestion.

On the other hand, in wireless networks during fading, the available bandwidth is reduced and the outgoing rate will be lowered. Consequently, input and output buffers will accumulate the incoming traffic indicating the onset of congestion. The channel fading is estimated by using the feedback information provided by the DPC protocol (Zawodniok and Jagannathan 2004) for the next packet transmission. The DPC algorithm predicts the channel state for the subsequent time interval and calculates the required power. If this power exceeds the maximum threshold, then the channel is considered to be unsuitable for transmission, and the proposed congestion control scheme can initiate the backoff process by reducing incoming traffic. Hence, information from DPC can be utilized to predict the onset of congestion due to fading channels.

Once the onset of congestion has been assessed, different strategies can be applied to prevent it from occurring. We propose a scheme, with a goal to prevent and minimize the effect of congestion while ensuring weighted fairness. When applied at every node in the wireless sensor network, it will render a fair and distributed congestion control scheme. The employed algorithms minimize queue overflows at a particular node by regulating the incoming flow. The admissible incoming traffic is calculated based on three factors:

Predicted outgoing flow — The outgoing flow is periodically measured, and an adaptive scheme is used to accurately predict the outgoing flow in the next period; moreover, the next hop node can reduce the outgoing flow assessment by applying a control over its incoming flows.

Wireless link state — The predicted outgoing flow rate is further reduced when the DPC protocol predicts a severe channel fading, which will disrupt communication on the link.

Queue utilization — The algorithm restricts the incoming flow based on the current queue utilizations and predicted outgoing flow, thus reducing buffer overflows.

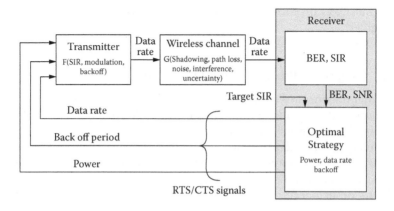

FIGURE 9.2
DPC with rate adaptation.

Weights are utilized in the proposed scheme to provide fair allocation of resources to all traffic flows based on the initial values. The weighted fairness guarantees that data from each source will be delivered to the destination in a timely and orderly manner. The weights associated with the packets in transit are used to ensure fair service.

9.2.2 Overview

A novel scheme, shown in Figure 9.2, is then derived based on the channel state, transmitter intended rate, and backlog. The scheme can be summarized in the following steps:

1. The buffer occupancies at the transmitter and receiver nodes along with the transmitter power required to overcome the channel state at the subsequent time interval will be used to detect an onset of congestion. The rate selection algorithm is then executed at the receiver to determine the appropriate rate (or available bandwidth) for the predicted channel state.

2. The available bandwidth (or rate) is allocated for the flows according to the flow weights. This ensures weighted fairness in terms of bandwidth allocation among the neighboring nodes. The weights can be selected initially and held subsequently or updated over time.

3. The DPC and rate information is communicated by the receiver to the transmitter for every link. At the transmitter node, a backoff interval is selected by using the proposed scheme based on the *assigned* outgoing rate.

The dynamic weight adaptation scheme can be used to further enhance the throughput while ensuring fairness. Packets at each node can be scheduled by using the adaptive and distributed fair scheduling (ADFS) scheme (Regatte and Jagannathan 2004) presented in Chapter 7 via flow-assigned weights that are updated based on the network state to ensure the fair handling of the packets.

REMARK 1

The feedback information is piggybacked to the ACK frame of the medium access control (MAC) protocol. This ensures that the feedback is successfully received by the node in the previous hop. In contrast, the CODA scheme transmits an indication of the congestion using a broadcast message without looking for any acknowledgment. Consequently, the delivery of such messages to all relevant nodes is not guaranteed.

REMARK 2

In the scheme proposed in this chapter, a single MAC data rate is considered without addressing the interlayer coordination and routing protocols. However, the mathematical analysis suggests that changes in routes and MAC data rates (bandwidth) will be accommodated by the outgoing traffic estimation algorithm. Any temporary fluctuation of the MAC data rate will be filtered out by averaging the outgoing flow measurements over the update period. Also, the persistent changes of the MAC data rate or establishment of a new route will be closely tracked by the outgoing flow estimation algorithm. Some insight into these issues is presented later in this chapter using a simulation scenario. This congestion control methodology utilizes both rate-based control and backoff interval selection schemes along with DPC. The embedded channel estimator in DPC indicates the state of the wireless channel, which is utilized to assess the onset of congestion. An overview of DPC is given in Chapter 6.

Next, the metrics that will decide the performance of the available congestion control protocols will be described. These performance measures are used to evaluate the proposed protocol and compare it with CODA and the standard IEEE 802.11 protocol.

9.2.3 Performance Metrics

The onset of congestion causes packets to be dropped because of buffer overflows. Packets dropped at the intermediate nodes will cause low network throughput and decreases energy efficiency because of retransmissions. Consequently, the total number of packets dropped at the intermediate nodes will be considered as a metric for the designed protocol. Energy efficiency measured as the number of bits transmitted per joule will be used as the second metric. The network efficiency measured as the total throughput at the base station will be taken as an additional metric. Weighted fairness will be used as a metric, because congestion can cause unfair handling of flows. Formally, the weighted fairness is defined in terms of fair allocation of resources as

$$\left| W_f(t_1, t_2)/\varphi_f - W_m(t_1, t_2)/\varphi_m \right| = 0 \qquad (9.1)$$

where f and m are considered flows, φ_f is the weight of flow f, and $W_f(t_1, t_2)$ is the aggregate service (in bits) received by it in the interval $[t_1, t_2]$. Finally, *fairness index* (*FI*) (Vaidya et al. 2000), which is defined as $FI = (\sum_f T_f/\phi_f)^2/(\eta * \sum_f (T_f/\phi_f)^2)$, where T_f is the throughput of flow f and η is the number of flows, will be utilized as a metric.

The proposed congestion control scheme ensures stability and performance, analytically. The proposed congestion control scheme based on buffer occupancy and backoff interval selection is summarized next.

9.3   Adaptive Congestion Control

The proposed adaptive congestion control scheme consists of adaptive rate and backoff interval selection schemes. The adaptive rate selection scheme will minimize the effect of congestion on a hop-by-hop basis by estimating the outgoing traffic flow. This scheme when implemented at each node will become a backpressure signal to the sources. Consequently, the congestion is alleviated (1) by designing suitable backoff intervals for each node based on channel state and current traffic and (2) by controlling the flow rates of all nodes, including the source nodes to prevent buffer overflowing. Next, we describe the rate and backoff selection algorithms in detail. Then, the data dissemination and fair scheduling are presented. Finally, the fairness guarantee of the proposed scheme is shown.

9.3.1   Rate Selection Using Buffer Occupancy

The rate selection scheme takes into account the buffer occupancy and a target outgoing rate. The target rate at the next hop node indicates what the incoming rate should be. The selection of the incoming rate is described next.

Consider buffer occupancy at a particular node i shown in Figure 9.3. The change in buffer occupancy in terms of incoming and outgoing traffic at this node is given as

$$q_i(k+1) = Sat_p[q_i(k) + T \cdot u_i(k) - f_i(u_{i+1}(k)) + d(k)] \qquad (9.2)$$

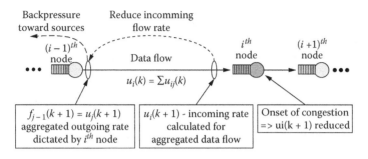

FIGURE 9.3
Rate selection overview.

where T is the measurement interval, $q_i(k)$ is the buffer occupancy of node i at time instant k, $u_i(k)$ is a regulated (incoming) traffic rate, $d(k)$ is an unknown disturbance in traffic, $f_i(\bullet)$ represents an outgoing traffic that is dictated by the next hop node $i+1$ and is disturbed by changes in channel state, and Sat_p is the saturation function that represents the finite-size queue behavior. The regulated incoming traffic rates $u_i(k)$ have to be calculated and propagated as a feedback to the node $i-1$ located on the path to the source, which is then used to estimate the outgoing traffic for this upstream node $f_{i-1}(\cdot)$.

Select the desired buffer occupancy at node i to be q_{id}. Then, buffer occupancy error defined as $e_{bi}(k) = q_i(k) - q_{id}$ can be expressed using Equation 9.2 as $e_{bi}(k+1) = q_i(k) + T \cdot u_i(k) - f_i(u_{i+1}(k)) + d(k) - q_{id}$. Next, the controller is introduced and its stability analysis is presented by using two different scenarios.

In the simple case, where the objective is to show that the scheme works, it is assumed that the outgoing traffic $f_i(\cdot)$ value is known. Theorem 9.3.1 shows the asymptotic stability of the system. Consequently, the queue level, $q_i(\cdot)$, will closely track the ideal level, q_{id}. Moreover, if the queue level exceeds the ideal level at any time instance, the feedback controller will quickly force the queue level to the target value. The second case presented in Theorem 9.3.2 relaxes the assumption of full knowledge about the outgoing flow $f_i(\cdot)$. The stability will hold even when the full knowledge of the outgoing flow is unknown as long as the traffic flow estimation error does not exceed the maximum value f_M. On the other hand, Theorem 9.3.3 shows that an adaptive scheme is capable of predicting the outgoing traffic $\hat{f}_i(\cdot)$ with error bounded by the maximum value f_M. In consequence, the proposed controller with adaptive scheme will ensure tracking of the ideal queue level even in the presence of bounded estimation errors in traffic flow.

CASE 1   The outgoing traffic $f_i(\cdot)$ is known. Now, define the traffic rate input, $u_i(k)$, as

$$u_i(k) = Sat_p(T^{-1}[f_i(u_{i+1}(k)) + (1 - k_{bv})e_{bi}(k)]) \tag{9.3}$$

where k_{bv} is a gain parameter. In this case, the buffer occupancy error at the time $k+1$ becomes

$$e_{bi}(k+1) = Sat_p[k_{bv}e_{bi}(k) + d(k)] \tag{9.4}$$

The buffer occupancy error will become zero as $k \to \infty$, provided $0 < k_{bv} < 1$.

CASE 2   The outgoing traffic $f_i(\cdot)$ is unknown and has to be estimated. In such a case, we define the traffic rate input, $u_i(k)$, as

$$u_i(k) = Sat_p[T^{-1}(\hat{f}_i(u_{i+1}(k)) + (1 - k_{bv})e_{bi}(k))] \tag{9.5}$$

where $\hat{f}_i(u_{i+1}(k))$ is an estimate of the unknown outgoing traffic $f_i(u_{i+1}(k))$.

In this case, the buffer occupancy error at the time instant $k+1$ becomes $e_{bi}(k+1) = Sat_p(k_{bv}e_{bi}(k) + \tilde{f}_i(u_{i+1}(k)) + d(k))$, where $\tilde{f}_i(u_{i+1}(k)) = f_i(u_{i+1}(k)) - \hat{f}_i(u_{i+1}(k))$ represents the estimation error of the outgoing traffic.

THEOREM 9.3.1 (IDEAL CASE)
Consider the desired buffer length, q_{id}, to be finite, and the disturbance bound, d_M, to be equal to zero. Let the virtual-source rate for Equation 9.2 be given by Equation 9.3. Then, the buffer occupancy feedback system is globally asymptotically stable provided $0 < k_{bv\,max}^2 < 1$.

PROOF   Let us consider the Lyapunov function candidate $J = [e_{bi}(k)]^2$. Then, the first difference is

$$\Delta J = [e_{bi}(k+1)]^2 - [e_{bi}(k)]^2 \tag{9.6}$$

Substituting error at time $k+1$ from Equation 9.4 in Equation 9.8 yields

$$\Delta J = \left(k_{bv}^2 - 1\right)[e_{bi}(k)]^2 \leq -\left(1 - k_{bv\,max}^2\right)\|e_{bi}(k)\|^2 \tag{9.7}$$

The first difference of the Lyapunov function candidate is negative for any time instance k. Hence, the closed-loop system is globally asymptotically stable.

REMARK 3

The preceding theorem using the Lyapunov method shows that under the ideal case of no errors in traffic estimation and with no disturbances, the control scheme will ensure that the actual queue level converges to the target value asymptotically.

THEOREM 9.3.2 (GENERAL CASE)

Consider the desired buffer length, q_{id}, to be finite, and the disturbance bound, d_M, to be a known constant. Let the virtual-source rate for Equation 9.2 be given by Equation 9.5 with the network traffic being estimated properly such that the approximation error $\tilde{f}_i(\cdot)$ is bounded above by f_M. Then, the buffer occupancy feedback system is globally uniformly bounded provided $0 < k_{bv} < 1$.

PROOF   Let us consider the Lyapunov function candidate $J = [e_{bi}(k)]^2$. Then, the first difference is

$$\Delta J = [k_{bv}e_{bi}(k) + \tilde{f}_i(u_{i+1}(k)) + d(k)]^2 - [e_{bi}(k)]^2 \qquad (9.8)$$

The stability condition $\Delta J \leq 0$ is satisfied if and only if

$$\|e\| > (f_M + d_M)/(1 - k_{bv\max}) \qquad (9.9)$$

When this condition is satisfied, the first difference of the Lyapunov function candidate is negative for any time instance k. Hence, the closed-loop system is globally uniformly bounded.

REMARK 4

The preceding theorem using the Lyapunov method shows that under the general case of where errors in traffic estimation are upper-bounded and with bounded disturbances, the control scheme will ensure that the actual queue level converges close to the target value.

Next the outgoing traffic function is estimated, using a vector of traffic parameters θ, by $f_i(u_{i+1}(k)) = \theta \cdot f_i(k-1) + \varepsilon(k)$, where $f_i(k-1)$ is the past value of the outgoing traffic and the approximation error $\varepsilon(k)$ is assumed bounded by the known constant ε_N. Now, define traffic estimate in the controller as $\hat{f}_i(u_{i+1}(k)) = \hat{\theta}(k)f_i(k-1)$, where $\hat{\theta}_i(k)$ is the actual vector of traffic parameters, $\hat{f}_i(u_{i+1}(k))$ is an estimate of the unknown outgoing traffic $f_i(u_{i+1}(k))$, and $f_i(k-1)$ is the past value of the outgoing traffic.

THEOREM 9.3.3 (NO TRAFFIC ESTIMATION ERROR)

Given the aforementioned incoming rate selection scheme with variable θ_i estimated accurately (no estimation error) and the backoff interval updated as in

Equation 9.5, then the mean estimation error of the variable θ_i along with the mean error in queue utilization converges to zero asymptotically, if the parameter θ_i is updated as

$$\hat{\theta}_i(k+1) = \hat{\theta}_i(k) + \lambda \cdot u_i(k) \cdot e_{fi}(k+1) \qquad (9.10)$$

provided(a) $\lambda \|u_i(k)\|^2 < 1$ and(b) $K_{f vmax} < 1/\sqrt{\delta}$, where $\delta = 1 / [1 - \lambda * \|u_i(k)\|^2]$, $K_{f vmax}$ is the maximum singular value of K_{fv}, λ, the adaptation gain, and $e_{fi}(k) = f_i(k) - \hat{f}_i(k)$, the error between the estimated value and the actual one.

The rate selected by the preceding algorithm in Equation 9.5 does not take into account the fading channels whether an adaptive scheme is employed or not for traffic estimation. It only detects the onset of congestion by monitoring the buffer occupancy. Under the fading wireless channels, the transmitted packets will not be decoded and dropped at the receiver, thereby requiring retransmissions as the effects of fading channels are not explicitly considered. To mitigate congestion due to channel fading, the selected rate from Equation 9.5 has to be reduced when the transmission power calculated by the DPC scheme exceeds the transmitter node's capability (greater than maximum transmission power). This is accomplished by using virtual rates and backoff interval selection. Selecting the backoff interval for a given node is a difficult task because it depends upon the backoff intervals of all neighboring nodes, which are normally unknown. Therefore an adaptive scheme is proposed to estimate the backoff interval of a given node.

9.3.2   Backoff Interval Selection

Because multiple nodes in a wireless sensor network compete to access the shared channel, backoff interval selection for nodes plays a critical role in deciding which node gains access to the channel. Thus, the proposed rate selection is implemented by suitably modifying the backoff intervals of the nodes around the congested node to achieve the desired rate control. In the case of contention-based protocols, it is difficult to select an appropriate backoff interval because of multiple nodes competing for the channel. For a given node, a relationship between transmission rate and backoff interval depends upon the backoff intervals of all nodes within a sensing range of a transmitting node in the CSMA or CA paradigm. To calculate this relationship, a node needs to know the backoff intervals of all its neighbors, which is not feasible in a wireless network because of a large traffic overhead resulting from communication.

Therefore, we propose using a distributed and predictive algorithm to estimate backoff intervals, such that a target rate is achieved. The main goal is to select the backoff interval, BO_i, at the ith transmitting node such that the actual throughput meets the desired outgoing rate $f_i(k)$. To simplify calculations, we consider the inverse of the backoff interval, which is denoted as $VR_i = 1/BO_i$, where VR_i is the virtual rate at the ith node and BO_i is the corresponding backoff interval. The fair scheduling algorithm schedules the packet transmissions according to the calculated node's backoff interval; the fair scheduling scheme is discussed in the next subsection. The interval is counted down when a node does not detect any transmission, and pauses otherwise. Consequently, a node will gain access to the channel proportional to its virtual rate and inversely proportional to the sum of virtual rates of its neighbors. The actual rate of the ith node is a fraction of the channel bandwidth, $B(t)$, defined as

$$R_i(t) = B(t) \cdot VR_i(t) \Big/ \sum_{l \in S_i} VR_l(t) = B(t) \cdot VR_i(t) / TVR_i(t) \qquad (9.11)$$

where TVR_i is the sum of all virtual rates for all neighbor nodes.

Because the scheme considers only a single modulation scheme, the bandwidth, B, is assumed time-invariant until the backoff interval is selected. It is assumed that the total bandwidth is constant as long as communication is possible on a link (when the received power is above a certain threshold). However, when severe fading occurs, the bandwidth will drop to zero. In such a case, backoff intervals are set at a large value, *lar*, to prevent unnecessary transmissions when a suitable signal-to-noise ratio (SNR) cannot be achieved at a destination node because of power constraints. Additionally, under normal circumstances, the algorithm presented as follows is used to calculate the backoff interval BO_i, which is then randomized to minimize probability of collision on access between nodes. Consequently, the MAC layer backoff timer BT_i value is defined as

$$BT_i = \begin{cases} \rho * BO_i(k), & \text{for } B(k) = 1 \\ lar, & \text{for } B(k) = 0 \end{cases} \qquad (9.12)$$

where ρ is a random variable with mean 1, *lar*, a large value of the backoff interval, and $B(k)$, the variable used to identify whether there is an onset of channel fading or not.

Equation 9.11 represents the relationship between the backoff intervals and the outgoing flow rate. To design a controller that will track the target value of that rate, the system equation is differentiated and then transformed into discrete-time domain. This allows the design of a feedback controller for the selection of the appropriate backoff interval.

Theorem 9.3.4 and Theorem 9.3.5 state that the proposed backoff selection scheme ensures convergence of the actual value to its target of outgoing traffic in both (1) the ideal case, in which the throughput dynamics are known, and (2) the general case, In which the dynamics are estimated by an adaptive scheme. In the latter case, the estimation error is bounded by a known value, ε_N. The proofs guarantee stability in the sense of Lyapunov (Jagannathan 2006, Zawodniok and Jagannathan 2006) as the backoff interval selection scheme is posed as a feedback dynamic system.

9.3.2.1   Adaptive Backoff Interval Selection

Differentiate Equation 9.11 to get

$$\dot{R}_i(t) = B \cdot TVR_i^{-2}(t)[\dot{V}R_i(t) \cdot TVR_i(t) - VR_i(t) \cdot T\dot{V}R_i(t)] \qquad (9.13)$$

To transform the differential equation into the discrete-time domain, Euler's formula is used:

$$R_i(k+1) - R_i(k) = B \cdot TVR_i^{-2}(k)$$
$$\times \left[(VR_i(k+1) - VR_i(k))TVR_i(k) - VR_i(k)(TVR_i(k+1) - TVR_i(k)) \right] \qquad (9.14)$$

After applying Equation 9.11, we can transform Equation 9.14 to get

$$R_i(k+1) = [R_i(k)VR_i(k+1)/VR_i(k)] + R_i(k)[1 - TVR_i(k+1)/TVR_i(k)] \quad (9.15)$$

Now, define $\alpha_i(k) = 1 - TVR_i(k+1)/TVR_i(k)$, $\beta_i(k) = R_i(k)/VR_i(k)$,   and $v_i(k) = VR_i(k+1) = 1/BO_i(k+1)$. The variable α_i describes a variation of backoff intervals of flows at the neighboring nodes from the time instant k to $k+1$. This variation is caused because of congestion resulting from traffic and fading channels. Because this information is not available locally, it is considered an unknown parameter and thus estimated by the algorithm. The parameter β_i is the ratio between the actual and the used virtual rate at time instant k and can be easily calculated. The term v_i is the backoff interval that needs to be calculated for the node under consideration.

Now, Equation 9.15 can be written as

$$R_i(k+1) = R_i(k)\alpha_i(k) + \beta_i(k)v_i(k) \qquad (9.16)$$

Equation 9.16 indicates that the achieved rate at the instant $k+1$ depends on the variations of backoff intervals in the neighboring nodes. Now, select the backoff interval as

$$v_i(k) = (\beta_i(k))^{-1}[f_i(k) - R_{ij}(k)\hat{\alpha}_i(k) + \kappa_v e_i(k)] \qquad (9.17)$$

where $\hat{\alpha}_i(k)$ is the estimate of $\alpha_i(k)$, $e_i(k) = R_i(k) - f_i(k)$, defined as throughput error, and κ_v is the feedback gain parameter. In this case, the throughput error is expressed as

$$e_i(k+1) = K_v e_i(k) + \alpha_i(k) R_i(k) - \hat{\alpha}_i(k) R_i(k) = K_v e_i(k) + \tilde{\alpha}_i(k) R_i(k) \quad (9.18)$$

where $\tilde{\alpha}_i(k) = \alpha_i(k) - \hat{\alpha}_i(k)$ is the error in estimation.

The throughput error of the closed-loop system for a given link is driven by the error in backoff intervals of the neighbors, which are typically unknown. If these uncertainties are properly estimated, a suitable backoff interval is selected for the node under consideration such that a suitable rate is selected to mitigate potential congestion. If the error in uncertainties tends to zero, Equation 9.18 reduces to $e_i(k+1) = \kappa_v \cdot e_i(k)$. In the presence of backoff interval variations of neighboring nodes, only a bound on the error in backoff interval selection for the node under consideration can be shown. In other words, the congestion control scheme will ensure that the actual throughput is close to its target value, but it will not guarantee convergence of actual backoff interval to its ideal target for all the nodes. It is very important to note that unless suitable backoff intervals are selected for all the nodes, congestion cannot be prevented.

THEOREM 9.3.4 (BACKOFF SELECTION UNDER IDEAL CIRCUMSTANCES)
Given the aforementioned backoff selection scheme with variable α_i estimated accurately (no estimation error), and the backoff interval updated as in Equation 9.17, then the mean estimation error of the variable α_i along with the mean error in throughput converges to zero asymptotically, if the parameter α_i is updated as

$$\hat{\alpha}_i(k+1) = \hat{\alpha}_i(k) + \sigma \cdot R_i(k) \cdot e_i(k+1) \quad (9.19)$$

Provided

$$\text{(a) } \sigma \|R_i(k)\|^2 < 1 \quad \text{and} \quad \text{(b) } K_{vmax} < 1/\sqrt{\delta} \quad (9.1)$$

where $\delta = 1/[1 - \sigma * \|R_i(k)\|^2]$, K_{vmax} is the maximum singular value of K_v, and σ, the adaptation gain.

PROOF Define the Lyapunov function candidate

$$J = e_i^2(k) + \sigma^{-1} \tilde{\alpha}_i^2(k) \quad (9.21)$$

whose first difference is

$$\Delta J = \Delta J_1 + \Delta J_2 = e_i^2(k+1) - e_i^2(k) + \sigma^{-1}\left[\tilde{\alpha}_i^2(k+1) - \tilde{\alpha}_i^2(k)\right] \quad (9.22)$$

Consider ΔJ_1 from Equation 9.22 and substitute Equation 9.18 to get

$$\Delta J_1 = e_i^2(k+1) - e_i^2(k) = \left(k_v e_i(k) + \tilde{\alpha}_i(k)R_i(k)\right)^2 - e_i^2(k) \tag{9.23}$$

Taking the second term of the first difference from Equation 9.22 and substituting Equation 9.19 yields

$$\Delta J_2 = \sigma^{-1}\left[\tilde{\alpha}_i^2(k+1) - \tilde{\alpha}_i^2(k)\right] = -2\left[k_v e_i(k)\right]\tilde{\alpha}_i(k)R_i(k)$$
$$- 2\left[\tilde{\alpha}_i(k)R_i(k)\right]^2 + \sigma R_i^2(k)\left[k_v e_i(k) + \tilde{\alpha}_i(k)R_i(k)\right]^2 \tag{9.24}$$

Combine Equation 9.23 and Equation 9.24 to get

$$\Delta J = -\left[1 - \left(1 + \sigma R_i^2(k)k_v^2\right)\right]e_i^2(k) + 2\sigma R_i^2(k)[k_v e_i(k)][\tilde{\alpha}_i(k)R_i(k)]$$
$$-\left(1 - \sigma R_i^2(k)\right)[\tilde{\alpha}_i(k)R_i(k)]^2 \leq -\left(1 - \delta k_{v\max}^2\right)\|e_i(k)\|^2$$
$$-\left(1 - \sigma\|R_i(k)\|^2\right)\cdot\left\|\tilde{\alpha}_i(k)R_i(k) - \left(\sigma\|R_i(k)\|^2 \Big/\left[1 - \sigma\|R_i(k)\|^2\right]\right)k_v e_i(k)\right\|^2 \tag{9.25}$$

where δ is given after (9.20). Now, taking expectations on both sides yields

$$E(\Delta J) \leq -E\left\{\left(1 - \delta k_{v\max}^2\right)\|e_i(k)\|^2\right.$$
$$\left.-\left(1 - \sigma\|R_i(k)\|^2\right)\left\|\tilde{\alpha}_i(k)R_i(k) + \sigma\|R_i(k)\|^2\Big/\left[1 - \sigma\|R_i(k)\|^2\right]k_v e_i(k)\right\|^2\right\} \tag{9.26}$$

Because $E(J) > 0$ and $E(\Delta J) \leq 0$, this shows the stability in the mean, in the sense of Lyapunov (Jagannathan 2006) provided the conditions in Equation 9.20 hold, so $E[e_i(k)]$ and $E[\tilde{\alpha}_i(k)]$ (and hence $E[\hat{\alpha}_i(k)]$) are bounded in the mean if $E[e_i(k_0)]$ and $E[\tilde{\alpha}_i(k_0)]$ are bounded. Summing both sides of Equation 9.26 and taking limits $\lim_{l\to\infty}E(\Delta J)$, the error $E[\|e_i(k)\|] \to 0$.

Consider the closed-loop throughput error with estimation error, $\varepsilon(k)$, as

$$e_i(k+1) = K_v e_i(k) + \alpha_i(k)R_i(k) + \varepsilon(k). \tag{9.27}$$

THEOREM 9.3.5 (BACKOFF SELECTION IN GENERAL CASE)
Assume the hypothesis as given in Theorem 9.3.4, and let the uncertain param-
eter α_i be estimated using Equation 9.18 with $\varepsilon(k)$, the error in estimation
considered bounded above such that $\|\varepsilon(k)\| \leq \varepsilon_N$, where ε_N is a known constant.
Then, the mean error in throughput and the estimated parameters are bounded
provided Equation 9.20 and Equation 9.20 hold.

PROOF Define a Lyapunov function candidate as in Equation 9.21 whose first difference is given by Equation 9.22. The first term ΔJ_1 and the second term ΔJ_2 can be obtained respectively as

$$\Delta J_1 = e_i^2(k)k_v^2 + 2[k_v e_i(k)][\tilde{\alpha}_i(k)R_i(k)] + [\tilde{\alpha}_i(k)R_i(k)]^2$$
$$+ \varepsilon^2(k) + 2[k_v e_i(k)]\varepsilon(k) + 2\varepsilon(k)e_i(k) - e_i(k)^2 \tag{9.28}$$

$$\Delta J_2 = -2[k_v e_i(k)][\tilde{\alpha}_i(k)R_i(k)]$$
$$- 2[\tilde{\alpha}_i(k)R_i(k)]^2 + \sigma R_i^2(k)[k_v e_i(k) + \tilde{\alpha}_i(k)R_i(k)]^2$$
$$- 2\left[1 - \sigma R_i^2(k)\right]e_i(k)\varepsilon(k) \tag{9.29}$$
$$+ 2\sigma R_i^2(k)[k_v e_i(k)]\varepsilon(k) + \sigma R_i^2(k)\varepsilon^2(k)$$

Following Equation 9.25 and completing the squares for $\tilde{\alpha}_i(k)R_i(k)$ yields

$$\Delta J \leq -\left(1 - \delta k_{v\max}^2\right)\left(\|e_i(k)\|^2 - \frac{2\sigma k_{v\max}\|R_i(k)\|^2}{1 - \delta k_{v\max}^2}\varepsilon_N\|e_i(k)\| - \frac{\delta}{1 - \delta k_{v\max}^2}\varepsilon_N^2\right)$$
$$- \left(1 - \sigma\|R_i(l)\|^2\right)\left\|\tilde{\alpha}_i(k)R_i(k) - \frac{\sigma\|R_i(k)\|^2}{1 - \sigma\|R_i(k)\|^2}(k_v e_i(k) + \varepsilon(k))\right\|^2 \tag{9.30}$$

with δ as given after Equation 9.20. Taking expectations on both sides,

$$E(\Delta J) \leq -E\left\{\left(1 - \delta k_{v\max}^2\right)\left(\|e_i(k)\|^2 - \frac{2k_{v\max}\sigma\|R_i(k)\|^2}{1 - \delta k_{v\max}^2}\varepsilon_N\|e_i(k)\| - \frac{\delta}{1 - \delta k_{v\max}^2}\varepsilon_N^2\right)\right.$$
$$\left. + \left(1 - \sigma\|R_i(k)\|^2\right)\left\|\tilde{\alpha}_i(k)R_i(k) - \frac{\sigma\|R_i(k)\|^2}{1 - \sigma\|R_i(k)\|^2}(k_v e_i(k) + \varepsilon(k))\right\|^2\right\} \tag{9.31}$$

as long as Equation 9.20 holds, and $E[\|e_i(k)\|] > (1 - \sigma k_{v\max}^2)^{-1} \varepsilon_N (\sigma k_{v\max} + \sqrt{\sigma})$. This demonstrates that $E(\Delta J)$ is negative outside a compact set U. According to a standard Lyapunov extension (Jagannathan 2006, Zawodniok and Jagannathan 2006), the throughput error $E[e_i(k)]$ is bounded for all $l \geq 0$. It is required to show that $\hat{\alpha}_i(k)$ or, equivalently, $\tilde{\alpha}_i(k)$ is bounded. The dynamics in error in the parameters estimates are

$$\tilde{\alpha}_i(k+1) = \left[1 - \sigma R_i^2(k)\right]\tilde{\alpha}_i(k) - \sigma R_i(k)[k_v e_i(k) + \varepsilon(k)] \qquad (9.32)$$

where the error, $e_i(k)$, is bounded and estimation error, $\varepsilon(l)$, is bounded. Applying the persistency of excitation condition (Jagannathan 2006), one can show that $\tilde{\alpha}_i(k)$ is bounded.

9.3.2.2 Rate Propagation

This total incoming rate is then divided among the upstream nodes proportionally to the sum of flow weights passing through a given node j as $u_{ij}(k) = u_i(k) \cdot \sum_n^{flows\,at\,j^{th}\,node}[\varphi_n] / \sum_m^{flows\,at\,i^{th}\,node}[\varphi_m]$, where $u_{ij}(k)$ is the rate allocated for a transmitting node j at receiving node i, $u_i(k)$ is the rate selected for all incoming flows at the ith node, given by Equation 9.3, and φ_n and φ_m are preassigned weights of the nth and mth flows, respectively. Next, the selected rate, $u_{ij}(k)$, is communicated to the upstream node j to mitigate congestion. This feedback continues recursively to the nodes upstream from the congested link, so that they will also reduce transmission rates and thus prevent overflowing buffers. One can update the preassigned weights to guarantee weighted fairness and to improve throughput as discussed next.

9.3.3 Fair Scheduling

Data packets at a receiver are first scheduled using the adaptive dynamic fair scheduling (ADFS) scheme (Regatte and Jagannathan 2004). Weights that correspond to the packet flows are used to build a schedule for transmission. This algorithm ensures weighted fairness among the flows passing a given node. The proposed scheme offers an additional feature of dynamic weight adaptation that further boosts the fairness and guarantees performance analytically as presented in this chapter.

This ADFS feature increases throughput while ensuring fairness of the flows by adjusting per-packet weight during congestion. This feature, though utilized here, can be optional in the congestion control scheme because it introduces additional overhead, although shown to be low in Chapter 7, in the form of (1) extra bits in each packet to carry the weight and (2) additional calculations performed to evaluate fairness and update the packet weight at each hop. However, this algorithm is necessary in a dynamic wireless networking environment.

Dynamic weight adaptation given in Equation 9.33 is utilized in the ADFS scheme. ADFS follows the weighted fairness criterion defined in Equation 9.1. The initial weights are selected by using the user-defined QoS criteria. Then, the packet weights are dynamically adapted with network state defined as a function of delay experienced, number of packets in the queue, and the previous weight of the packet. In fact, analytical results are included in Chapter 7 to demonstrate the throughput and end-to-end delay bounds in contrast with the existing literature. The NS simulation results indicate that the proposed scheme renders a fair protocol for WSN even in the presence of congestion due to fading channels. The weights are updated as follows.

9.3.3.1 Dynamic Weight Adaptation

To account for the changing traffic and channel conditions that affect the fairness and end-to-end delay, the weights for the flows are updated dynamically as

$$\hat{\varphi}_{ij}(k+1) = \xi\hat{\varphi}_{ij}(k) + \varsigma E_{ij} \qquad (9.33)$$

where $\hat{\varphi}_{ij}(k)$ is the actual weight for the ith flow, jth packet at time k, ξ and ς are design constants, $\{\xi, \varsigma\} \in [-1,1]$, and E_{ij} is defined as $E_{ij} = e_{bi} + 1/e_{ij,delay}$ where e_{bi} is the error between the expected length of the queue and the actual size of the queue and $e_{ij,delay}$ is the error between the expected delay and the delay experienced by the packet so far.

According to E_{ij}, when queues build up, the packet weights will be increased to clear the backlog. Similarly, with large end-to-end delays, the packet weights will be assigned high, so that the nodes can service these packets sooner. Note that the value of E_{ij} is bounded because of finite queue length and delay, as packets experiencing delay greater than the delay error limit will be dropped. The updated weights are utilized to schedule packets. Then, the packet to be transmitted is sent to the MAC layer, where the backoff interval scheme is implemented. Next, the proposed backoff selection algorithm is presented, followed by the convergence of the backoff interval selection scheme and the fairness and throughput guarantee of the scheduling scheme.

9.3.3.2 Fairness and Throughput Guarantee

To prove that the dynamic weight adaptation is fair, we need to show a bound on $|W_f(t_1,t_2)/\phi_f - W_m(t_1,t_2)/\phi_m|$ for a sufficiently long interval $[t_1,t_2]$ in which both flows, f and m, are backlogged. Next, three theorems are given to show the performance guarantee. Theorem 9.3.6 guarantees proportional fairness for all flows, whereas Theorem 9.3.7 guarantees that the minimal throughput is achieved by each flow. Finally, Theorem 9.3.8 ties both provisions from Theorem 9.6 and Theorem 9.7, thus guaranteeing overall performance of the proposed scheme. The proofs closely follow those given by Regatte and Jagannathan (2004).

REMARK 5
In fact, the weight update (Equation 9.13) ensures that the actual weight assigned to a packet at each node converges close to its target value.

REMARK 6
The value ϕ_{ij} is finite for each flow at a given node. Let $\hat{\phi}_{ij}$ be the weight error defined as $\tilde{\phi}_{ij} = \phi_{ij} - \hat{\phi}_{ij}$. Note that from now on, the weight of a packet for flow f at node l is denoted as $\phi_{f,l}$ and it can be related to ϕ_f at each node as $\phi_{f,l} = \sigma_f \phi_f$.

THEOREM 9.3.6
For any interval $[t_1, t_2]$ in which flows f and m are backlogged during the entire interval, the difference in the service received by two flows at a cluster head is given as

$$\left| W_f(t_1, t_2) / \phi_{f,l} - W_m(t_1, t_2) / \phi_{m,l} \right| \leq l_f^{\max} / \phi_{f,l} + l_m^{\max} / \phi_{m,l} \qquad (9.34)$$

REMARK 7
Notice that Theorem 9.3.6 holds regardless of the service rate of the cluster head. This demonstrates that the algorithm achieves fair allocation of bandwidth and thus meets a fundamental requirement of a fair scheduling algorithm for integrated services networks.

REMARK 8
Weights can be selected initially and may not get updated. If this occurs, it becomes the self-clocked fair queuing (SFQ) scheme (Golestani 1994). One can still use the proposed backoff interval selection scheme. However, it is demonstrated in Chapter 7 that the ADFS increases throughput while ensuring fairness compared with SFQ.

THEOREM 9.3.7 (THROUGHPUT GUARANTEE)
If Q is the set of flows served by an ADFS node following FC service model with parameters $(\lambda(t_1, t_2), \psi(\lambda))$, and $\sum_{n \in Q} \phi_{n,l} \leq \lambda(t_1, t_2)$, then for all intervals

$[t_1, t_2]$ *in which flow f is backlogged throughout the interval,* $W_f(t_1, t_2)$ *is given as*

$$W_f(t_1, t_2) \geq \phi_{f,l}(t_2 - t_1) - \phi_{f,l} \sum_{n \in Q} l_n^{max} \bigg/ \lambda(t_1, t_2) - \phi_{f,l} \, \psi(\lambda) \bigg/ \lambda(t_1, t_2) - l_f^{max}$$

(9.35)

PROOF The proof follows in similar lines to ad hoc wireless networks (Regatte and Jagannathan 2004).

THEOREM 9.3.8 (THROUGHPUT AND FAIRNESS GUARANTEE)
Assume the hypothesis given in Theorem 9.6 and Theorem 9.7. If the backoff intervals are selected as according to Equation 9.17 and packet weights are updated by using Equation 9.33, then the proposed scheme will deliver throughput with a bounded error and ensure fairness even in the presence of congestion.

PROOF Follows from the proofs of Theorem 9.3.6 and Theorem 9.3.7.

9.4 Simulation

The proposed scheme was implemented in NS-2 simulator and evaluated against DPC (Zawodniok and Jagannathan 2004) and the 802.11 protocols as well as CODA. Information for backoff calculation is available locally. Hence, only the rate information has to be fed back to upstream nodes and is incorporated into the MAC frames. The calculations of the rate and backoff intervals are performed periodically at every 0.5 sec.

The parameters that were used include a 2-Mbps channel with path loss, shadowing, and Rayleigh fading with the AODV routing protocol. The queue limit is set to 50 with the packet size taken as 512 bytes. The SFQ algorithm (Bennett and Zhang 1996) was used to ensure fairness among the flows passing at a given node. It uses the assigned weights to schedule packets for transmission, and the weights are not updated. The proposed scheme is compared with others. The CODA scheme has been implemented in NS-2 by carefully following the description in Wan et al. (2003). Additionally, performance of the scheme in case of variations in outgoing flow rate was assessed using MATLAB. Next, the results are discussed.

Example 9.4.1: Tree Topology Results
The tree topology, typical of a sensor network, is used for the simulations in which the sources represent cluster heads, and the sink is a base station collecting the sensor data. Traffic accumulates near the destination node

thus causing congestion at the intermediate nodes. In the simulations, the traffic consists of five flows, which had been simulated for two cases: (1) with the same weights, equal to 0.2, and (2) with weights equal to 0.4, 0.1, 0.2, 0.2, and 0.1, respectively. All the sources generate equal traffic, which exceeds channel capacity, so that congestion can be created. The initial rates for each flow have been assigned proportional to the weight, such that they saturate the radio channel. It has to be noted that in such a heavily congested network, the RTS-CTS-DATA-ACK handshake and the standard windows contention mechanism of the 802.11 protocol are unable to prevent collisions. Moreover, the high number of collisions occurring leads to underutilization of the wireless channel in case of the 802.11 protocol. The DPC protocol improves the channel utilization in the presence of collisions, as described by Zawodniok and Jagannathan (2004). However, the imbalance between incoming and outgoing flows due to congestion is not addressed by the DPC, thus still resulting in buffer overflows and a significant number of dropped packets.

Figure 9.4 and Figure 9.5 depict the throughput/weight (normalized weights) ratio for each flow. Ideally, the throughput over the initial weight ratio plot should be a straight line parallel to the x-axis for fair scheduling schemes. The proposed protocol results in a fair allocation of bandwidth when compared to the DPC and to the 802.11 MAC protocols, even for the case of variable weights assigned to the flows, as observed in Figure 9.5. The proposed protocol achieves this by taking weight into account during a packet scheduling and flow control. Consequently, flows with different

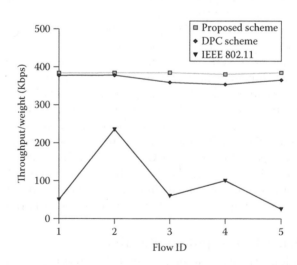

FIGURE 9.4
Performance for the tree topology.

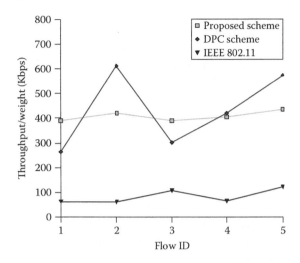

FIGURE 9.5
Performance for tree topology with varying flow weights.

weights will get proportional service rates. The DPC protocol achieves very good fairness in the case of identical weights. However, the DPC fails when the weights are unequal because there is no mechanism to vary allocation of the channel resources. Overall, the proposed scheme achieves better fairness compared to the DPC and 802.11 protocols.

Table 9.1 summarizes the overall performance of the protocols. The fairness index (FI) in Table 9.1 shows the fairness in case of varying flow weights. An ideal protocol will have the FI equal to 1.00. Both the 802.11 and DPC protocols have FI smaller than 1.00, indicating unfair handling of flows, whereas the proposed scheme achieves fairness index equal to 1.00, indicating fair allocation of resources to the flows. The proposed protocol achieves an end-to-end fairness by recursively applying the proposed scheme at every node, which, in turn, guarantees the hop-by-hop fairness.

TABLE 9.1

Protocol Performance Comparison

Protocol	Average Delay (sec)	Fairness Index (FI) (mix weights)	Network Efficiency (kbps)	Energy Efficiency (kb/J)
Proposed	0.8	1.00	400.99	13.05
802.11	—	0.91	77.86	3.23
DPC	1.06	0.91	368.55	11.79

In terms of throughput, the 802.11 protocol performs unsatisfactorily because it cannot handle the increased number of collisions that occur in a heavily congested network as it was unable to predict the onset of congestion to prevent it. The network becomes deadlocked because frequent collisions prevent the 802.11 protocol from successfully delivering packets to the destination node, as observed in Figure 9.5. As a result, the average delay for 802.11 is significantly high compared to other schemes. On the other hand, the DPC detects the idle channel after collisions and resumes transmission sooner than the 802.11 protocol (Zawodniok and Jagannathan 2004), thus improving throughput in a congested network. In the case of the proposed scheme, the throughput is further increased because it prevents the onset of congestion. By applying feedback in the form of backpressure signal, the proposed scheme prevents packet losses; therefore, it minimizes energy and bandwidth wastage, and maximizes end-to-end throughput. Consequently, the proposed algorithm outperforms other protocols.

Figure 9.6 and Figure 9.7 present throughput and drop rate for flow 1, for instance. It can be noticed that, during the time interval of 21 to 31 sec, the link from relay node to destination experiences severe fading and no transmission is possible. The proposed protocol prevents overflowing buffer at the relay node by sending backpressure indication to the sources and preventing them from overflowing the relay node. Thus, no increase in drop rate is observed for the rate-based protocol.

Figure 9.8 and Figure 9.9 illustrate the "weight * delay" metric, which describes a weighted fairness in terms of end-to-end delay. The desired

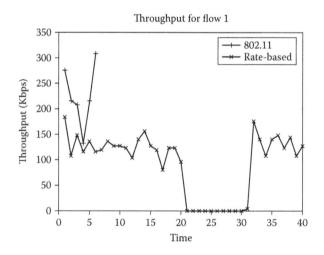

FIGURE 9.6
Throughput for flow 1.

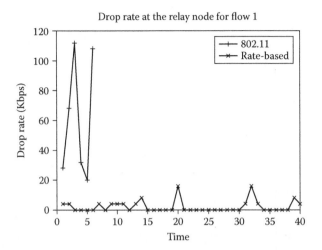

FIGURE 9.7
Drop rate at the relay node.

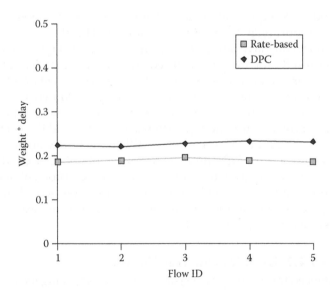

FIGURE 9.8
Weighted delay with equal flow weights (const = 0.2).

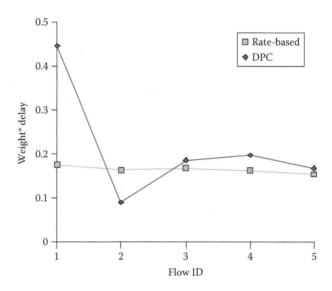

FIGURE 9.9
Weighted delay with varying flow weights.

plot should have equal values for all flows because the delay will be inversely proportional to the QoS parameter. The flows with higher weight value will have lower delay, and *vice versa*. Figure 9.8 shows simulation results for the case with equal flow weights. The proposed protocol is fair for all flows, because it schedules packets and adjusts feedback information using corresponding weights. Moreover, the end-to-end delays are smaller for the proposed protocol when compared to the DPC protocol, because the congestion is mitigated and the packets are transmitted without unnecessary delays. When the flow weights are varied, as in Figure 9.9, the DPC protocol becomes unfair because it does not differentiate between flows. By contrast, the proposed protocol maintains fairness, because it can provide different service rates according to the flow weights. In general, the results for end-to-end delay are consistent with the earlier throughput results.

Example 9.4.2: Unbalanced Tree Topology

The unbalanced tree topology consists of one flow (#1) being located closer to the destination and four flows (#2 to #5) located farther away from the destination. Consequently, without adaptive weights, the first flow is favored at the expense of the other four flows. In Figure 9.10, the throughput/weight (normalized weights) ratio for each flow is depicted. The proposed protocol without weight adaptation performs better than the

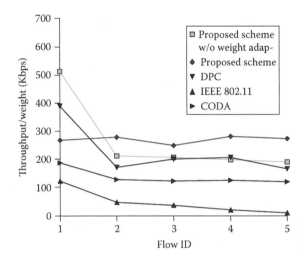

FIGURE 9.10
Performance for unbalanced tree topology.

DPC because, by alleviating congestions, it does not waste bandwidth for collisions and retransmissions. However, it is not able to provide fair service for all flows as it cannot identify flows hampered by congestion on preceding hops. In contrast, when the weighted fairness part is added to the proposed congestion control scheme, it identifies that flows 2 through 5 are hindered because of congestion and network topology; thus, their weights are adjusted at the next hops to meet fairness criteria. As a result, at the destination, all the flows achieve the same weighted throughput and end-to-end delay.

The 802.11 protocol performs unsatisfactorily for all flows as it stalls because of severe congestion. In comparison, the CODA scheme improves the performance of the network over the 802.11 protocol as it restricts network traffic during the congestion by using the backpressure mechanism. However, CODA is not able to achieve throughput comparable to the proposed protocol because the proposed scheme can mitigate onset of congestion by precisely controlling queue utilization. In contrast, CODA uses a binary bit to identify the onset of congestion and has no precise control over the incoming flows; thus, it cannot completely prevent buffer overflows. Additionally, the CODA scheme does not consider weighted fairness, thus unfairly handling different flows. Overall, the proposed protocol improves the performance of the network by 93 to 98% when compared with the CODA scheme, thus justifying increased processing requirements of the proposed scheme.

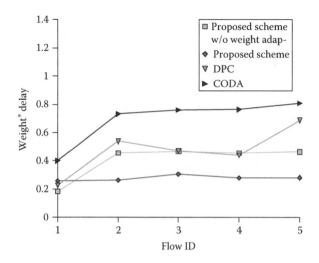

FIGURE 9.11
Packet weight and delay product with equal flow weights (const = 0.2).

Figure 9.11 presents the weight and delay product for all flows and protocols. This figure does not contain the 802.11 protocol performance results as the protocol is quickly stalled because of congestion and only very few packets are received at the destination. Consequently, the observed delay cannot be compared with the other protocols. In case of the CODA scheme, the node's transmission is halted for a random period of time as stated by the congestion policy (Wan et al. 2003). In consequence, the end-to-end delay is increased when passing though the congested area. Also, in this case, the proposed protocol with weight adaptation outperforms other protocols as it assigns radio resources proportionally to the weights associated with flows. Additionally, the hop-by-hop weight adaptation takes into account problems that packets encountered hitherto. Hence, the local unfairness is amended on the later hops.

Example 9.4.3: Performance Evaluation with Outgoing Flow Variation

Additional simulation has been performed using MATLAB to evaluate performance of the outgoing traffic estimation and buffer occupancy controller under varying outgoing flow rate. The algorithm presented before has been implemented in MATLAB and executed for varying outgoing traffic. The queue size is set to 20 packets, the ideal queue level is set to

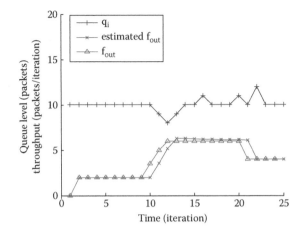

FIGURE 9.12
Queue utilization and estimation of the outgoing flow.

10 packets, the controller parameters $k_{vb} = 0.1$ and $\lambda = 0.001$. The actual outgoing flow is set to 0 packets per iteration at the beginning of simulation, then increased to 2, and later to 6 packets per iteration. Next, the outgoing flow is decreased to 4 packets per iteration. These outgoing flow variations can also be viewed as MAC data rate changes, thus providing an indication how the proposed protocol performs in networks that support multiple modulation rates.

Figure 9.12 illustrates the actual and estimated value of the outgoing flow, together with the queue utilization. The outgoing flow estimation is able to track the actual value, f_{out}. The queue utilization, q_i, varies from the ideal value when the sudden change in the outgoing flow occurs, because the outgoing traffic estimation could not predict such abrupt changes. However, after the sudden change, in just a few iterations of the algorithm, the queue level has converged to the ideal value as the estimation scheme quickly detected and accommodated the changed bandwidth. Additionally, Figure 9.13 presents the error in estimation of the outgoing flow, e_f, and the error of queue utilization, e_{bi}. The errors are bounded and quickly converge to zero because the scheme adapts to the changed outgoing flow rate. Moreover, in a gradual change in the outgoing traffic rate, for example during time interval of 10 to 14 iterations, the outgoing flow estimation quickly adapts to the change, and the estimation error decreases even though the outgoing flow rate continues to increase.

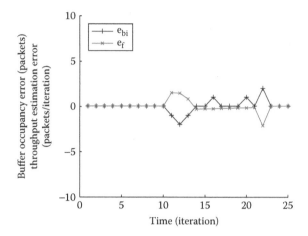

FIGURE 9.13
Queue utilization error and outgoing traffic estimation error.

9.5 Conclusions

This chapter presents a novel predictive congestion control scheme whereby the congestion is mitigated by suitably predicting the backoff interval of all the nodes based on the current network conditions. The network conditions include the traffic flow through a given region and channel state. Simulation results show that the proposed scheme increases throughput, network efficiency, and energy conservation. With the addition of a fair scheduling algorithm, the scheme guarantees desired QoS and weighted fairness for all flows even during congestion and fading channels. Finally, the proposed scheme provides a hop-by-hop mechanism for throttling packet flow rate, which will help in mitigating congestion. The convergence analysis is demonstrated by using a Lyapunov-based analysis.

References

Bennett, J.C.R. and Zhang, H., WF2Q: worst-case fair weighted fair queuing, *Proceedings of the IEEE INFOCOM*, Vol. 1, March 1996, pp. 120–128.

Golestani, S.J., A self-clocked fair queuing scheme for broadband applications, *Proceedings of the IEEE INFOCOM*, Vol. 2, June 1994, pp. 636–646.

Hull, B., Jamieson, K., and Balakrishnan, H., Mitigating congestion in wireless sensor networks, *Proceedings of the ACM SenSys*, 2004.

Jagannathan, S., End to end congestion control in high-speed networks, *Proceedings of the IEEE LCN*, 2002, pp. 547–556.

Jagannathan, S., *Neural Network Control of Nonlinear Discrete-Time Systems*, Taylor and Francis (CRC), Boca Raton, FL, 2006.

Kuo, W.-K. and Kuo, C.-C.J., Enhanced backoff scheme in CSMA/CA for IEEE 802.11, *Proceedings of the IEEE Vehicular Technology Conference*, Vol. 5, October 2003, pp. 2809–2813.

Peng, M., Jagannathan, S., and Subramanya, S., End to end congestion control of multimedia high speed Internet, *Journal of High Speed Networks*, to appear in 2006.

Regatte, N. and Jagannathan, S., Adaptive and distributed fair scheduling scheme for wireless ad hoc networks, *Proceedings of the World Wireless Congress*, May 2004, pp. 101–106.

Vaidya, N.H., Bahl, P., and Gupta, S., Distributed fair scheduling in a wireless LAN, *Proceedings of the 6th Annual International Conference on Mobile Computing and Networking*, August 2000, pp. 167–178.

Wan, C.-Y., Eisenman, S., and Campbell, A., CODA: congestion detection and avoidance in sensor networks, *Proceedings of the ACM SenSys'03*, Los Angeles, CA, November 2003, pp. 266–279.

Wang, S.-C., Chen, Y.-M., Lee, T.H., Helmy, A., Performance evaluations for hybrid IEEE 802.11b and 802.11g wireless networks, *Proceedings of the 24th IEEE IPCCC*, April 2005, pp. 111–118.

Yi, Y. and Shakkottai, S., Hop-by-hop congestion control over wireless multi-hop networks, *Proceedings of IEEE INFOCOM*, Vol. 4, Hong-Kong, March 2004, pp. 2548–2558.

Zawodniok, M. and Jagannathan, S., A distributed power control MAC protocol for wireless ad hoc networks, *Proceedings of the IEEE WCNC*, Vol. 3, March 2004, pp. 1915–1920.

Zawodniok, M. and Jagannathan, S., Dynamic programming-based rate adaptation and congestion control MAC protocol for wireless ad hoc networks, *Proceedings of the 2006 IEEE Conference on Local Computer Networks*, to appear in November 2006.

Zawodniok, M. and Jagannathan, S., Predictive congestion control MAC protocol for wireless sensor networks, *Proceedings of the International Conference on Control and Automation (ICCA)*, Vol. 1, June 2005, pp. 185–190.

Problems

Section 9.3

Problem 9.3.1: Simulate the queue utilization state equation (Equation 9.2) with the congestion control scheme, using Equation 9.5

for the selection of incoming flow rate and estimating the outgoing rate with the neural network equation for traffic parameter updates (Equation 9.10). Assume that the outgoing flow rate changes from the initial 5 packets per interval to 7 packets per interval at iteration 10. Calculate estimated value of outgoing flow, estimation error, queue level, and the difference from ideal queue utilization. The queue size is equal to 20 packets, the ideal queue utilization is equal to 10 packets, $\lambda = 0.001$, and $k_{bv} = 0.1$. Assume that the disturbance, $d(k)$, is equal to zero.

Problem 9.3.2: Repeat simulation in Problem 9.3.1 with disturbance, $d(k)$, altering between +1 and 1 for each iteration ($d(k) = k \bmod 2$).

Section 9.4

Problem 9.4.1: Select a high-density WSN with several cluster heads. Redo Example 9.4.1.

Problem 9.4.2: Redo Example 9.4.2 using a high-density ESN consisting of 150 nodes.

Problem 9.4.3: Redo Example 9.4.3 by varying the packet size at the buffers. Plot end-to-end delay, queue utilization, and energy efficiency.

10

Adaptive and Probabilistic Power Control Scheme for RFID Reader Networks

In Chapter 6, a power control scheme for wireless ad hoc and sensor networks is presented. In this chapter, the distributed power control (DPC) for a different type of network is discussed wherein by controlling power, detection range and read rates can be enhanced.

In radio frequency identification (RFID) systems (Finkenzeller and Waddington 2000), the detection range and read rates will suffer from interference among high-power reading devices. This problem becomes severe and degrades system performance in dense RFID networks. Consequently, DPC schemes and associated medium access protocols (MACs) are needed for such networks to assess and provide access to the channel so that tags can be read accurately. In this chapter, we investigate a suite of feasible power control schemes (Cha et al. 2006) to ensure overall coverage area of the system while maintaining a desired read rate. The power control scheme and the MAC protocol dynamically adjust the RFID reader power output in response to the interference level seen during tag reading and acceptable signal-to-noise ratio (SNR). The distributed adaptive power control (DAPC) and probabilistic power control (PPC) from Cha et al. (2006) are introduced as two possible solutions. A suitable backoff scheme is also added with DAPC to improve coverage. Both the methodology and implementation of the schemes are presented, simulated, compared, and discussed for further work.

10.1 Introduction

The advent of radio frequency identification (RFID) technology has brought with it increased visibility into manufacturing process and industry (Finkenzeller and Waddington 2000). From supply chain logistics to enhanced shop floor control, this technology presents many opportunities for process improvement or reengineering. The underlying principle of RFID technology is to obtain information from tags by using readers through radio

461

frequency (RF) links. The RFID technology basics and current standards can be found at the EPC Global Web site (http://www.epcglobalinc.org/).

In passive RFID systems, tags harvest energy from the carrier signal, which is obtained from the reader to power internal circuits. Moreover, passive tags do not initiate any communication but only decode modulated command signals from the readers and respond accordingly through backscatter communication (Rao 1999). The nature of RF backscatter requires high power ouput at the reader, and theoretically higher output power offers farther detection range with a desirable bit error rate (BER). For 915 MHz ISM bands, the output power is limited to 1 W (FCC Code 2000). When multiple readers are deployed in a working environment, signals from one reader may reach others and cause interference. This RFID interference problem was explained in Engels (2002) as reader collision.

The work by Engels (2002) suggested that RFID frequency interference occurs when a signal transmitted from one reader reaches another and jams its ongoing communication with tags in range. Studies also show that interrogation zones among readers need not overlap for frequency interference to occur, the reason being that power radiated from one reader needs to be at the level of the tag backscatter signal (μW) (Karthaus and Fischer 2003) to cause interference when reaching others. For a desired coverage area, readers must be placed relatively close to one another, forming a dense reader network. Consequently, frequency interference normally occurs, which results in limited read range, inaccurate reads, and long reading intervals. Placement of readers to mimize the interference and maximize the read range is an open problem.

To date, frequency interference has been described as "collision," as in a *yes* or *no* case in which a reader in the same channel at a certain distance causes another reader not to read any of its tags in its range. In fact, higher interference only implies that the read range is reduced significantly, but not to zero. This result is mathematically given in Section 10.2. Previous attempts (Waldrop et al. 2003, Tech Report 2005) to solve this channel access problem were based on either spectral or temporal separation of readers. Colorwave (Waldrop et al. 2003) and "listen before talk" implemented as per CEPT regulations (Tech Report 2005) rely on time-based separation, whereas frequency hopping spread spectrum (FHSS) implemented as per the FCC regulations (FCC Code 2000) utilizes multiple frequency channels. The former strategy is inefficient in terms of reader time and average read range, whereas the latter is not universally permitted by regulations. The work from Cha et al. (2006) is specifically targeted for RFID networks to overcome these limitations.

In this chapter, two power control schemes from Cha et al. (2006) that employ reader transmission power as the system control variable to achieve a desired read range and read rates are discussed. Degree of interference measured at each reader is used as a local feedback parameter to dynamically adjust its transmission power. With the same underlying concept, decentralized adaptive power control uses SNR to adapt power at discrete time steps, whereas PPC adapts the transmission power based on a certain probability distribution. A Lyapunov-based approach is used to show the convergence of the proposed DAPC scheme. Simulation results demonstrate theoretical conclusions.

10.2 Problem Formulation

The frequency interference problem needs to be fully understood before a solution can be evolved. In this section, we present an analysis of this problem and the assumptions made.

10.2.1 Mathematical Relations

In a backscatter communication system, SNR must meet a required threshold $R_{required}$, which can be expressed as

$$R_{required} = (E_b/N_0)/(W/D) \tag{10.1}$$

where E_b is the energy per bit of the received signal in watts, N_0 is the noise power in watts per Hertz, D is the bit rate in bits per second, and W is the radio channel bandwidth in Hertz. For a known modulation method and BER, E_b/N_0 can be calculated. Hence, $R_{required}$ can be selected based on desired a data rate and BER.

For any reader i, the following must hold for successful tag detection:

$$\frac{P_{bs}}{I_i} = R_i \geq R_{required} \tag{10.2}$$

where P_{bs} is the backscatter power from a tag, I_i is the interference at the tag backscatter frequency, and R_i is the SNR at a given reader.

In general, P_{bs} can be evaluated in terms of the reader transmission power P_i and tag distance r_{i-t}. Other variables such as reader and tag antenna gains, modulation indexing, and wavelength, derived in Rappaport (1999), can be considered as constants and simplified in Equation 10.3 as K_1. Then,

$$P_{bs} = K_1 \cdot \frac{P_i}{r_{i-t}^{4q}} = g_{ii} \cdot P_i \tag{10.3}$$

where q is an environment-dependent variable considering path loss, and g_{ii} represents the channel loss from reader i to tag and back. Communication channel between the reader and interrogated tag should be in a relatively short range; for this reason Rayleigh fading and shadowing effects are not considered for the reader–tag link. Influence by reflection can also be considered as a constant merging into g_{ii} assuming the environment is relatively stable. Hence, P_{bs} can be evaluated using path loss alone and by ignoring other channel uncertainties. However, the channel uncertainites are considered during the calculation of interference as reader locations are relatively farther away compared to a reader and a tag, and readers are power sources.

Interference caused by reader j at reader i is given as

$$I_{ij} = K_2 \cdot \frac{P_j}{r_{ij}^{2q}} \cdot 10^{0.1\zeta} \cdot X_{ij}^2 = g_{ij} \cdot P_j \qquad (10.4)$$

where P_j is the transmission power of reader j, r_{ij} is the distance between the two readers, K_2 represents all other constant properties, $10^{0.1\zeta}$ corresponds to the effect of shadowing, and X is a random variable with Rayleigh distribution (Rappaport 1999) to account for Rayleigh fading loss in the channel between reader j and reader i. After simplification, g_{ij} represents the channel loss from reader j to reader i. Note that because the interference actually occurs at the tag backscatter sideband, only power at that particular frequency needs to be considered. This factor is also accounted for in K_2 and g_{ij}.

Cumulative interference I_i at any given reader i is essentially the sum of interferences introduced by all other readers plus the variance of the noise η:

$$I_i = \sum_{j \neq i} g_{ij} P_j + \eta \qquad (10.5)$$

Given the transmission power and interference, the actual detection range of a reader is given by

$$r_{actual}^{4q} = \frac{K_1 \cdot P_i}{R_{required} \cdot I_i} \qquad (10.6)$$

Received SNR for a tag at a desired range r_d can be calculated as

$$R_{rd} = \frac{K_1 \cdot P_i}{r_d^{4q} \cdot I_i} \qquad (10.7)$$

Merging Equation 10.6 and Equation 10.7, we can calculate the actual detection range r_{actual} in terms of R_{rd} as

$$r_{actual} = r_d \left(\frac{R_{rd}}{R_{required}} \right)^{1/4q} \tag{10.8}$$

For analysis purposes, we assume that any tag within such a range can be successfully detected by the reader on account of BER specification. If a reader is completely isolated, meaning no interference, a maximum range r_{max} can be achieved by using it at the maximum power P_{max} of a given reader. In a practical application, it is not possible to expect this maximum range owing to interference. It is important to note from Equation 10.8 that the detection range and SNR are interchangeable and, therefore, our proposed algorithms target the required SNR. By viable power control both read rate and coverage can be achieved.

By substituting Equation 10.3 and Equation 10.4 into Equation 10.2, the SNR as a time-varying function is given by

$$R_i(t) = \frac{P_{bs}(t)}{I_i(t)} = g_{ii} \cdot P_i(t) \left/ \left(\sum_{j \neq i} g_{ij}(t) P_j(t) + u_i(t) \right) \right. \tag{10.9}$$

Note that g_{ii} is constant for a particular reader–tag link if we assume that the tag is stationary. If the desired range for the reader is defined as r_d, which is less than r_{max}, then we can define the SNR for the backscatter signal from a tag placed at a distance r_d to a reader as

$$R_{i-rd}(t) = \frac{P_{bs-rd}(t)}{I_i(t)} = g_{ii-rd} \cdot P_i(t) \left/ \left(\sum_{j \neq i} g_{ij}(t) P_j(t) + u_i(t) \right) \right. \tag{10.10}$$

where

$$g_{ii-rd} = \frac{K_1}{r_d^{4q}} \tag{10.11}$$

Equation 10.10 shows the basic relationship between the SNR and the output power of all readers through interference experienced at a particular in the network. This relationship is used throughout this chapter to derive the power control algorithms.

10.2.2 Two-Reader Model

To better understand the problem, a simple two-reader model is considered first. Two readers i and j spaced $D(i, j)$ apart, each with the desired range R_{i_1} and R_{j_1}, respectively, are shown in Figure 10.1. Readers must

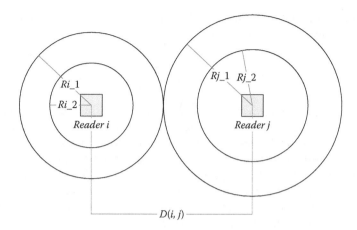

FIGURE 10.1
Two-reader model.

provide transmission powers P_i and P_j to achieve their respective desired range without considering interference. However, because of the interference introduced by each other, the actual detection range in fact decreases to R_{i_2} and R_{j_2}, respectively.

As a result of not achieving the SNR at a desired detection range owing to interference, readers must attempt to increase their transmission power. If both readers increase their powers greedily, they will eventually reach the maximum power without achieving the desired range, on account of increased interferences. Further, the SNR target is not met and as a result the tags are not read even by those that are in range. One could solve this problem by operating them in mutually exclusive time slots. However, as the number of readers increases, this strategy severely degrades each reader's average read time and detection range and eventually increases reading intervals.

A more appropriate solution is to balance the transmission power between the two readers in order to reach the equilibrium where multiple readers can achieve their respective read ranges. In this model, if reader i transmits at P_{max} and reader j is off, a read range greater than the targeted value of R_{i_1} can be achieved. On the other hand, there exists a power level at which reader j can transmit and still allow i to achieve read range R_{i_1}. This process can be applied in reverse to enable reader j to achieve its targeted range. Under such circumstances, the average read range of both readers is improved over the typical on and off cycle. Such a yielding strategy is required in dense reader networks in which the desired range may not be achieved by all the readers simultaneously. The effect of the improvement due to this strategy will be significant in dense networks. The next section details such a decentralized strategy.

10.2.3 Distributed Solution

In this chapter, two schemes of DPC are introduced — DAPC and PPC. DAPC involves systematic power updates based on local interference measurements at each reader. It also uses embedded channel prediction to account for the time-varying fading channel state for the next cycle. In Section 10.3, it is shown that the proposed DAPC scheme will converge to any target SNR value in the presence of channel uncertainties. For dense networks in which the target SNR cannot be reached by all readers simultaneously, an additional selective backoff method is incorporated besides power updates, introducing a degree of yielding to ensure that all readers achieve their desired range.

By contrast, in the PPC scheme, a probability distribution is specified for each reader to select output power from. Statistical distribution for the desired read range can be specified as the target. To achieve the target, the output power distribution on each reader is altered based on interference measurements. The relationship between the two distributions is analytically derived in Section 10.4.

10.2.4 Standards

Implementing FHSS on readers has been explored in the past as a solution to the interference problem. Although FHSS reduces the probablity of interference, it is not a universal solution because of the differing spectral regulations over the world. In this work, frequency hopping is not considered. New standards (EPC 2003) have been designed in dense reader networks by spectrally separating the reader and tag modulation frequencies. However, subject to the Transmit Mask specifications and hardware implementations, substantial interference will still exist at the sideband frequencies of a tag in a highly dense reader network. The from Cha et al. (2006) work is not dependent on any existing RFID standards or implementations and can be easily adapted to improve the performances of RFID reader networks.

10.3 Decentralized Adaptive Power Control

Decentralized, also known as DPC, protocols have been extensively studied in the field of wireless communication, including in ad hoc networks (Zawodniok and Jagannathan 2004) and cellular networks (Jagannathan et al. 2006). Conceptually, power control in an RFID reader network is similar to these protocols. However, there are several fundamental differences between them because of the unique nature of the communication interface and RFID application. Moreover, a tag is not smart compared to a cell phone or a sensor node, and therefore such schemes have to be modified for RFID applications.

First, the main goal of DPC in wireless communication is to conserve energy while maintaining desired quality-of-service (QoS) requirements. In Park and Sivakumar (2002), Jung and Vaidya (2002), Jagannathan et al. (2006), and Zawodniok and Jagannathan (2004), authors propose different power-updating schemes to maintain a target SNR threshold for successful communication. By contrast, the work proposed for RFID systems is to reduce interference introduced by others while maintaining read range requirements at each reader, thereby achieving optimal coverage for all readers and read rates. Second, DPC for ad hoc and cellular networks requires a feedback signal between the transmitter and receiver. In RFID reader networks, the reader acts both as a transmitter and receiver. Hence, the feedback is internal to the reader and does not result in any communication overhead. Third, in contrast to low-power wireless networks run on battery power, RFID readers in dense networks may not achieve the target SNR even at maximum power owing to the high levels of interference. Finally, in contrast with a connection-oriented network in which each node transmits only when it is needed most, RFID readers are required to be always on and transmitting in order to read the tags. Therefore, it is more difficult to distribute channel access among all readers.

This DAPC algorithm consists of two building blocks: adaptive power update and selective backoff. The goal of adaptive power update is to achieve required SNR with an appropriate output power by correctly estimating the interference and any channel uncertainties. In dense networks, selective backoff forces high-power readers to yield so that other readers can achieve required SNR. We now discuss these two building blocks of DAPC in depth.

10.3.1 Power Update

The development and performance of DAPC are now demonstrated analytically. Differentiating the SNR from Equation 10.10 as the channel interference follows the time-varying nature of the channel, we get

$$R_{i-rd'}(t) = g_{ii-rd} \cdot \frac{P_i'(t)I_i(t) - P_i(t)I_i'(t)}{I_i^2(t)} \qquad (10.12)$$

where $R_{i-rd'}(t)$, $P_i'(t)$, and $I_i'(t)$ are the derivatives of $R_{i-rd}(t)$, $P_i(t)$, and $I_i(t)$, respectively.

Applying Euler's formula and following similar development as that of Chapter 5, $x'(t)$ can be expressed as $\frac{x(l+1)-x(l)}{T}$ in the discrete time domain, where T is the sampling interval. Equation 10.12 can be transformed into the discrete time domain as

$$\frac{R_{i-rd}(l+1)-R_{i-rd}(l)}{T} = \frac{g_{ii-rd} \cdot P_i(l+1)}{I_i(l)T}$$

$$-\frac{g_{ii-rd} \cdot P_i(l)}{I_i^2(l)T} \cdot \sum_{j \neq i} \begin{pmatrix} [g_{ij}(l+1)-g_{ij}(l)]P_j(l) \\ +g_{ij}(l)[P_j(l+1)-P_j(l)] \end{pmatrix} \quad (10.13)$$

After the transformation, Equation 10.13 can be expressed as

$$R_{i-rd}(l+1) = \alpha_i(l)R_{i-rd}(l) + \beta_i v_i(l) \quad (10.14)$$

where

$$\alpha_i(l) = 1 - \frac{\displaystyle\sum_{j \neq i} \Delta g_{ij}(l)P_j(l) + \Delta P_j(l)g_{ij}(l)}{I_i(l)} \quad (10.15)$$

$$\beta_i = g_{ii-rd} \quad (10.16)$$

and

$$v_i(l) = P_i(l+1)/I_i(l) \quad (10.17)$$

With the inclusion of noise, Equation 10.14 is written as

$$R_{i-rd}(l+1) = \alpha_i(l)R_{i-rd}(l) + \beta_i v_i(l) + r_i(l)\omega_i(l) \quad (10.18)$$

where $\omega(l)$ is the zero mean stationary stochastic channel noise with $r_i(l)$ being its coefficient.

From Equation 10.18, we can obtain the SNR at time instant $l+1$ as a function of channel variation from time instant l to $l+1$. The difficulty in designing the DAPC is that channel variation is not known beforehand. Therefore, α must be estimated for calculating the feedback control. Now, define; $y_i(k) = R_{i-rd}(k)$, then Equation 10.18 can be expressed as

$$y_i(l+1) = \alpha_i(l)y_i(l) + \beta_i v_i(l) + r_i(l)\omega_i(l) \quad (10.19)$$

As α_i, r_i are unknown, Equation 10.19 can be transformed into

$$y_i(l+1) = [\alpha_i(l) \quad r_i(l)]\begin{bmatrix} y_i(l) \\ \omega_i(l) \end{bmatrix} + \beta_i v_i(l) = \theta_i^T(l)\psi_i(l) + \beta_i v_i(l) \quad (10.20)$$

where $\theta_i^T(l) = [\alpha_i(l) \quad r_i(l)]$ is a vector of unknown parameters, and $\psi_i(l) = \left[\begin{smallmatrix} y_i(l) \\ \omega_i(l) \end{smallmatrix}\right]$ is the regression vector. Now, selecting feedback control for DAPC as

$$v_i(l) = \beta_i^{-1}[-\hat{\theta}_i(l)\psi_i(l) + \gamma + k_v e_i(l)] \tag{10.21}$$

where $\hat{\theta}_i(l)$ is the estimate of $\theta_i(l)$, the SNR error system is expressed as

$$e_i(l+1) = k_v e_i(l) + \theta_i^T(l)\psi_i(l) - \hat{\theta}_i^T(l)\psi_i(l) = k_v e_i(l) + \tilde{\theta}_i^T(l)\psi_i(l) \tag{10.22}$$

where $\tilde{\theta}_i(l) = \theta_i(l) - \hat{\theta}_i(l)$ is the error in estimation.

From Equation 10.22, it is clear that the closed-loop SNR error system is driven by the channel estimation error. If the channel uncertainties are properly estimated, then SNR estimation error tends to be zero; therefore, the actual SNR approaches the target value. In the presence of error in estimation, only the boundedness of error in SNR can be shown. Given the closed-loop feedback control and error system, we can now advance to the channel estimation algorithms.

Consider now the closed-loop SNR error system with channel estimation error, $\varepsilon(l)$, as

$$e_i(l+1) = k_v e_i(l) + \tilde{\theta}_i^T(l)\psi_i(l) + \varepsilon(l) \tag{10.23}$$

where $\varepsilon(l)$ is the error in estimation, which is considered bounded above $\|\varepsilon(l)\| \leq \varepsilon_N$, with ε_N a known constant.

THEOREM 10.3.1
Given the DPC scheme above with channel uncertainties, if the feedback for the DPC scheme is selected as in Equation 10.21, then the mean channel estimation error along with the mean SNR error converges to zero asymptotically, if the parameter updates are taken as

$$\hat{\theta}_i(l+1) = \hat{\theta}_i(l) + \sigma\psi_i(l)e_i^T(l+1) - \Gamma\|I - \psi_i^T(l)\psi_i(l)\|\hat{\theta}_i(l) \tag{10.24}$$

where $\varepsilon(l)$ is the error in estimation, which is considered bounded above $\|\varepsilon(l)\| \leq \varepsilon_N$, ε with ε_N a known constant. Then the mean error in SNR and the estimated parameters are bounded.

10.3.2 Selective Backoff

In a dense reader environment, it is inconceivable that all readers are able to achieve their target SNR together, owing to severe congestion that affects both read rates and coverage. These readers will eventually reach maximum power as a result of the adaptive power update. This demands a time-based yielding strategy, with some readers allowing others to achieve their target SNR.

Whenever the reader finds the target SNR is not achievable at maximum power, implying that the interference level is too high in the network, it should back off to a low output power for a period of time. As interference is a locally experienced phenomenon, multiple readers will face this situation and they will all be forced to back off. The rapid reduction of power will result in significant improvement of SNR at other readers. After waiting for the backoff period, a reader will return to normal operation and attempt to achieve the target SNR. The process is repeated for every reader in the network. To fairly distribute the channel access among all congested readers, certain quality measurements must be ensured for all readers in the backoff scheme. The selective backoff scheme uses the percentage of time a reader has achieved desired range as the quality control parameter to ensure fairness.

After backing off, each reader must wait for a time duration τ_w. To illustrate the effect of backoff, τ_w is defined as a logarithmic function of the percentage of time ρ a reader has attained the required SNR. A neglected reader will exit backoff mode quickly and attain the required SNR, whereas other readers in the vicinity fall back. The calculation of τ_w is given by

$$\tau_w = 10 \cdot [\log_{10}(\rho + 0.01) + 2] \tag{10.25}$$

Using the preceding equation, a reader with a ρ of 10% will wait for 10 time intervals, whereas the waiting time for ρ of 100% equals 20. A plot of waiting time τ_w vs. ρ is presented in Figure 10.2.

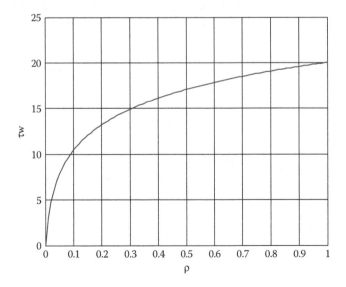

FIGURE 10.2
Selective backoff function plot.

TABLE 10.1

Selective Backoff Pseudocode

If reader is not in backoff mode
   If $P_{next} == P_{max}$
      change reader to backoff mode
      initialize wait time τ_w
If reader is in backoff mode
      Set $P_{next} = P_{min}$
      decrease τ_w
      If $\tau_w == 0$
         reader exit backoff mode

The backoff policy will cause negative changes in interference, and hence does not adversely affect the performance of the adaptive power update. A detailed pseudocode for implementing selective backoff is given in Table 10.1.

10.3.3 DAPC Implementation

DAPC can be easily implemented at the MAC layer of the RFID reader; MAC implementation is not covered in detail here. The algorithm requires two parameters to be known initially. These are the desired range, r_d, and required SNR, $R_{required}$.

Proposed DAPC can be seen as a feedback between the transmitter and receiver units of a reader. A block diagram of the implementation is shown in Figure 10.3. The detailed description of the algorithm implementation follows:

1. Power update block at the receiver unit of a reader obtains sensed interference $I(l)$.

2. In the power update block, based on r_d, $R_{required}$, and current power $P(l)$, the current SNR $R_{i-rd}(l)$ is calculated.

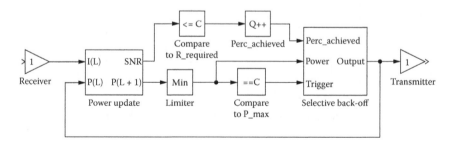

FIGURE 10.3
Block diagram for DAPC implementation.

3. $R_{i-rd}(l)$ is compared to R_{required}, and percentage of time required for achieving required SNR, ρ is calculated and recorded.

4. Based on Equation 10.24, the channel is estimated for the next time step $l + 1$, and the power for $P(l + 1)$ is also calculated using the feedback control from Equation 10.21.

5. $P(l + 1)$ is then limited to maximum power P_{max}; if $P(l + 1)$ is greater than P_{max}, the selective backoff scheme is triggered, otherwise $P(l + 1)$ is used as the output power for the next cycle.

6. The selective backoff block follows the algorithm provided in the preceding subsection and restricts the final output power for the next cycle.

Simulation and results of the preceding implementation are discussed in Section 10.5 and Section 10.6, respectively, along with those of PPC. Next, the PPC is discussed.

10.4 Probabilistic Power Control

The idea of PPC comes from simple TDM algorithms. If a reader is assigned a time slot to transmit in full power while others are turned off, it will achieve its maximum range. A round-robin assignment of time slots can ensure that all readers operate with no interference. However, this is inefficient in terms of average read range, reader utilization, and waiting periods. It is obvious that more than one reader can operate in the same time slot but at different power levels to accomplish better overall read range. If the power levels at all readers change in each time slot following a certain distribution, over time, every reader will be able to achieve its peak range while maintaining a good average.

For a distributed solution, this would involve setting a probability distribution for power to be selected for each time step. Such a distribution would need to be adapted based on the density and other parameters of the reader network.

10.4.1 Power Distribution

Equation 10.9 states that the read range of a particular reader is dependent on its transmission power and the interference experienced, which is a function of powers of all other readers. If reader powers follow a certain probability distribution, the distribution of read ranges for each reader is a function of these power distributions

$$F(r_i) = f_i(F(P_1), \ldots, F(P_n)) \tag{10.26}$$

where $F(r_i)$ is the cumulative density function of the read range of the reader i, and $F(P_i)$ is the cumulative power density function of reader i. Performance metrics, including mean read range μ_r and percentage of time ρ the desired range r_d was achieved, characterized the read range distribution $F(r_i)$.

$$F(r_i) = g_i(\mu_r, \rho) \tag{10.27}$$

To achieve targeted characteristics on the read range distribution, we need to modify the power distribution freely. Beta distribution, demonstrated in Figure 10.4, is specifically chosen for this reason: by specifying the shape variables α and β, one can change the cumulative density function in the domain from 0 to 1 (0 to 100% power). By changing these two parameters, we can control the power distribution and thus attempt to achieve desired targets on the read range distribution in Equation 10.26. Power using Beta distribution can be represented as

$$F(P_i) = H(P_i : \alpha, \beta) \tag{10.28}$$

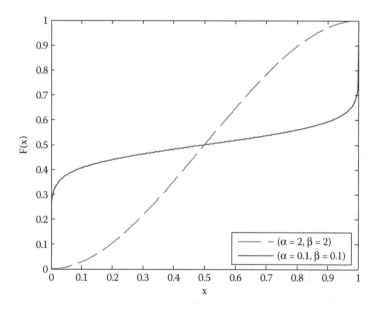

FIGURE 10.4
Cumulative density function of read range.

As shown in Figure 10.4, *Beta*(0.1, 0.1) renders 30% probability in selecting either high or low power. On an average, a third of the total readers will not operate in each time slot, and therefore the interference levels will be reduced. Such a distribution is expected to perform well in dense networks, because it works similar to a time-slotting method. For sparser networks in which the target SNR is achievable for all readers, power distribution *Beta*(0.1, 0.1) will degrade the performance as readers will be off 30% of the time. Meanwhile, the distribution generated by *Beta*(2, 2) will result in higher probability being in the medium-power range, and it will achieve better results as higher output power can overcome the interference produced in sparser networks. It is important to note that dense RFID networks involve 30 to 40 readers whereas sparse networks may involve 5 to 10 readers unlike in wireless ad hoc and sensor networks, where dense networks may involve several hundred to thousand sensor nodes.

10.4.2 Distribution Adaptation

Equation 10.26 represents the relationship between the cumulative density function of read range and output power of all readers. However, in a distributed implementation, operation parameters such as the power distribution and location of a reader are not known to the other readers. Hence, these parameters have to be reflected in a measurable quantity; Equation 10.5 provides such a representative quantity in the form of interference, which leads to Equation 10.28 as

$$F(r_i) = l_i(F(P_1), F(I_i)) \qquad (10.29)$$

Substituting Equation 10.27 and Equation 10.28 into Equation 10.29,

$$g_i(\mu_r, \tau_r) = l_i(H(\alpha, \beta), F(I_i)) \qquad (10.30)$$

Transforming Equation 10.30, we can represent α and β in terms of μ_r, ρ, and $F(I_i)$ as

$$[\alpha, \beta] = h_i(\mu_r, \rho, F(I_i)) \qquad (10.31)$$

where $F(I_i)$, the cumulative density function of interference, can be statistically evaluated by observing the interference level at each reader over time. It can also be interpreted as the local density around the reader.

The function represented by Equation 10.31 involves joint distributions of multiple random variables, and it is complex and difficult to extract. However, it is easy to obtain numerical data sets of this function from simulation. Such data sets can be used potentially to train a neural network that could provide a model of the function. In this chapter, we do not attempt to provide the interference-based adaptive distribution tuning scheme for the PPC. We only implement PPC using fixed power distributions for all scenarios to observe the overall performance patterns and to understand the differences between the DAPC and PPC. The two distributions, $Beta(0.1, 0.1)$ and $Beta(2, 2)$ used in Figure 10.4 are chosen for the simulation and compared for performance evaluation.

In terms of implementation, PPC only requires a power control block that selects output power based on predefined probability distributions. However, a more complex model of PPC can be generated provided the relationship in Equation 10.31 can be obtained. This PPC requires interference measurement and dynamically adjusts the power distribution based on interference to maximize μ_r and ρ.

10.5 Simulation Results

The simulation environment is set up in MATLAB. Full model of DAPC and PPC are implemented for comparison. Both algorithms are tested under the same configuration.

10.5.1 Reader Design

Reader power is implemented as a floating point number varying from 0 to 30 dBm (1 W) as per FCC regulations. For error-free detection, the reader should maintain a target SNR of 14 (~11 dB). Other system constants are designed so that the maximum read range of a reader in an isolated environment is 3 m. Interference experienced at any reader is calculated based on a matrix consisting of power and positions of all other readers plus the channel variation g_{ij}. A desired range of 2 m is specified based on the worst case analysis.

For the proposed DAPC, power update parameters Kv and σ are both set to 0.001. For proposed PPC, both $Beta(0.1, 0.1)$ and $Beta(2, 2)$ are implemented.

10.5.2 Simulation Parameters

For both models, random topologies are generated in order to emulate denser networks with a suitable number of readers. The RFID network with a suitable density for a given scenario is created by placing the readers with the minimum distance between them and the maximum area under test. The minimum distance between any two readers is varied from 4 to 14 m and the maximum size of the coordinate is adjusted accordingly. The number of readers is changed from 5 to 60 to create denser network and to test the scalability of the proposed schemes. Each simulation scenario is executed for 10,000 iterations.

10.5.3 Evaluation Metrics

To demonstrate the typical performance of the reader network, the cumulative range distribution of a reader can be plotted. In Figure 10.5, the cumulative density function $F(x)$ of read range x for a reader using DAPC is plotted. From this plot, we can observe the minimum and maximum detection range as well as the percentile of attainment of certain ranges.

To evaluate the performances of the proposed algorithms, the metrics average read range, percentage of time the desired range is attained, average output power, and average interference experienced are evaluated across all readers for each scenario and simulation results are given.

10.5.4 Results and Analysis

In Figure 10.6, the output power, interference level, and detection range at a particular reader are plotted vs. time for the DAPC in a dense network. It is seen that the DAPC attempts to achieve the desired range by increasing

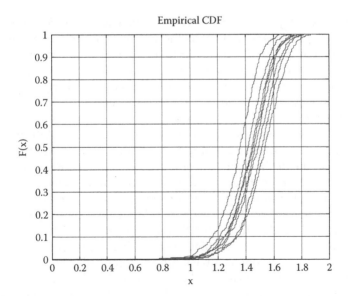

FIGURE 10.5
Cumulative density function of the read range for a reader using DAPC.

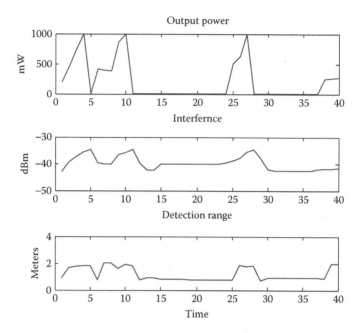

FIGURE 10.6
Output power, interferences, and detection range as a function of time in seconds.

power; however, the interference level is too high and, therefore, the reader reaches maximum power and enters the selective backoff scheme. It is also observed that as the reader backs off to low power value, the interference level increases, meaning that other readers are taking the advantage and accessing the channel. This plot also demonstrates the changes in backoff time corresponding to the desired range achievement; for example, time interval 12 to 24 sec and 28 to 37 sec.

The performance in sparse networks is discussed first. With the minimum distance of 9 m between any two readers, the average percentage of time ρ that the desired range is attained across all readers is presented in Figure 10.7. Note that each reader has a maximum detection range of 3 m without interference, and the desired range is set to 2 m in the presence of multiple readers. The DAPC is observed to have superior performance over the two PPC algorithms for this sparse network. The DAPC converges to 100% desired range achievement with the appropriate parameter estimation and closed-loop feedback control described in Section 10.3. The results justify the theoretical conclusions. It is also shown that *Beta*(2, 2) performs better than *Beta*(0.1, 0.1) in terms of ρ. With a *Beta*(2, 2) distribution, every reader will be on and transmitting at medium power most of the time. With sparse networks and small interferences, the medium power

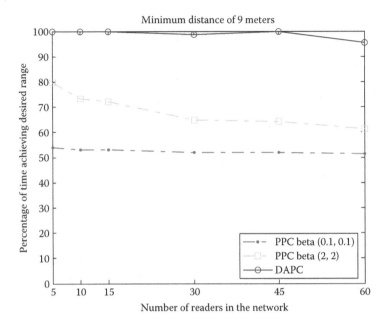

FIGURE 10.7
Percentage of time the desired range is achieved with number of readers.

overcomes the interference produced, thereby achieving the desired range. In contrast, *Beta*(0.1, 0.1) has a 30% probability of being off; therefore, the probability of attaining the desired range will be low.

In Figure 10.8, considering the average detection range for the same scenario, the DAPC converges to the 2 m desired range and outperforms both PPC algorithms. We can also observe the average power level used for each algorithm in Figure 10.9. As the mean for both *Beta*(2, 2) and *Beta*(0.1, 0.1) is 0.5, the average reader output power stays at 500 mW, which is half of the maximum power. Meanwhile, DAPC is able to dynamically adjust its output power to find the optimal level for which the desired range can be achieved as the size of the network varies.

Performance of the power control schemes in denser networks is now analyzed. For network with a minimum distance of 6 m, the desired range is not attainable by all readers as the transmission power is not able to overcome the interference, forcing the yielding strategy of each algorithm to test. The detection range and percentile vs. number of readers are presented in Figure 10.10 and Figure 10.11 respectively. As the number of readers increases, the overall interference in the network will also increase. Consequently, the percentage of time ρ a reader attains its desired range will drop, as shown in Figure 10.10. It is observed that PPC with *Beta*(0.1, 0.1) offers the best performance in terms of ρ. This is because on

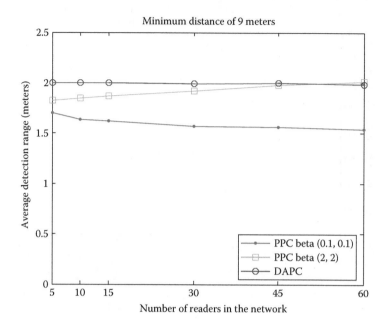

FIGURE 10.8
Average detection range in meters with number of readers.

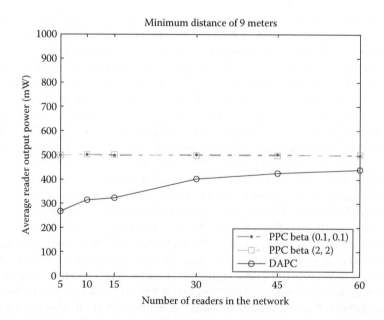

FIGURE 10.9
Average output power per reader with number of readers.

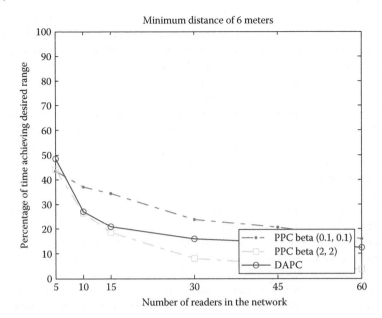

FIGURE 10.10
Percentage of time desired range is achieved with number of readers.

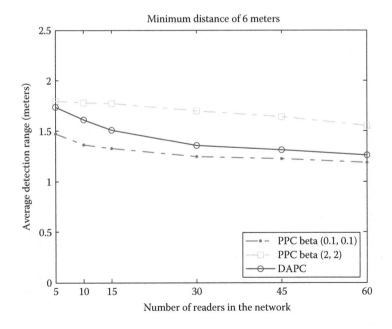

FIGURE 10.11
Average detection range with number of readers.

average 30% of the readers will be switched off for each time interval whereas the other 30% transmit at full power. Hence, readers in full power have a greater probability of attaining the desired range, whereas the average detection range is sacrificed for this achievement. The relatively poor performance in average detection range compared to DPC and PPC $Beta(2,2)$ can be observed in Figure 10.11.

Although the percentage of time a target range is achieved is low for $Beta(2,2)$, it provides the best average detection range of all three algorithms. DAPC with the selective backoff scheme finds a balance between the two evaluation metrics. These show that there is a trade-off between percentage times the target range and average detection range are achieved.

The average detection range and percentile plots can also be produced by fixing the number of readers and varying the minimum distance between any two readers. Shown in Figure 10.12 and Figure 10.13, DAPC is seen to converge as the minimum distance between any two readers decreases, which again verifies the theoretical conclusions for the power update scheme. With the same explanation as discussed previously, PPC with $Beta(0.1, 0.1)$ performs better in achieving desired range, whereas $Beta(2,2)$ gives better average detection range.

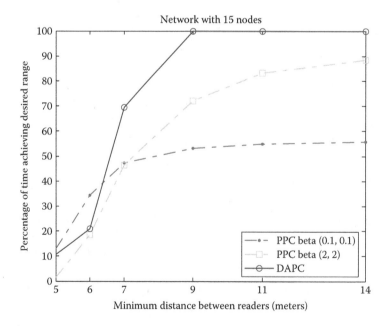

FIGURE 10.12
Percentage of time the target range is achieved with minimum distance.

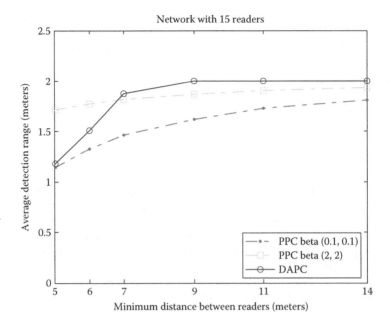

FIGURE 10.13
Average detection range with minimum distance.

10.6 Conclusions

Two algorithms for RFID reader read range and interference management based on DAPC are explored and analyzed. Both algorithms can be implemented as power control MAC protocols for MATLAB-based RFID reader network simulation. DAPC is seen to converge at a fast rate to the required SNR if it is achievable within power limitations. Selective backoff algorithm in DAPC enhances the channel utilization in denser networks. PPC is not fully implemented to tune in with the network density; however, it still shows advantages in scalability and fairness of channel assessment. Furthermore, implementation details of both algorithms are discussed.

This chapter details a novel interpretation of the reader collision problem that can be applied to other similar RF systems also. We have demonstrated that high-power RFID networks suffer from severe interferences and cause problems on other lower-power RF devices. These problems may not be resolved easily at the RF communication level and, therefore, two power control algorithms, DAPC and PPC, are introduced. Further work on DAPC would involve automatically tuning the selective backoff implementations based on interference and quality measurements.

Further work on PPC would concentrate on developing a method to internally adapt the power distribution based on interference measurements to achieve specified statistical goals for the read range.

References

Cha, K., Ramachandran, A., and Jagannathan, S., Adaptive and probabilistic power control schemes and hardware implementation for dense RFID networks, *Proceedings of the IEEE International Conference of Decision and Control*, pp. 1858–1863, 2006.

Engels, D.W., The Reader Collision Problem, MIT Auto ID Center, MIT-AUTOID-WH-007, 2002.

EPC Radio-Frequency Identity Protocols Generation 2 Identity Tag (Class 1): Protocol for Communications at 860MHz-960MHz, *EPC Global Hardware Action Group (HAG), EPC Identity Tag (Class 1) Generation 2*, Last-Call Working Draft Version 1.0.2, November 24, 2003.

EPCGlobal Inc publications, http://www.epcglobalinc.org/.

FCC Code of Federal Regulations, Title 47, Vol. 1, Part 15, Sections 245–249, 47CFR15, October 2000.

Finkenzeller, K. and Waddington, R., *RFID Handbook: Radio-Frequency Identification Fundamentals and Applications*, John Wiley & Sons, January 2000.

Jagannathan, S., *Neural Network Control of Nonlinear Discrete-Time Systems*, Taylor and Francis (CRC Press), Boca Raton, FL, 2006.

Jagannathan, S., Zawodniok, M., and Shang, Q., Distributed power control of cellular networks in the presence of channel uncertainties, *IEEE Transactions on Wireless Communications*, Vol. 5, No. 3, 540–549, March 2006.

Jung, E.-S. and Vaidya, N.H., A power control MAC protocol for ad hoc networks, *Proceedings of the ACM MOBICOM*, 2002.

Karthaus, U. and Fischer, M., Fully Integrated passive UHF RFID Transponder IC With 16.7-uW Minimum RF Input Power, *IEEE Journal of Solid-State Circuits*, Vol. 38, No. 10, October 2003.

Park, S.-J. and Sivakumar, R., Quantitative analysis of transmission power control in wireless ad-hoc networks, *Proceedings of the ICPPW'02*, August 2002.

Rao, K.V.S., An overview of back scattered radio frequency identification systems (RFID), *Proceedings of the IEEE Microwave Conference*, Vol. 3, 1999, pp. 746–749.

Rappaport, T.S., *Wireless Communications, Principles and Practices*, Prentice Hall, NJ, 1999.

TR (Technical Report) on LBT (listen-before-talk) for Adaptive Frequency Agile SRD's as Implemented in the Draft EN 302 288, ETSI TR 102 378 V1.1, October 2005.

Waldrop, J., Engels, D.W., and Sharma, S.E., Colorwave: an anti-collision algorithm for the reader collision problem, *Proceedings of the IEEE ICC*, Vol. 2, 2003, pp. 1206–1210.

Zawodniok, M. and Jagannathan, S., A distributed power control MAC protocol for wireless ad hoc networks, *Proceedings of the IEEE WCNC*, Vol. 3, March 2004, pp. 1915–1920.

Problems

Section 10.4

Problem 10.4.1: Show that the probabilistic-based power control converges to a target value.

Section 10.6

Problem 10.6.1: Derive the minimum number of readers and their placement equations in order to achieve a target SNR.

Index

T - #0152 - 101024 - C0 - 229/152/29 [31] - CB - 9780824726751 - Gloss Lamination